T0182042

# Optimization by GRASP

Mauricio G.C. Resende • Celso C. Ribeiro

# Optimization by GRASP

Greedy Randomized Adaptive Search
Procedures

 Springer

Mauricio G.C. Resende
Modeling and Optimization Group (MOP)
Amazon.com, Inc.
Seattle, WA, USA

Celso C. Ribeiro
Instituto de Ciência da Computação
Universidade Federal Fluminense
Niterói, Rio de Janeiro, Brazil

ISBN 978-1-4939-8227-1     ISBN 978-1-4939-6530-4     (eBook)
DOI 10.1007/978-1-4939-6530-4

Printed on acid-free paper

This Springer imprint is published by Springer Nature
The registered company is Springer Science+Business Media LLC
The registered company address is: 233 Spring Street, New York, NY 10013, U.S.A

*In memory of*
*David Stifler Johnson*

# Foreword

In recent years, advances in metaheuristics have given practitioners a powerful framework for making key decisions in problems as diverse as telecommunications network design and supply chain planning to scheduling in transportation systems. GRASP is a metaheuristic that has enjoyed wide success in practice, with an extraordinarily broad range of applications to real-world optimization problems. Starting from the seminal 1989 paper by Feo and Resende, over the past 25 years, a large body of work on greedy randomized adaptive search procedures has emerged. A vast array of papers on GRASP have been published in the open literature, and numerous MSc and PhD theses have been written on the subject. This book is a timely and welcome addition to the metaheuristics literature, bringing together this body of work in a single volume.

The account of GRASP in this book is especially commendable for its readability, covering many facets of this metaheuristic, such as solution construction, local search, hybridizations, and extensions. It is organized into four main sections: introduction to combinatorial optimization, fundamentals of heuristic search, basic GRASP, and advanced topics. The book can be used as an introductory text, not only to GRASP but also to combinatorial optimization, local search, path-relinking, and metaheuristics in general. For the more advanced reader, chapters on hybridization with path-relinking and parallel and continuous GRASP present these topics in a clear and concise fashion. The book additionally offers a very complete annotated bibliography of GRASP and combinatorial optimization.

For the practitioner who needs to solve combinatorial optimization problems, the book provides implementable templates for all algorithms covered in the text.

This book, with its excellent overview of the state of the art of GRASP, should appeal to researchers and practitioners of combinatorial optimization who have a need to find optimal or near-optimal solutions to hard optimization problems.

Boulder, CO, USA
May 2016

Fred Glover

# Preface

Greedy randomized adaptive search procedures, or GRASP, were introduced by T. Feo and M. Resende in 1989 as a probabilistic heuristic for solving hard set covering problems. Soon after its introduction, it was recognized as a general purpose metaheuristic and was applied to a number of other combinatorial optimization problems, including scheduling problems, the quadratic assignment problem, the satisfiability problem, and graph planarization. At the Spring 1991 ORSA/TIMS meeting in Nashville, T. Feo and M. Resende presented the first tutorial on GRASP as a metaheuristic, which was followed by their tutorial published in the *Journal of Global Optimization* in 1995. Since then, GRASP has gained wide acceptance as an effective and easy-to-implement metaheuristic for finding optimal or near-optimal solutions to combinatorial optimization problems.

This book has been many years in planning. Though many books have been written about other metaheuristics, including genetic algorithms, tabu search, simulated annealing, and ant colony optimization, a book on GRASP had yet to be published. Since the subject has had 25 years to mature, we feel that this is the right time for such a book. After Springer agreed to publish this book, we began the task of writing it in 2010.

We have been collaborating on the design and implementation of GRASP heuristics since 1994 when we decided, at the TIMS XXXII International Meeting in Anchorage, Alaska, to partner in designing a GRASP for graph planarization. Since then, we have worked together on a number of papers on GRASP, including three highly cited surveys.

This book is aimed at students, engineers, scientists, operations researchers, application developers, and other specialists who are looking for the most appropriate and recent GRASP-based optimization tools to solve particular problems. It focuses on algorithmic and computational aspects of applied optimization with GRASP. Emphasis is given to the end user, providing sufficient information on the broad spectrum of advances in applied optimization with GRASP.

The book grew from talks and short courses that we gave at many universities, companies, and conferences. *Optimization by GRASP* turned out to be not only a book on GRASP but also a pedagogical book on heuristics, metaheuristics and its basics, foundations, extensions, and applications. We motivate the subject with a number of hard combinatorial optimization problems expressed in simple descriptions in the first chapter. This is followed by an overview of complexity theory that makes the case for heuristics and metaheuristics as very effective strategies for solving hard or large instances of the so-called intractable *NP*-hard optimization problems. In our view, most metaheuristics share a number of common building blocks that are combined following different strategies to overcome premature local optimality. Such building blocks are explored, for example, in the chapters or sections on greedy algorithms, randomization, local search, cost updates and candidate lists, solution perturbations and ejection chains, adaptive memory and elite sets, path-relinking, runtime distributions and probabilistic analysis tools, parallelization strategies, and implementation tricks, among other topics. As such, preliminary versions of this text have been used in the last three years as a textbook for the course on metaheuristics at the graduate program in computer science at Universidade Federal Fluminense, Brazil, complemented with specific reading material about other metaheuristics, where it has matured and was exposed to criticisms and suggestions from many students and colleagues.

The book begins in Chapter 1 with an introduction to optimization and a discussion about solution methods for discrete optimization, such as exact and approximate methods, including heuristics and metaheuristics.

We then provide in Chapter 2 a short tour of combinatorial optimization and computational complexity, in which we introduce metaheuristics as a very effective tool for approximately solving hard optimization problems.

This is followed in Chapter 3 with solution construction methods, including greedy algorithms and their relation to matroids, adaptive greedy and semi-greedy algorithms, and solution repair procedures.

Chapter 4 focuses on local search. We discuss solution representation, neighborhoods, and the solution space graph. We then focus on local search methods, covering neighborhood search strategies, cost function updates, and candidate list strategies. Ejection chains and perturbations as well as other strategies to escape from local optima are discussed.

Chapter 5 introduces the basic GRASP as a semi-greedy multistart procedure with local search. Techniques for accelerating the basic procedure are pointed out. Probabilistic stopping criteria for GRASP are also discussed. The chapter concludes with a short introduction to the application of GRASP as a heuristic for multiobjective optimization.

Chapter 6 focuses on time-to-target plots (or runtime distributions) for comparing exponentially distributed runtimes, such as those for GRASP heuristics, and runtimes with general distributions, such as those for GRASP with path-relinking. Runtime distribution will be extensively used throughout this book to assess the performance of stochastic search algorithms.

Extended GRASP construction heuristics are covered in Chapter 7. The chapter begins with reactive GRASP and then covers topics such as probabilistic choice of the construction parameter, random plus greedy and sampled greedy constructions, construction by cost perturbation, and the use of bias functions in construction.

The chapter continues with the use of memory, learning, and the proximate optimality principle in construction, pattern-based construction, and Lagrangean GRASP.

Path-relinking is introduced in Chapter 8. The chapter provides a template for path-relinking and discusses its mechanics and implementation strategies. Other topics related to path-relinking are also discussed in this chapter. This includes how to deal with infeasibilities in path-relinking, how to randomize path-relinking, and external path-relinking and its relation to diversification.

The hybridization of GRASP with path-relinking is covered in Chapter 9. The chapter begins by providing motivation for hybridizing path-relinking with GRASP to provide GRASP with a memory mechanism. It then goes on to discuss elite sets and how they can be used as a way to connect GRASP and path-relinking. The chapter ends with a discussion of evolutionary path-relinking and restart mechanisms for GRASP with path-relinking heuristics.

The implementation of GRASP on parallel machines is the topic of Chapter 10. The chapter introduces two types of strategies for parallel implementation of GRASP: multiple-walk independent-thread and multiple-walk cooperative-thread strategies. It then goes on to illustrate these implementation strategies with three examples: the three-index assignment problem, the job shop scheduling problem, and the 2-path network design problem.

Continuous GRASP extends GRASP heuristics from discrete optimization to continuous global optimization. This is the topic of Chapter 11. After establishing the similarities and differences between GRASP for discrete optimization and continuous GRASP (or simply C-GRASP), the chapter describes the construction and local search phases of C-GRASP and concludes with several examples applying C-GRASP to multimodal box-constrained optimization.

The book concludes with Chapter 12 where four implementations of GRASP and GRASP with path-relinking are described in detail. These implementations are for the 2-path network design problem, the graph planarization problem, the unsplittable network flow problem, and the maximum cut problem.

Each chapter concludes with bibliographical notes.

Writing this book was certainly a long and arduous task, but most of all it has been an amazing experience. The many trips between Holmdel, Seattle, and Rio de Janeiro and the periods the authors spent visiting each other along the last six years have been gratifying and contributed much to fortify an already strong friendship. We had a lot of fun and we are very happy with the outcome of this project. We will be even happier if the readers appreciate reading and using this book as much as we enjoyed writing it.

Seattle, WA, USA                                        Mauricio G.C. Resende
Rio de Janeiro, RJ, Brazil                              Celso C. Ribeiro
May 2016

# Acknowledgments

Over the years, we have collaborated with many people on research related to topics covered in this book. We make an attempt to acknowledge all of them below, in alphabetical order, and apologize in case someone was omitted from this long list: James Abello, Vaneet Aggarwal, Renata Aiex, Daniel Aloise, Dario Aloise, Adriana Alvim, Diogo Andrade, Alexandre Andreatta, David Applegate, Aletéia Araújo, Aaron Archer, Silvio Binato, Ernesto Birgin, Isabelle Bloch, Maria Claudia Boeres, Maria Cristina Boeres, Julliany Brandão, Luciana Buriol, Vicente Campos, Suzana Canuto, Sergio Carvalho, W.A. Chaovalitwongse, Bruno Chiarini, Clayton Commander, Abraham Duarte, Alexandre Duarte, Christophe Duhamel, Sandra Duni-Ekişoğlu, João Lauro Facó, Djalma Falcão, Haroldo Faria Jr., Tom Feo, Eraldo Fernandes, Daniele Ferone, Paola Festa, Rafael Frinhani, Micael Gallego, Fred Glover, Fernando Carvalho-Gomes, José Fernando Gonçalves, José Luis González-Velarde, Erico Gozzi, Allison Guedes, William Hery, Michael Hirsch, Rubén Interian, David Johnson, Howard Karloff, Yong Li, X. Liu, David Loewenstern, Irene Loiseau, Abilio Lucena, Rafael Martí, Cristian Martínez, Simone Martins, Geraldo Mateus, Thelma Mavridou, Marcelo Medeiros, Rafael Melo, Claudio Meneses, Renato Moraes, Luis Morán-Mirabal, Leonardo Musmanno, Fernanda Nakamura, Mariá Nascimento, Thiago Noronha, Carlos Oliveira, Panos Pardalos, Luciana Pessoa, Alexandre Plastino, Leonidas Pitsoulis, Marcelo Prais, Fábio Protti, Tianbing Qian, Michelle Ragle, Sanguthevar Rajasekaran, Martin Ravetti, Vinod Rebello, Lucia Resende, Alberto Reyes, Caroline Rocha, Noemi Rodriguez, Isabel Rosseti, Jesus Sánchez-Oro, Andréa dos Santos, Ricardo Silva, Stuart Smith, Cid de Souza, Mauricio de Souza, Reinaldo Souza, Fernando Stefanello, Sandra Sudarksy, Franklina Toledo, Giorgio de Tomi, Gerardo Toraldo, Marco Tsitselis, Eduardo Uchoa, Osman Ulular, Sebastián Urrutia, Reinaldo Vallejos, Álvaro Veiga, Ana Viana, Dalessandro Vianna, Carlos Eduardo Vieira, Eugene Vinod, and Renato Werneck.

The authors are particularly indebted to Simone Martins for her careful revision of this manuscript. We are also very thankful to Fred Glover for kindly agreeing to write the foreword of this book.

The second author is grateful to Jiosef Fainberg, Julieta Guevara, Nelson Macu-lan, and Segyu Rinpoche for their friendship and support throughout the preparation of this book.

Finally, a special thanks goes to the artist Frances Stark for agreeing to let us use a reproduction of an image of her collage *I must explain, specify, rationalize, classify, etc.* that appears on page xv of this book. The text in this piece is taken from Witold Gombrowicz's novel *Ferdydurke*. As in the commentary by art historian Alex Kitnick (Kitnick, 2013), the work "has to do with beginnings, blank pages, and the question of artistic labor, but it is also concerned with how one arranges oneself in relation to language. The question here is less how to make a first mark than how to organize a set of information and desires in relation to one's own person." We feel that it perfectly illustrates the effort we made to collect, organize, explain, and convey as clearly as possible the fundamentals, principles, and applications of optimization by GRASP.

Seattle, WA, USA                                        Mauricio G.C. Resende
Rio de Janeiro, RJ, Brazil                                  Celso C. Ribeiro
May 2016

Another preface....without a preface I cannot possibly go on. I must explain, specify, rationalize, classify, bring out the root idea underlying all other ideas in the book, demonstrate and make plain the essential griefs and hierarchy of ideas which are here isolated and exposed... thus enabling the reader to find the work's head, legs nose, fingers and to prevent him from co___ __g a_ __telling me that I don't know what I'm __riving __nd that instead of marching forward __traig__ __rect like the great writers of all __ __s, I ___ ___ly revolving ridiculously on my ow___ ___ then shall the fundamental overall ang___ ___here art thou great-grandmother of all___ ___The deeper I dig, the more I explore an___ ___, the more clearly do I see that in re__ ___rimary, the fundamental grief is pure___ ___mply, in my opinion, the agony of bad ___ ___form, defective appearance, th___ ___phraseology, grimaces, faces...yes, thi___ ___igin, the source, the fount from wh___ ___ flow harmoniously all the other___ ___ts, follies, and afflictions without any exc___ ___ __whatever. Or perhaps it would be as well ___ ___asize that the primary and fundamenta___ ___ __ that born of the constraint of man by ___ ___ from the fact that we suffocate and s___ ___ __e narrow and rigid idea of ourselve___ ___rs have of us.

**Frances Stark**
b. 1967; Newport Beach, CA
*I must explain, specify, rationalize, classify, etc.*, 2008.
Acrylic, fiber-tipped pen, graphite pencil, inset laser print, and paper collage on paper

# Contents

# Chapter 1
# Introduction

In this first chapter, we introduce general optimization problems and the class of combinatorial optimization problems. As a motivation, we present a number of fundamental combinatorial optimization problems that will be revisited along the next chapters of this book. We also contrast exact and approximate solution methods and trace a brief history of approximate algorithms (or heuristics) from $A^*$ to meta-heuristics, going through greedy algorithms and local search. We motivate the reader and outline, chapter by chapter, the material in the book. Finally, we introduce basic notation and definitions that will be used throughout the book.

## 1.1 Optimization problems

In its most general form, an *optimization problem* can be cast as

$$\text{optimize } f(S) \tag{1.1}$$

subject to

$$S \in F, \tag{1.2}$$

where $F$ is a *feasible set* of solutions and $f$ is a real-valued *objective function* that associates each *feasible solution* $S \in F$ to its cost or value $f(S)$. In the case of a *minimization problem* we seek a solution minimizing $f(S)$, while in the case of a *maximization problem* we search for a solution that maximizes $f(S)$ over the entire domain $F$ of feasible solutions.

A *global optimum* of a minimization problem is a solution $S^* \in F$ such that $f(S^*) \leq f(S)$, $\forall S \in F$. Similarly, the global optimum of a maximization problem is a solution $S^* \in F$ such that $f(S^*) \geq f(S)$, $\forall S \in F$.

Optimization problems are commonly classified into two groups: those with continuous variables, that in principle can take any real value, and those represented by discrete variables, that can take only a finite or a countably infinite set of values.

© Springer Science+Business Media New York 2016
M.G.C. Resende, C.C. Ribeiro, *Optimization by GRASP*,
DOI 10.1007/978-1-4939-6530-4_1

The latter are called *combinatorial optimization problems* and reduce to the search for a solution in a finite (or, alternatively, countably infinite) set, which can typically be formed by binary or integer variables, permutations, paths, trees, cycles, or graphs. Most of this book is concerned with combinatorial optimization problems with a single objective function, although Section 5.6 briefly addresses approaches for multiobjective problems and Chapter 11 describes a method for continuous problems.

Combinatorial optimization problems and their applications abound in the literature and in real-life, as will be illustrated later in this book. As a motivation, some fundamental combinatorial optimization problems will be presented in the next section, together with examples of their basic applications. These problems will be revisited many times along the next chapters of this book. In particular, they will be formalized and discussed in detail in the next chapter, where we show that some combinatorial optimization problems are intrinsically harder to solve than others. By harder, we mean that state-of-the-art algorithms to solve them can be very expensive in terms of the computation time needed to find a global optimum or that only small problems can be solved in a reasonable amount of time.

Understanding the inner computational complexity of each problem is an absolute prerequisite for the identification and development of an appropriate, effective, and efficient algorithm for its solution.

## 1.2 Motivation

We motivate our introductory discussion with the description of six typical and fundamental combinatorial optimization problems.

### Shortest path problem

Suppose a number of cities are distributed in a region and we want to travel from city $s$ to city $t$. The distances between each pair of cities are known beforehand. We can either go directly from $s$ to $t$ if there is a road directly connecting these two cities, or start in $s$, traverse one or more cities, and end up in $t$. A path from $s$ to $t$ is defined to be a sequence of two or more cities that starts in $s$ and ends in $t$. The length of a path is defined to be the sum of the distances between consecutive cities in this path. In the *shortest path problem*, we seek, among all paths from $s$ to $t$, one which has minimum length.                                                                   ∎

### Minimum spanning tree problem

Suppose that a number of points spread out on the plane have to be interconnected. Once again, the distances between each pair of points are known beforehand. Some points have to be made pairwise connected, so as to establish a unique path between any two points. In the *minimum spanning tree problem*, we seek to determine which pairs of points will be directly connected such that the sum of the distances between the selected pairs is minimum. ∎

## Steiner tree problem in graphs

Assume that a number of terminals (or clients) have to be connected by optical fibers. The terminals can be connected either directly or using a set of previously located hubs at fixed positions. The distances between each pair of points (be them terminals or hubs) are known beforehand. In the *Steiner tree problem in graphs*, we look for a network connecting terminals and hubs such that there is exactly one unique path between any two terminals and the total distance spanned by the optical fibers is minimum. ∎

## Maximum clique problem

Consider the global friendship network where pairs of people are considered to be either friends or not. In the *maximum clique problem*, we seek to determine the largest set of people for which each pair of people in the set are mutual friends. ∎

## Knapsack problem

Consider a hiker who needs to pack a knapsack with a number of items to take along on a hike. The knapsack has a maximum weight capacity. Each item has a given weight and some utility to the hiker. If all of the items fit in the knapsack, the hiker packs them and goes off. However, the entire set of items may not fit in the knapsack and the hiker will need to determine which items to take. The *knapsack problem* consists in finding a subset of items with maximum total utility, among all sets of items that fit in the knapsack. ∎

## Traveling salesman problem

Consider a traveling salesman who needs to visit all cities in a given sales territory. The salesman must begin and end the tour in a given city and visit each other city in the territory exactly once. Since each city must be visited only once, a tour can be represented by a circular permutation of the cities. Assuming the distances between each pair of cities are known beforehand, the objective of the *traveling salesman problem* is to determine a permutation of the cities that minimizes the total distance traveled. ∎

## 1.3 Exact vs. approximate methods

An *exact method* or *optimization method* for solving an optimization problem is one that is guaranteed to produce, in finite time, a global optimum for this problem and a proof of its optimality, in case one exists, or otherwise show that no feasible solution exists. Globally optimal solutions are often referred to as exact optimal solutions. Among the many exact methods for solving combinatorial optimization problems, we find algorithmic paradigms such as cutting planes, dynamic programming, backtracking, branch-and-bound (together with its variants and extensions, such as branch-and-cut and branch-and-price), and implicit enumeration. Some of these paradigms can be viewed as tree search procedures, in the sense that they start from a feasible solution (which corresponds to the root of the tree) and carry out the search for the optimal solution by generating and scanning the nodes of a subtree of the solution space (whose nodes correspond to problem solutions).

Chapter 2 shows that efficient exact algorithms are not known (and are unlikely to exist) for a broad class of optimization problems classified as *NP*-hard. These problems are often referred to as *intractable*. Even though the size of the problems that can be solved to optimality (exactly) has been always increasing due to algorithmic and technological developments, there are problems (or problem instances) that are not amenable to be solved by exact methods. Other approaches, based on different paradigms, are needed to tackle such hard and large optimization problems.

As opposed to exact methods, *approximate methods* are those that provide feasible solutions that, however, are not necessarily optimal. Approximate methods usually run faster than exact methods. As a consequence, approximate methods are capable of handling larger problem instances than are exact methods. In this book, we use the terms *heuristic* and approximate method interchangeably.

Relevant work on heuristics or approximate algorithms for combinatorial optimization problems can be traced back to the origins of the field of Artificial Intelligence in the 1960s, with the development and applications of A$^*$ search.

*Constructive heuristics* are those that build a feasible solution from scratch. They are often based on greedy algorithms and their connections with matroid theory. Greedy algorithms and their extensions will be thoroughly studied in Chapter 3.

*Local search* procedures start from a feasible solution and improve it by successive small modifications until a solution that cannot be further improved is encountered. Although they often provide high-quality solutions whose values are close to those of optimal solutions, in some situations they can become prematurely trapped in low-quality solutions. Local search heuristics are explored in Chapter 4.

*Metaheuristics* are general high-level procedures that coordinate simple heuristics and rules to find high-quality solutions to computationally difficult optimization problems. Metaheuristics are often based on distinct paradigms and offer different mechanisms to go beyond the first solution obtained that cannot be improved by local search. They are among the most effective solution strategies for solving combinatorial optimization problems in practice and very frequently produce much better solutions than those obtained by the simple heuristics and rules they coordinate.

## 1.4 Metaheuristics

In this section, we recall the principles of some of the most used metaheuristics. These solution approaches have been instrumental and contributed to most developments and applications in the field of metaheuristics. Among them, we can cite genetic algorithms, simulated annealing, tabu search, variable neighborhood search (VNS), and greedy randomized adaptive search procedures (GRASP), with the last one being the focus of this book.

*Genetic algorithms* are search procedures based on the mechanics of evolution and natural selection. These algorithms evolve populations of solutions that are pairwise combined to generate offspring. Elements of the solution population are also submitted to mutation processes that create individuals with new characteristics. The most fit individuals in each generation are those that most likely survive and pass their characteristics to the individuals of the next generation. Several operators and strategies can be used to create the initial population and to implement the mechanisms of mating, reproduction, mutation, and selection. In general, the evolution process ends after a number of generations without improvement of the best individual. Although the original implementations of genetic algorithms were based almost exclusively on probabilistic choices for mating, reproduction, and mutation, modern versions incorporate developments from the field of optimization and heuristics, such as the use of greedy randomized algorithms to generate the initial population and local search to improve the characteristics of the offspring.

Annealing is the physical process of heating up a solid until it melts, followed by its cooling down until crystallization. The free energy of the solid is minimized in this process. Practical experiments show that slow cooling schemes lead to final states with lower energy levels. By establishing associations between the physical states of the solid submitted to the annealing process and the feasible solutions of an optimization problem, and between the free energy of the solid and the cost function to be optimized, the *simulated annealing* metaheuristic mimics this process for the solution of a combinatorial optimization problem. It can be seen as a form of controlled random walk in the space of feasible solutions. The method is very general, can be easily implemented and applied, and has the ability to find good approximate solutions whose value is often close to the optimal. It can even be proved to converge to the optimal solution under certain circumstances, although at the cost of unlimited computation time. In practice, it can require large computation times and its memoryless characteristic does not contribute to more effective iterations and, consequently, to the efficiency of the approach.

*Tabu search* is a metaheuristic that guides a local search procedure to explore the solution space beyond local optimality. Its basic principle consists in pursuing local search whenever it encounters a local optimum. At such points, instead of moving to an improving solution, the algorithm moves to the least deteriorating solution in the neighborhood of that local optimum, under the expectation that after some steps a better solution will be found. However, moving to a worse solution can lead to cycling, since the algorithm can return to the previous local optimum at the next or in a later iteration. To avoid cycling back to solutions already visited, tabu search makes use of a short-term memory which contains recently visited solutions or, more often and in more clever implementations, the attributes of the current solution which should not be changed in order to prevent cycling. This short-term memory of forbidden solutions or attributes is called a tabu list. More complete or sophisticated variants of the algorithm also make use of medium-term and long-term memories which are used to intensify the search in promising regions of the solution space, or to diversity the search towards new regions that have not been properly explored.

*Variable neighborhood search* (VNS) is a metaheuristic based on the exploration of multiple neighborhood definitions imposed on the same solution space. Each of its iterations has two main steps: shaking and local search. With shaking, a neighbor of the current solution is randomly generated. Local search is applied to the solution obtained by the shaking step. VNS systematically exploits the idea of neighborhood change, both in the search for local optima, as in the process of escaping from the valleys that contain them.

GRASP is an acronym for *greedy randomized adaptive procedures* and is among the most effective metaheuristics for solving combinatorial optimization problems. It is a multistart procedure, in which each iteration consists basically of two phases: construction and local search. The construction phase builds a feasible solution, whose neighborhood is investigated until a local minimum is found during the local search phase. The best overall solution is kept as the result. GRASP and VNS are somehow complementary, in the sense that randomization is applied in GRASP at the construction phase, while in VNS it is applied in the local search phase.

The remainder of this book is entirely devoted to the presentation of the main building blocks, algorithms, performance evaluation tools, case studies, and strategies for sequential and parallel implementations of GRASP for solving optimization problems.

## 1.5 Graphs: basic notation and definitions

As many of the combinatorial optimization problems studied in this book come from graph theory and its applications, we introduce in this section some basic notation and definitions that will be used throughout the book.

Given a set $V = \{1, \ldots, n\}$, we denote by $|V| = n$ the cardinality (i.e., the number of elements) of $V$. We define $2^V$ as the set formed by all subsets of $V$, including the empty set $\varnothing$ and the set $V$ itself.

A *graph* $G = (V, U)$ is defined by a set $V = \{1, \ldots, n\}$ of nodes and a set $U \subseteq V \times V$ of unordered pairs $(i, j)$ of nodes $i, j \in V$ called edges. Therefore, both pairs $(i, j)$ or $(j, i)$ can be used to represent the same edge between $i, j \in V$ in $U$. A graph is said to be *complete* if there is an edge in $U$ between any two distinct nodes $i, j \in V$. A graph can also be referred to as an *undirected graph*. A *path* $P_{st}(G)$ in an undirected graph $G$ from $s \in V$ to $t \in V$ is defined as a sequence of nodes $i_1, i_2, \ldots, i_{q-1}, i_q \in V$, where $i_1 = s$, $i_q = t$, and each edge $(i_k, i_{k+1}) \in U$, for any $k = 1, \ldots, q - 1$. The number of edges in this path is given by $q - 1$. A graph $G = (V, U)$ is said to be *connected* if there is at least one path $P_{st}(G)$ connecting every pair of nodes $s, t \in V$. A *subgraph* $G' = (V', U')$ of $G = (V, U)$ is such that for any pair of nodes $i, j \in V'$, edge $(i, j) \in U'$ if and only if $(i, j) \in U$, and therefore $V' \subseteq V$ and $U' \subseteq U$.

A *spanning tree* of a graph $G = (V, U)$ is a connected subgraph of $G$ with the same node set $V$ and whose edge set $U' \subseteq U$ has exactly $n - 1$ edges.

Given a graph $G = (V, U)$ and a subset $V'$ of its node set $V$, the graph $G(V') = (V', U')$ *induced* in $G$ by $V'$ has $U' = \{(i, j) \in U : i, j \in V'\}$ as its edge set.

A *clique* of a graph $G = (V, U)$ is a subset of nodes $C \subseteq V$ such that $(i, j) \in U$ for every pair of nodes $i, j \in C$, with $i \neq j$. Alternatively, we can say that $C$ is a clique if the graph $G(C)$ induced in $G$ by $C$ is complete. The size of a clique is defined to be its cardinality $|C|$. A subset $I \subseteq V$ of the nodes in $G$ is said to be an *independent set* or a *stable set* if every two vertices in $I$ are not directly connected by an edge, i.e., if $(i, j) \notin U$ for all $i, j \in I$ such that $i \neq j$.

A *directed graph* $G = (V, A)$ is defined by a set $V = \{1, \ldots, n\}$ of nodes and a set $A \subseteq V \times V$ of ordered pairs $(i, j)$ of nodes $i, j \in V$ called arcs. A *path* $P_{st}(G)$ in a directed graph $G$ from $s \in V$ to $t \in V$ is defined as a sequence of nodes $i_1, i_2, \ldots, i_{q-1}, i_q \in V$, where $i_1 = s$, $i_q = t$, and each arc $(i_k, i_{k+1}) \in A$, for any $k = 1, \ldots, q - 1$. A directed graph $G = (V, A)$ is said to be *strongly connected* if there is at least one path $P_{st}(G)$ connecting node $s$ to node $t$ and another path $P_{ts}(G)$ connecting node $t$ to node $s$, for every pair of nodes $s, t \in V$.

A *Hamiltonian path* in a directed or undirected graph is a path between two nodes that visits each node of the graph exactly once. A *Hamiltonian cycle* in a directed or undirected graph is a Hamiltonian path that is also a cycle, i.e., its extremities coincide. Every Hamiltonian cycle corresponds to a circular permutation of the nodes of the graph. A Hamiltonian cycle is also known as a *Hamiltonian tour* or, simply, as a *tour*.

## 1.6 Organization

In addition to this introductory chapter, this book contains another 11 chapters. Each chapter concludes with a section with bibliographical notes.

Chapter 2 introduces combinatorial optimization problems and their computational complexity. First, some fundamental problems are formulated and then basic concepts of the theory of computational complexity are introduced, with special emphasis on decision problems, polynomial-time algorithms, and *NP*-complete problems. The chapter concludes with a discussion of solution approaches for *NP*-hard problems, introducing constructive heuristics, local search, and, finally, metaheuristics.

Chapter 3 addresses the construction of feasible solutions. We begin by considering greedy algorithms and showing how they are related with matroid theory. We then consider adaptive greedy algorithms, which are a generalization of greedy algorithms. Next, we present semi-greedy algorithms that are obtained by randomizing greedy or adaptive greedy algorithms. The chapter concludes with a discussion of solution repair procedures.

Chapter 4 deals with local search. A local search method is one that starts from any feasible solution and visits a sequence of other (feasible or infeasible) solutions, until a feasible solution that cannot be further improved is found. Local improvements are evaluated with respect to neighbor solutions that can be obtained by slight modifications applied to the solution currently being visited. We introduce in this chapter the concept of solution representation, which is instrumental in the design and implementation of local search methods. We also define neighborhoods of combinatorial optimization problems and moves between neighbor solutions. We illustrate the definition of a neighborhood by a number of examples for different problems. Local search methods are introduced and different implementation issues are discussed, such as neighborhood search strategies, quick cost updates, and candidate list strategies.

Chapter 5 presents the basic structure of a greedy randomized adaptive search procedure (or, more simply, GRASP). We first introduce random and semi-greedy multistart procedures and show how solutions produced by these procedures differ. The hybridization of a semi-greedy procedure with a local search method within an iterative procedure constitutes a GRASP heuristic. Efficient implementation strategies are also discussed in this chapter, as well as probabilistic stopping criteria. The chapter concludes with a short introduction to the application of GRASP as a heuristic for multiobjective optimization.

Chapter 6 covers runtime distributions. Also called time-to-target plots, runtime distributions display on the ordinate axis the probability that an algorithm will find a solution at least as good as a given target value within a given running time, shown on the abscissa axis. They provide a very useful tool to characterize the running times of stochastic algorithms for combinatorial optimization problems and to compare different algorithms or strategies for solving a given problem. Accordingly, they have been widely used as a tool for algorithm design and comparison.

Chapter 7 considers enhancements, extensions, and variants of greedy randomized adaptive construction procedures such as Reactive GRASP, the probabilistic choice of the parameter used in the construction of restricted candidate lists, random plus greedy and sampled greedy constructions, cost perturbations, bias functions, using principles of intelligent construction based on memory and learning, proximate optimality and local search applied to partially constructed solutions, and pattern-based construction strategies using vocabulary building or data mining.

Chapter 8 introduces path-relinking, an important search intensification strategy. Being a major enhancement to heuristic search methods for solving combinatorial optimization problems, its hybridization with other metaheuristics has led to significant improvements in both solution quality and running times. In this chapter, we review the fundamentals of path-relinking, implementation issues and strategies, and the use of randomization in path-relinking.

Chapter 9 covers the hybridization of GRASP with path-relinking. Path-relinking is a major enhancement that adds a long-term memory mechanism to the otherwise memoryless GRASP heuristics. GRASP with path-relinking implements a long-term memory with an elite set of diverse high-quality solutions found during the search. In its most basic implementation, at each GRASP iteration the path-relinking operator is applied between the solution found by local search and a randomly selected solution from the elite set. The solution resulting from path-relinking is a candidate for inclusion in the elite set. In this chapter we examine elite sets, their integration with GRASP, the basic GRASP with path-relinking procedure, several variants of the basic scheme (including evolutionary path-relinking), and restart strategies for GRASP with path-relinking heuristics.

Chapter 10 introduces parallel GRASP heuristics. Parallel computers and parallel algorithms have been increasingly finding their way into metaheuristics. Most of the parallel implementations of GRASP found in the literature consist in either partitioning the search space or the iterations and assigning each partition to a processor. These implementations can be categorized as following the multiple-walk independent-thread approach, with the communication among processors during GRASP iterations being limited to the detection of program termination and gathering the best solution found over all processors. Parallel strategies for the parallelization of GRASP with path-relinking can follow not only the multiple-walk independent-thread but also the multiple-walk cooperative-thread approach, in which the processors share the information about the elite solutions they visited at previous iterations. This chapter covers multiple-walk independent-thread strategies, multiple-walk cooperative-thread strategies, and some applications of parallel GRASP.

Chapter 11 considers Continuous GRASP, or C-GRASP, which extends GRASP to the domain of continuous box-constrained global optimization. The algorithm searches the solution space over a dynamic grid. Each iteration of C-GRASP consists of two phases. In the construction (or diversification) phase, a greedy randomized solution is constructed. In the local search (or intensification) phase, a local search algorithm is applied, starting from the first phase solution, and a locally optimal solution is produced. A deterministic rule triggers a restart after each C-GRASP iteration. This chapter addresses the construction phase and the restart strategy and also presents a local search procedure. The chapter concludes with examples of continuous functions optimized with an implementation of C-GRASP.

The book concludes with Chapter 12, in which we consider four case studies, 2-path network design, graph planarization, unsplittable multicommodity flow, and maximum cut, to illustrate the application and the implementation of GRASP heuristics. The key point here is not to show numerical results or comparative statistics with other approaches but, instead, to show how to customize the GRASP metaheuristic for each particular problem.

## 1.7 Bibliographical notes

The shortest path problem and the minimum spanning tree problem that were used to motivate Section 1.2 have been addressed in many papers and textbooks (see Cormen et al. (2009) in particular). Although many references exist for the other problems discussed in this chapter, we refer the reader to Pardalos and Xue (1994) for the maximum clique problem, Martello and Toth (1990) for the knapsack problem, and Lawler et al. (1985), Gutin and Punnen (2002), and Applegate et al. (2006) for the traveling salesman problem. The Steiner tree problem in graphs appeared first in Hakimi (1971) and Dreyfus and Wagner (1972). See also Maculan (1987), Winter (1987), Goemans and Myung (1993), Hwang et al. (1992), Voss (1992), and Ribeiro et al. (2002).

The textbooks by Nilsson (1971; 1982) and Pearl (1985) are fundamental references on the origins, principles, and applications of A* search and heuristic search methods introduced in Section 1.3. Cormen et al. (2009) present a good coverage of greedy algorithms and an introduction to matroid theory. Pitsoulis (2014) offers a more in-depth coverage of matroids.

Yagiura and Ibaraki (2002) trace back the history of local search since the work of Croes (1958). Kernighan and Lin (1970) and Lin and Kernighan (1973) were among the first to propose local search algorithms for the graph partitioning and the traveling salesman problem, respectively. The book by Hoos and Stützle (2005) is a thorough study of the foundations and applications of stochastic local search.

Genetic algorithms and research in metaheuristics were pioneered in the book of Holland (1975). Other developments in genetic algorithms appeared in textbooks by Reeves and Rowe (2002), Goldberg (1989), and Michalewicz (1996), among others. The work on optimization by simulated annealing was pioneered by Kirkpatrick et al. (1983), with accounts of later developments and applications being found in textbooks by van Laarhoven and Aarts (1987) and Aarts and Korst (1989). The seminal papers of Glover (1989; 1990) established the fundamentals, extensions, and uses of tabu search. They provided solid foundations and originated most of the developments from where the field of metaheuristics flourished. An alternative approach based on virtually the same principle of using a short-term memory was independently proposed by Hansen (1986), in a method named steepest-ascent mildest-descent. The reader is also referred to the textbook by Glover and Laguna (1997). Variable neighborhood search (VNS) was proposed by Mladenović and Hansen (1997), followed by other reviews by Hansen and Mladenović (1999; 2002; 2003).

The fundamentals of GRASP were originally proposed by Feo and Resende (1989). This article was followed by many others proposing variants, extensions, hybridizations, and applications of GRASP. Extensive literature reviews containing late developments and applications appeared in papers by Feo and Resende (1995), Festa and Resende (2002; 2009a;b), Resende and Ribeiro (2003b; 2005a; 2010; 2014), Pitsoulis and Resende (2002), Ribeiro (2002), Resende and González-Velarde (2003), Resende et al. (2012), and Resende and Silva (2013).

The reader interested in other relevant, but less explored approaches such as ant colony optimization, iterated local search, scatter search, and particle swarm optimization, can look into the broad and ever evolving literature on the subject, in particular the handbooks edited by Reeves (1993), Glover and Kochenberger (2003), Gendreau and Potvin (2010), and Burke and Kendall (2005; 2014). Sörensen (2015) offers a critical view of the explosion of metaheuristic methods based on metaphors of some natural or man-made processes and concludes by pointing out some of the most promising research avenues for the field of metaheuristics.

The reader is referred to the books by Bondy and Murty (1976), West (2001), and Diestel (2010), among others, for notation, definitions, theoretical results, and algorithms in graphs.

# Chapter 2
# A short tour of combinatorial optimization and computational complexity

This chapter introduces combinatorial optimization problems and their computational complexity. We first formulate some fundamental problems already introduced in the previous chapter and then consider basic concepts of the theory of computational complexity, with special emphasis on decision problems, polynomial-time algorithms, and *NP*-complete problems. The chapter concludes with a discussion of solution approaches for *NP*-hard problems, introducing constructive heuristics, local search or improvement procedures and, finally, metaheuristics.

## 2.1 Problem formulation

An *instance* of a combinatorial optimization problem is defined by a finite *ground set* $E = \{1, \ldots, n\}$, a set of feasible solutions $F \subseteq 2^E$, and an objective function $f : 2^E \rightarrow \mathbb{R}$. In the case of a minimization problem, we seek a *global optimal solution* $S^* \in F$ such that $f(S^*) \leq f(S)$, $\forall S \in F$. The ground set $E$, the cost function $f$, and the set of feasible solutions $F$ are defined for each specific problem. Similarly, in the case of a maximization problem, we seek an optimal solution $S^* \in F$ such that $f(S^*) \geq f(S)$, $\forall S \in F$.

Each of the six problems considered in Section 1.2 is an example of a combinatorial optimization problem that can be formulated as described below.

### Shortest path problem – Revisited

Let $G = (V, A)$ be a directed graph, where $V$ is its set of nodes and $A$ its set of arcs. Each city corresponds to a node of this graph. The origin $s$ and destination $t$ are two special nodes in $V$. For every pair of cities $i, j \in V$ that are directly connected, let

© Springer Science+Business Media New York 2016
M.G.C. Resende, C.C. Ribeiro, *Optimization by GRASP*,
DOI 10.1007/978-1-4939-6530-4_2

$d_{ij}$ be the length of arc $(i, j) \in A$. Furthermore, let $P_{st}(G)$ be a path from $s$ to $t$ in $G$, defined as a sequence of nodes $i_1, i_2, \ldots, i_{q-1}, i_q \in V$, with $i_1 = s$ and $i_q = t$. The length of path $P_{st}(G)$ is given by $f(P_{st}(G)) = \sum_{k=1}^{q-1} d_{i_k, i_{k+1}}$.

Therefore, in the case of the shortest path problem, the ground set $E$ consists of the arc set $A$. The set of feasible solutions $F \subseteq 2^E$ is formed by all subsets of $E$ that correspond to paths from $s$ to $t$ in $G$. The objective of the shortest path problem is to find a path $P^* \in F$ that minimizes the objective function $f(P)$ over all paths $P \in F$ from $s$ to $t$ in $G$.

Consider the example in Figure 2.1, not drawn to scale. The shortest path from node 1 to node 6 is $1 - 2 - 3 - 6$ and is shown in red. The length of this path is $55 + 20 + 25 = 100$.     ∎

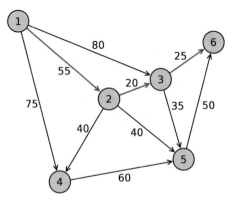

**Fig. 2.1** The shortest path from node 1 to node 6 is $1 - 2 - 3 - 6$ and is indicated by red arcs. This path has length 100.

## Minimum spanning tree problem – Revisited

Let $G = (V, U)$ be a graph, where the node set $V$ corresponds to points to be connected and its edge set $U$ is formed by unordered pairs of points $i, j \in V$, with $i \neq j$. Let $d_{ij}$ be the length (or weight) of edge $(i, j) \in U$. In addition, let $T(G) = (V, U')$ be a spanning tree of graph $G$, i.e., a connected subgraph of $G$ with the same node set $V$ and whose edge set $U' \subseteq U$ has exactly $n - 1 = |V| - 1$ edges. The total weight of tree $T(G)$ is given by $f(T(G)) = \sum_{(i,j) \in U'} d_{ij}$.

Therefore, in the case of the minimum spanning tree problem, the ground set $E$ consists of the set $U$ of edges. The set of feasible solutions $F \subseteq 2^E$ is formed by all subsets of edges that correspond to spanning trees of $G$. The objective of the minimum spanning tree problem is to find a spanning tree $T^* \in F$ such that $f(T^*) \leq f(T)$ for all $T \in F$.

Consider the example in Figure 2.2, not drawn to scale. The minimum spanning tree of the graph in this figure is shown in red and has five edges: $(1, 2), (2, 3), (2, 4), (3, 5)$, and $(3, 6)$. Its total weight is $55 + 20 + 40 + 35 + 25 = 175$.     ∎

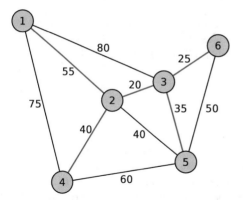

**Fig. 2.2** A minimum spanning tree is shown in red and has total length 175.

## Steiner tree problem in graphs – Revisited

Let $G = (V,U)$ be a graph, where the node set is $V = \{1,\ldots,n\}$ and the edge set $U$ is formed by unordered pairs of points $i, j \in V$, with $i \neq j$. Let $d_{ij}$ be the length of edge $(i, j) \in U$. Furthermore, let $T \subseteq V$ be a subset of terminal nodes that have to be connected. A *Steiner tree* $S = (V',U')$ of $G$ is a subtree of $G$ that connects all nodes in $T$, i.e., $T \subseteq V' \subseteq V$. The total length of the Steiner tree $S$ is given by $f(S) = \sum_{(i,j)\in U'} d_{ij}$.

Therefore, in the case of the Steiner tree problem in graphs, the ground set $E$ once again consists of the set $U$ of edges. The set of feasible solutions $F \subseteq 2^E$ is formed by all subsets of edges that correspond to Steiner trees of $G$. The objective of the Steiner tree problem in graphs is to find a Steiner tree $S^* \in F$ such that $f(S^*) \leq f(S)$ for all $S \in F$.

Consider the graph in the example in Figure 2.3, not drawn to scale. The terminal nodes are represented by circles, while the optional nodes correspond to squares. The minimum Steiner tree is shown in red and makes use of the optional nodes 5 and 6. Its total length is $5 + 5 + 5 + 5 + 5 = 25$.

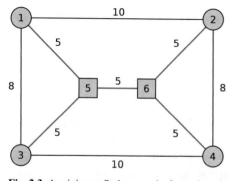

**Fig. 2.3** A minimum Steiner tree is shown in red and has total length 25.

The nonterminal, optional nodes in $V \setminus T$ that are effectively used to connect the terminal nodes in $T$ are called *Steiner nodes*. The Steiner tree problem in graphs reduces to a shortest path problem whenever $|T| = 2$. Furthermore, it reduces to a minimum spanning tree problem whenever $T = V$.    ■

### Maximum clique problem – Revisited

Let $G = (V, U)$ be a graph, where the node set $V = \{1, \ldots, n\}$ corresponds to the set of people in the world. For every two people $i, j \in V$, the edge $(i, j) \in U$ if and only if $i$ and $j$ are friends. A clique is a subset $C \subseteq V$ such that $(i, j) \in U$ for every pair $i, j \in C$ with $i \neq j$. The size of a clique is defined to be its cardinality, i.e., $f(C) = |C|$.

Therefore, in the case of the maximum clique problem, the ground set $E$ corresponds to the set of nodes $V$. The set of feasible solutions $F \subseteq 2^E$ is formed by all subsets of $V$ in which all nodes are pairwise adjacent. The objective of the maximum clique problem is to find a clique $C^* \in F$ such that $f(C^*) \geq f(C)$ for all $C \in F$.

Consider the example in Figure 2.4. The maximum clique is formed by the four nodes numbered 2, 3, 5, and 6. The edges connecting the nodes of this clique are illustrated in red.    ■

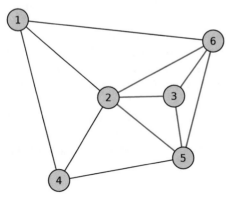

**Fig. 2.4** A maximum clique of size four formed by nodes 2, 3, 5, and 6 is illustrated with the edges connecting its nodes in red.

### Knapsack problem – Revisited

Let $b$ be an integer representing the maximum weight that can be taken in a hiker's knapsack and suppose the hiker has a set $I = \{1, \ldots, n\}$ of items to be placed in the knapsack. Let $a_i$ and $c_i$ be integer numbers representing, respectively, the weight and the utility of each item $i \in I$. Without loss of generality, we assume that each item fits in the knapsack by itself, i.e., $a_i \leq b$, for all $i \in I$. A subset of items $K \subseteq I$ is feasible if $\sum_{i \in K} a_i \leq b$. The utility of this subset is given by $f(K) = \sum_{i \in K} c_i$.

Therefore, in the case of the knapsack problem, the ground set $E$ consists of the set $I$ of items to be packed. The set of feasible solutions $F \subseteq 2^E$ is formed by all subsets of items $K \subseteq I$ for which $\sum_{i \in K} a_i \leq b$. The objective of the knapsack problem is to find a set of items $K^* \in F$ such that $f(K^*) \geq f(K)$ for all $K \in F$.

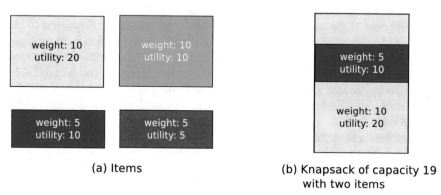

(a) Items

(b) Knapsack of capacity 19 with two items

**Fig. 2.5** Four items are candidates to be packed into a knapsack with maximum weight capacity of 19. The optimal solution packs the yellow and blue items with total weight of 15 and maximum total utility of 30.

Consider the example in Figure 2.5, where four items are available to a hiker to be placed in a knapsack of capacity 19. The weights of the yellow and green items are each equal to 10 and those of the blue and red items are both equal to 5. Therefore, only two of the four items fit together in the knapsack. The two heaviest items have utilities 20 and 10 to the hiker, while the two items with least weights have utilities 10 and 5. Since both large items (which combined have the highest utility to the hiker, but also the greatest weight) cannot be placed together in the knapsack, the hiker will need to select a large and a small item. Of each group, the hiker selects the one with maximum utility. The solution is shown on the right side of the figure. The yellow and blue items are placed in the knapsack and together they have a total weight of $5 + 10 = 15$ and a total maximum utility of $10 + 20 = 30$. ■

### Traveling salesman problem – Revisited

Let $V = \{1, \ldots, n\}$ be the set of cities a traveling salesman has to visit. If we consider the graph $G = (V, U)$ with non-negative lengths $d_{ij}$ associated with each existing edge $(i, j) \in U$, then any tour visiting each of the $n$ cities exactly once corresponds to a Hamiltonian cycle in $G$, i.e., a cycle in $G$ that visits every node exactly once. A feasible solution to the traveling salesman problem is a *tour* defined by a circular permutation $\pi = (i_1, i_2, \ldots, i_n, i_1)$ of the $n$ cities, with $i_j \neq i_k$ for every $j \neq k \in V$. This permutation is associated with the Hamiltonian cycle

$H = \{(i_1,i_2),(i_2,i_3),\dots,(i_{n-1},i_n),(i_n,i_1)\}$ in $G$, i.e., $(i_n,i_1) \in U$ and $(i_k,i_{k+1}) \in U$, for $k = 1,\dots,n-1$. The total length of this tour is given by $f(H) = \sum_{k=1}^{n-1} d_{i_k,i_{k+1}} + d_{i_n,i_1}$.

In case $G = (V,U)$ is not complete, for any pair of vertices $i, j \in V$ such that $(i,j) \notin E$, we can create a new edge $(i,j)$ with a sufficiently large length $d_{ij} = \infty$. Every Hamiltonian cycle in the original graph corresponds to a finite length Hamiltonian cycle in the resulting complete graph. Therefore, we can always assume, without loss of generality, that $G = (V,U)$ can be viewed as a complete graph.

In the case of the traveling salesman problem, the ground set $E$ consists of the edge set $U$. The set of feasible solutions $F \subseteq 2^E$ is formed by all subsets of edges corresponding to Hamiltonian cycles in $G$. The objective of the traveling salesman problem is to find a Hamiltonian cycle $H^* \in F$ such that $f(H^*) \le f(H)$ for all $H \in F$. Alternatively, we can view the ground set $E$ as formed by all vertices in $V$ and the set $F$ of feasible solutions formed by all circular permutations of the elements of the ground set.

Consider an instance of the traveling salesman problem, defined by the graph in Figure 2.6. Next, Figure 2.7 depicts in red a tour that visits cities $1 - 2 - 4 - 5 - 3 - 6 - 1$ in this order and has a total length of 325. The shortest tour, shown in red in Figure 2.8, visits cities $1 - 2 - 3 - 6 - 5 - 4 - 1$ in this order and has a total length of 285. ■

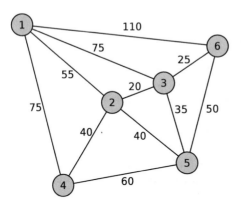

**Fig. 2.6** Instance of a traveling salesman problem with six cities.

## 2.2 Computational complexity

This book is mainly concerned with the solution of computationally difficult combinatorial optimization problems. In this section, we discuss the basics of the theory of computational complexity, which provides useful tools to differentiate between easy and hard combinatorial optimization problems.

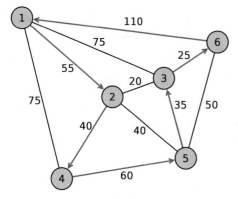

**Fig. 2.7** Example of a tour in this graph visiting cities $1 - 2 - 4 - 5 - 3 - 6 - 1$ in this order as illustrated in red, with a total length of 325.

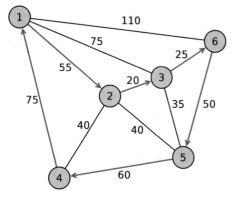

**Fig. 2.8** The shortest tour in this graph visits cities $1 - 2 - 3 - 6 - 5 - 4 - 1$ in this order as illustrated in red and has a total length of 285.

## 2.2.1 Polynomial-time algorithms

A *computational problem* is generally considered well-solved when there exists an efficient algorithm for its exact solution. Basically, *efficient algorithms* are those that are not too time-consuming and whose computation times do not grow excessively fast with the problem size. In fact, the rate of growth of the time taken by an algorithm is the main limitation for its use in practice. Algorithms with fast increasing computation times quickly become useless to solve real-world applications.

An algorithm is considered efficient and practically useful for solving a computational problem $\mathscr{P}$ whenever its time complexity (or its running time) grows as a polynomial function of the size of its input. In this context, the *input size* corresponds to the number of bits or to the amount of memory needed to store the data of any particular instance of $\mathscr{P}$. If we denote by $L$ the length of any reasonable encoding of the problem data, using an alphabet defined by at least two symbols, an algorithm $\mathscr{A}$ for this problem is said to run in polynomial time if there is a polynomial function $p$ such that the computation of $\mathscr{A}$ is bounded from above by $p(L)$. In this case, we say that algorithm $\mathscr{A}$ runs in time $O(p(L))$ if there exists an integer number $L_0$ and a real number $c$ such that the running time of algorithm $\mathscr{A}$ applied to any instance of problem $\mathscr{P}$ of size $L \geq L_0$ is less than or equal to $c \cdot p(L)$.

**Definition 2.1 (Polynomial algorithm).** An algorithm $\mathscr{A}$ for a problem $\mathscr{P}$ is said to be polynomial if there is a polynomial function $p$ such that $\mathscr{A}$ solves any instance of $\mathscr{P}$ in time bounded by $p(L)$, where $L$ is the length of a reasonable encoding of this instance.

Polynomial-time algorithms are known for two of the six combinatorial optimization problems used to motivate the discussion in this chapter. These are the shortest path problem and the minimum spanning tree problem. The other four – the Steiner tree problem in graphs, the maximum clique problem, the knapsack problem, and the traveling salesman problem – are typical examples of hard problems for which, to date, no polynomial-time algorithm is known. Hard optimization problems in this category are the main concern of this book and correspond to those that benefit from the solution methods presented here.

## 2.2.2 Characterization of problems and instances

In Section 2.1, we saw that an instance of a combinatorial optimization problem can be characterized by a finite ground set $E = \{1,\ldots,n\}$, a set $F$ of feasible solutions, and a cost function $f : F \to \mathbb{R}$ that associates a real value $f(S)$ with each feasible solution $S \in F$.

For each combinatorial optimization problem, we assume that its set $F$ of feasible solutions and its cost function $f(S)$ are implicitly given by two algorithms $\mathscr{A}_F$ and $\mathscr{A}_f$, respectively. Given an object $S \in 2^E$ and a set $P_F$ of parameters, the *recognition algorithm* $\mathscr{A}_F$ determines if the object $S$ belongs to $F$, the set of feasible solutions characterized by the parameters in $P_F$. Similarly, given a feasible solution $S \in F$ and a set of parameters $P_f$, the *cost function algorithm* $\mathscr{A}_f$ computes the cost function value $f(S)$. Therefore, we can say that each combinatorial optimization problem is characterized by the recognition algorithm $\mathscr{A}_F$ and the cost function algorithm $\mathscr{A}_f$, while each of its instances is associated with a pair of parameter sets $P_F$ and $P_f$. These concepts are illustrated below for some of the problems previously described in this chapter.

### Shortest path problem – Characterizing parameters and algorithms

In the case of the shortest path problem, the parameter set $P_F$ that establishes feasibility consists of the description of the directed graph $G = (V,A)$, where $V$ is its set of nodes and $A$ its set of arcs, together with the definition of the source and destination nodes $s,t \in V$. An object that is a candidate to be a feasible solution is defined by a subset $P$ of the arcs in $A$. The cost function parameter set $P_f$ is defined by the arc lengths $d_{ij}$, for every arc $(i,j) \in A$. The recognition algorithm $\mathscr{A}_F$ checks if $P$ corresponds to a path from $s$ to $t$ in $G$. Once the feasibility of a solution $P$ has been established (i.e., $P$ characterizes a path from $s$ to $t$), the cost function algorithm $\mathscr{A}_f$ adds up the lengths of all arcs in $P$ to compute the cost function value $f(P)$. ■

### Minimum spanning tree problem – Characterizing parameters and algorithms

In the minimum spanning tree problem, the parameter set $P_F$ that establishes feasibility consists exclusively of the description of the graph $G = (V,U)$ itself, where $V$ is its set of nodes and $U$ its set of edges. An object that is a candidate to be a feasible solution is defined by a subset $T$ of the edges in $U$. The cost function parameter set $P_f$ consists of the edge lengths $d_{ij}$, for every edge $(i,j) \in U$. The recognition algorithm $\mathscr{A}_F$ checks if $T$ corresponds to a spanning tree in $G$. Once the feasibility of a solution $T$ has been established (i.e., the graph induced in $G$ by the edge subset $T$ is connected and has exactly $|V| - 1$ edges), the cost function algorithm $\mathscr{A}_f$ adds up the lengths of all edges in $T$ to compute the cost function value $f(T)$. ■

### Maximum clique problem – Characterizing parameters and algorithms

In the case of the maximum clique problem, once again the parameter set $P_F$ that establishes feasibility consists exclusively of the description of the graph $G = (V,U)$, where $V$ is its set of nodes and $U$ its set of edges. An object that is a candidate to be a feasible solution is defined by a subset $C$ of the nodes in $V$. Since the cost of a feasible solution $C$ depends only on the number of nodes in $C$, no cost parameter exists and therefore $P_f = \varnothing$. The recognition algorithm $\mathscr{A}_F$ checks if $C$ corresponds to a clique in $G$. Once the feasibility of a solution $C$ has been established (i.e., every two nodes $i,j \in C$ are pairwise connected by an edge in $U$), the cost function algorithm $\mathscr{A}_f$ simply counts the number of nodes in $C$ to compute the cost function value $f(C)$. ■

### Knapsack problem – Characterizing parameters and algorithms

For the knapsack problem, the parameter set $P_F$ that establishes feasibility is defined by the maximum weight $b$ of the knapsack and by the weights $a_i$ of each item $i \in I$. An object that is a candidate to be a feasible solution is defined by a subset $B$ of the items in $I$. The cost function parameter set $P_f$ is defined by the utilities $c_i$, for each $i \in I$. The recognition algorithm $\mathscr{A}_F$ checks if $B$ corresponds to a feasible subset of items. Once the feasibility of a solution $B$ has been established (i.e., the sum of the weights of all items in $B$ is less than or equal to the maximum weight $b$), the cost function algorithm $\mathscr{A}_f$ adds up the utilities of all items in $B$ to compute the cost function value $f(B)$.                                                                                    ∎

**Traveling salesman problem – Characterizing parameters and algorithms**

In the case of the traveling salesman problem, the parameter set $P_F$ that establishes feasibility consists of the number $|V|$ of cities or vertices of the complete graph $G = (V, U)$. An object that is a candidate to be a feasible solution is characterized by a circular permutation of all cities in $V$ or by the set $H$ of edges in the associated Hamiltonian cycle in $G$. The cost function parameter set $P_f$ is defined by the distances $d_{ij}$, for every pair of cities $i, j \in V$, with $i \neq j$. The recognition algorithm $\mathscr{A}_F$ checks if $H$ corresponds to a Hamiltonian cycle in $G$. Once the feasibility of a solution $H$ has been established (i.e., $H$ characterizes a Hamiltonian cycle visiting every node of $G$ exactly once), the cost function algorithm $\mathscr{A}_f$ adds up the distances of all arcs in $H$ to compute the cost function value $f(H)$.                    ∎

### 2.2.3 One problem has three versions

A combinatorial optimization problem can therefore be alternatively stated in general as the following computational task:

**Definition 2.2 (Optimization problem).** Given representations for the parameter sets $P_F$ and $P_f$ for algorithms $\mathscr{A}_F$ and $\mathscr{A}_f$, respectively, find an optimal feasible solution.

The above formulation is usually referred to as the *optimization version* of the problem. However, if instead of finding an optimal solution itself, we are only interested in finding its value, then we have a more relaxed evaluation form of this problem:

**Definition 2.3 (Evaluation problem).** Given representations for the parameter sets $P_F$ and $P_f$ for algorithms $\mathscr{A}_F$ and $\mathscr{A}_f$, respectively, find the cost of an optimal feasible solution.

Under the reasonable assumption that $\mathscr{A}_f$ is a polynomial-time algorithm, which means that the cost of any solution can be efficiently computed, this *evaluation version* cannot be harder than the optimization version. This is so because once the optimization version of the problem has been solved and its optimal solution is known, its cost can be easily computed in polynomial time by the cost function algorithm $\mathscr{A}_f$.

A third version of a combinatorial optimization problem is particularly important in the context of complexity theory. The *decision version* (or *recognition version*) of a minimization problem amounts to a single question requiring a "yes" or "no" answer:

**Definition 2.4 (Decision version of a minimization problem).** Given representations for parameter sets $P_F$ and $P_f$ for algorithms $\mathscr{A}_F$ and $\mathscr{A}_f$, respectively, and an integer number $B$ that represents a bound, is there a feasible solution $S \in F$ such that $f(S) \leq B$?

Analogously, the decision version of a maximization problem asks for the existence of a feasible solution $S \in F$ whose cost $f(S)$ is greater than or equal to $B$. The decision version of a combinatorial optimization problem cannot be harder than its evaluation version. In fact, once the cost of an optimal solution has been obtained as the solution of the evaluation version, we can just compare it with the value of $B$ to give a "yes" or "no" answer to the decision version. We have therefore established a problem hierarchy, in which the decision version is not harder than the evaluation version that, in turn, is not harder than the optimization version.

**Maximum clique problem – Problem versions**

1. *Optimization version:* Given a graph $G = (V,U)$, find a maximum cardinality clique of $G$.
2. *Evaluation version:* Given a graph $G = (V,U)$, find the number of nodes in a maximum cardinality clique of $G$.
3. *Decision version:* Given a graph $G = (V,U)$ and an integer number $B$, is there a clique in $G$ with at least $B$ nodes? ∎

**Knapsack problem – Problem versions**

1. *Optimization version:* Given a set $I = \{1,\dots,n\}$ of items, integer weights $a_i$ and utilities $c_i$ associated with each item $i \in I$, and a maximum weight capacity $b$, find a subset $K^* \subseteq I$ of items such that $\sum_{i \in K^*} c_i = \max_{K \subseteq I}\{\sum_{i \in K} c_i : \sum_{i \in K} a_i \leq b\}$.
2. *Evaluation version:* Given a set $I = \{1,\dots,n\}$ of items, integer weights $a_i$ and utilities $c_i$ associated with each item $i \in I$, and a maximum weight capacity $b$, find $c^* = \max_{K \subseteq I}\{\sum_{i \in K} c_i : \sum_{i \in K} a_i \leq b\}$.
3. *Decision version:* Given a set $I = \{1,\dots,n\}$ of items, integer weights $a_i$ and utilities $c_i$ associated with each item $i \in I$, a maximum weight capacity $b$, and an integer $B$, is there $K \subseteq I$ such that $\sum_{i \in K} a_i \leq b$ and $\sum_{i \in K} c_i \geq B$? ∎

**Traveling salesman problem – Problem versions**

1. *Optimization version:* Given a complete graph $G = (V, U)$ with non-negative distances $d_{ij}$ between every pair of nodes $i, j \in V$, find a shortest Hamiltonian cycle in $G$.
2. *Evaluation version:* Given a complete graph $G = (V, U)$ with non-negative distances $d_{ij}$ between every pair of nodes $i, j \in V$, compute the length of a shortest Hamiltonian cycle in $G$.
3. *Decision version:* Given a complete graph $G = (V, U)$ with non-negative distances $d_{ij}$ between every pair of nodes $i, j \in V$ and an integer $B$, is there a Hamiltonian cycle in $G$ of length less than or equal to $B$? ∎

Under very reasonable assumptions, we can show that the three versions of any combinatorial problem have roughly the same computational complexity. If we have a polynomial-time algorithm to solve the decision version of a combinatorial problem, then in general we can also construct polynomial-time algorithms for solving the evaluation and the optimization versions.

As an example, consider the case of the asymmetric traveling salesman problem defined on a complete directed graph $G = (V, A)$, in which the distances $d_{ij}$ and $d_{ji}$ associated with a pair of arcs $(i, j) \in A$ and $(j, i) \in A$ are not necessarily the same. We first suppose that there exists an algorithm TSPDEC$(n, d, B)$ that solves the decision version of the traveling salesman problem. This algorithm provides the appropriate "yes" or "no" answer for any instance defined by $n$ cities, non-negative distances $d_{ij}$ for every pair of cities $i, j \in V = \{1, \dots, n\}$, with $i \neq j$, and an integer $B$. We also assume that algorithm TSPDEC$(n, d, B)$ runs in time $T(n)$.

Algorithm TSPOPT$(n, d)$, described in Figure 2.9, solves the optimization version of the asymmetric traveling salesman problem by repeatedly applying algorithm TSPDEC$(n, d, B)$ to slightly modified instances of its decision version. Lines 1 and 2 set initial values to $LB$ and $UB$ which are, respectively, trivial lower and upper bounds for the optimal solution value. Line 3 defines a sufficiently large value $BIG$ that will be used as a flag. The loop in lines 4 to 10 implements a binary search procedure that seeks the optimal solution value in the interval $[LB, UB]$. Line 4 interrupts the search as soon as an optimal solution is found, in which case both bounds $LB$ and $UB$ are equal to the optimal value. Line 5 asks for the existence of a solution whose length is less than or equal to $\lfloor (LB + UB)/2 \rfloor$. If there is one, then the upper bound $UB$ is reset to $\lfloor (LB + UB)/2 \rfloor$ in line 6. Otherwise, line 8 resets the lower bound $LB$ to $\lfloor (LB + UB)/2 \rfloor$. At the end of the execution of the loop in lines 4 to 10, the optimal solution value $LB = UB$ is saved to $OPT$ in line 11, providing a solution to the evaluation version.

Lines 12 to 24 compute the solution of the optimization version. The loop in lines 12 to 18 identifies the arcs that belong to an optimal Hamiltonian cycle. Lines 12 and 13 enumerate all ordered pairs of cities. In line 14, we save in $TMP$ the distance $d_{ij}$ associated with the arc $(i, j) \in A$, with $i, j \in V$ and $i \neq j$, and replace it with a sufficiently large value $BIG$ in line 15. Line 16 asks for the existence of a tour whose length is less than or equal to $OPT$ with the modified distance $d_{ij}$. If there is none, then arc $(i, j)$ must belong to the optimal solution and its length $d_{ij}$ is reset to the original value *TMP*. The arcs that have their lengths reset to *BIG* at the end of the loop in lines 12 to 18 are those that do not belong to an optimal solution. The optimal solution $S^*$ is initialized with the empty set in line 19. The loop in lines 20 to 24 builds an optimal tour. Lines 20 and 21 are used to enumerate all arcs or ordered pairs of cities. Line 22 inserts an arc $(i, j) \in A$ in the optimal solution if its length has not been reset to *BIG*, in which case it belongs to the optimal solution. Line 25 returns an optimal solution $S^*$ and its optimal value *OPT*, solving, respectively, the optimization and the evaluation versions.

```
begin TSPOPT(n, d);
1   LB ← 0;
2   UB ← n max_{i, j ∈ V : i ≠ j} {d_{ij}};
3   BIG ← UB + 1;
4   while UB ≠ LB do
5       if TSPDEC(n, d, ⌊(LB + UB)/2⌋) = "yes" then
6           UB ← ⌊(LB + UB)/2⌋;
7       else
8           LB ← ⌊(LB + UB)/2⌋;
9       end-if;
10  end-while;
11  OPT ← UB;
12  for j = 1, ..., n do
13      for i = 1, ..., n with i ≠ j do
14          TMP ← d_{ij};
15          d_{ij} ← BIG;
16          if TSPDEC(n, d, OPT) = "no" then d_{ij} ← TMP;
17      end-for;
18  end-for;
19  S* ← ∅;
20  for j = 1, ..., n do
21      for i = 1, ..., n with i ≠ j do
22          if d_{ij} ≠ BIG then S* ← S* ∪ {(i, j)};
23      end-for;
24  end-for;
25  return S*, OPT;
end TSPOPT.
```

**Fig. 2.9** Pseudo-code of algorithm TSPOPT$(n, d)$ for the optimization version of the traveling salesman problem based on the repeated execution of algorithm TSPDEC$(n, d, B)$ for the decision version.

Since algorithm TSPDEC$(n, d, B)$ runs in time $T(n)$, both the evaluation and the optimization versions can be solved in time $O(n^2 \cdot T(n))$, therefore within a polynomial factor of the time needed to solve the decision version. The binary search approach to solve the evaluation version can be extended to most problems under reasonable assumptions, while similar constructions are available for solving the optimization version by successive applications of an algorithm that solves the decision version.

### 2.2.4 The classes P and NP

We have seen in the previous section that the decision version of a combinatorial optimization problem amounts to a question that can be answered by either "yes" or "no":

SHORTEST PATH: Given a directed graph $G = (V,A)$, an origin node $s \in V$, a destination node $t \in V$, lengths $d_{ij}$ associated with every arc $(i, j) \in A$, and an integer $B$, is there a path from $s$ to $t$ in $G$ whose length is less than or equal to $B$?

SPANNING TREE: Given a graph $G = (V,U)$, a weight $d_{ij}$ associated with each edge $(i, j) \in U$, and an integer $B$, is there a spanning tree of $G$ such that the sum of the weights of its edges is less than or equal to $B$?

STEINER TREE IN GRAPHS: Given a graph $G = (V,U)$, lengths $d_{ij}$ associated with each edge $(i, j) \in U$, a subset $T \subseteq V$, and an integer $B$, is there a subtree of $G$ that connects all nodes in $T$ and such that the sum of its edge lengths is less than or equal to $B$?

CLIQUE: Given a graph $G = (V,U)$ and an integer $B$, is there a clique in $G$ with at least $B$ nodes?

KNAPSACK: Given a set $I = \{1,\ldots,n\}$ of items, integer weights $a_i$ and utilities $c_i$ associated with each item $i \in I$, a maximum weight capacity $b$, and an integer $B$, is there a subset of items $K \subseteq I$ such that $\sum_{i \in K} a_i \leq b$ and $\sum_{i \in K} c_i \geq B$?

TRAVELING SALESMAN PROBLEM (TSP): Given a set $V = \{1,\ldots,n\}$ of cities and non-negative distances $d_{ij}$ between every pair of cities $i, j \in V$, with $i \neq j$, and an integer $B$, is there a tour visiting every city of $V$ exactly once with length less than or equal to $B$?

Other examples of well-known computational problems that correspond to the decision versions of combinatorial optimization problems are

LINEAR PROGRAMMING: Given an $m \times n$ matrix $A$ of integer numbers, an integer $m$-vector $b$, an integer $n$-vector $c$, and an integer $B$, is there an $n$-vector $x \geq 0$ of rational numbers such that $A \cdot x = b$ and $c \cdot x \leq B$?

GRAPH COLORING: Given a graph $G = (V,U)$ and an integer $B$, is it possible to color the nodes of $G$ with at most $B$ colors, such that adjacent nodes receive different colors?

INDEPENDENT SET: Given a graph $G = (V,U)$ and an integer $B$, is there an independent set of nodes in $G$ (i.e., a subset of mutually nonadjacent nodes) with at least $B$ nodes?

INTEGER PROGRAMMING: Given an $m \times n$ matrix $A$ of integer numbers, an integer $m$-vector $b$, an integer $n$-vector $c$, and an integer $B$, is there an $n$-vector $x \geq 0$ of integer numbers such that $A \cdot x = b$ and $c \cdot x \leq B$?

In general, a *decision problem* is one that has only two possible solutions: either the answer "yes" or the answer "no." All the above decision versions of combinatorial optimization problems are decision problems. However, there are many other decision problems that have not been originally cast as optimization problems. Some examples are

HAMILTONIAN CYCLE: Given a graph $G = (V, U)$, is there a Hamiltonian cycle in $G$ visiting all its nodes exactly once?

GRAPH PLANARITY: Given a graph $G = (V, U)$, is it planar?

GRAPH CONNECTEDNESS: Given a graph $G = (V, U)$, is it connected?

SATISFIABILITY (SAT): Given $m$ disjunctive clauses $C_1, \ldots, C_m$ involving the Boolean variables $x_1, \ldots, x_n$ and their complements, is there a truth assignment of 0 (false) and 1 (true) values to these variables such that the formula $C_1 \wedge C_2 \wedge \cdots \wedge C_m$ is satisfiable?

Decision problems and the decision versions of optimization problems are closer to the prototype of computational problems studied by the theory of computation and play a very important role in complexity theory. Furthermore, since we have shown that the decision version of an optimization problem cannot be harder than the optimization version, if a decision problem cannot be solved in polynomial time, then its corresponding optimization version cannot be solved in polynomial time as well.

**Definition 2.5 (Class P).** A decision problem $\mathscr{P}$ belongs to the class $P$ if there exists an algorithm $\mathscr{A}$ that solves any of its instances in polynomial time.

In other words, the class $P$ is formed by "easy" decision problems that can be efficiently solved by polynomial-time algorithms. Some problems in this class among those we have already examined are SHORTEST PATH, SPANNING TREE, GRAPH CONNECTEDNESS, and LINEAR PROGRAMMING. For all of them, there are efficient algorithms that compute an exact "yes" or "no" answer in polynomial time.

Given a decision problem $\mathscr{P}$ and a "yes" instance $\mathscr{J}$, a *certificate* $c(\mathscr{J})$ is a string that encodes a solution and makes it possible to reach the "yes" answer for instance $\mathscr{J}$. A certificate is said to be *concise* if the length of its encoding is bounded from above by a polynomial in the amount of memory that is used to encode instance $\mathscr{J}$. With these definitions, we identify a broader class of decision problems.

**Definition 2.6 (Class NP).** A decision problem $\mathscr{P}$ belongs to the class $NP$ if there exists a certificate-checking algorithm $\mathscr{A}'$ such that, for any "yes" instance $\mathscr{J}$ of $\mathscr{P}$, there is a concise certificate $c(\mathscr{J})$ with the property that algorithm $\mathscr{A}'$ applied to instance $\mathscr{J}$ and certificate $c(\mathscr{J})$ reaches the answer "yes" in polynomial time.

For a problem to be in *NP*, it is not required that there exists an algorithm that computes an answer in polynomial time for every instance of this problem. All that is required is that there exists a concise certificate for any "yes" instance that can be checked for validity in polynomial time. The certificate-checking algorithm $\mathscr{A}'$ is usually a combination of the recognition algorithm $\mathscr{A}_F$ that checks for feasibility with the cost function algorithm $\mathscr{A}_f$ that computes the cost function value as defined in Section 2.2.1.

We next give examples of concise certificates and membership in *NP* for several combinatorial optimization problems.

### Maximum clique problem – Concise certificate and membership in *NP*

A certificate $c(\mathscr{J})$ for the maximum clique problem is an encoding of a possible list of nodes forming a clique. This certificate is concise, because it cannot have more than $|V|$ nodes. The certificate-checking algorithm is polynomial, since it amounts to first checking whether $c(\mathscr{J})$ corresponds to a subset of the nodes of the graph $G = (V, U)$, then verifying if there is an edge in $G$ for every pair of nodes in the certificate. This part corresponds to the application of the recognition algorithm $\mathscr{A}_F$ and is followed by the application of the cost function algorithm $\mathscr{A}_f$ that counts the number of nodes in the certificate and by the comparison with the integer parameter $B$. Therefore, the decision problem CLIQUE belongs to *NP*.     ∎

### Knapsack problem – Concise certificate and membership in *NP*

A certificate $c(\mathscr{J})$ for the knapsack problem is an encoding of a possible sequence of integer numbers representing a subset of the $n$ available items. Once again this certificate is concise, because it cannot have more than $n$ elements. The certificate-checking algorithm is polynomial and corresponds exactly to the recognition algorithm $\mathscr{A}_F$, since it amounts to adding up the weights of the items in $c(\mathscr{J})$ and comparing the total weight with the maximum weight capacity $b$. Next, the cost function algorithm $\mathscr{A}_f$ adds up the utilities of the items in $c(\mathscr{J})$ and their total utility is compared with the integer parameter $B$. Consequently, the decision problem KNAPSACK also belongs to *NP*.     ∎

### Traveling salesman problem – Concise certificate and membership in *NP*

A certificate $c(\mathscr{J})$ for the traveling salesman problem is an encoding of a possible permutation of the $n$ cities or nodes in the graph $G = (V, U)$. This certificate is also concise, because it must have exactly $|V|$ nodes. The certificate-checking algorithm is polynomial and corresponds to the recognition algorithm $\mathscr{A}_F$ since it amounts to checking if every city or node in the graph $G$ appears exactly once in the certificate. Finally, the cost function algorithm $\mathscr{A}_f$ adds up the lengths of the edges defined by the permutation established by certificate $c(\mathcal{J})$ and the total length of the tour is compared with the integer parameter $B$. Therefore, the decision problem TSP also belongs to *NP*. ∎

Examples of other decision problems in *NP* are STEINER TREE IN GRAPHS, GRAPH PLANARITY, GRAPH COLORING, INDEPENDENT SET, HAMILTO-NIAN CYCLE, SAT, and INTEGER PROGRAMMING.

To prove that a problem is in *NP*, one is not required to provide an efficient algorithm to compute the certificate $c(\mathcal{J})$ for any given instance $\mathcal{J}$. One has only to prove the existence of at least one concise certificate for each "yes" instance.

```
begin ND-01KSP(n, a, b, c, B);
1    for i = 1, . . . , n do
2        go to both X,Y;
3 X:     x_i ← 0;
4        go to Z;
5 Y:     x_i ← 1;
6 Z:     continue
7    end-for;
8    if ∑_{i=1}^{n} a_i x_i ≤ b and ∑_{i=1}^{n} c_i x_i ≥ B then return "yes";
end ND-01KSP.
```

**Fig. 2.10** Pseudo-code of the nondeterministic algorithm ND-01KSP$(n, a, b, c, B)$ that solves the decision version of the knapsack problem in polynomial time.

Nothing is required for the "no" instances: concise certificates should exist only for "yes" instances.

We now suppose that there exists a polynomial-time algorithm $\mathcal{A}$ for solving a decision problem $\mathcal{P}$ in P. In other words, algorithm $\mathcal{A}$ is able to provide the appropriate "yes" or "no" answer for every instance of $\mathcal{P}$. Therefore, the steps of algorithm $\mathcal{A}$ applied to a "yes" instance $\mathcal{J}$ can be represented as a string of polynomial size. This string is a concise certificate, since it can be checked in polynomial time to be a valid execution of $\mathcal{A}$. The existence of a concise certificate that can be checked in polynomial time for any "yes" instance $\mathcal{J}$ shows that $\mathcal{P}$ is also in *NP*. Therefore, whenever a decision problem $\mathcal{P} \in P$, it also holds that $\mathcal{P} \in NP$. In other words, $P \subseteq NP$.

We remark that the acronym *NP* stands for *nondeterministic polynomial*, and not for *nonpolynomial*, as it often appears erroneously in the literature. A *nondeterministic algorithm* can be seen as one that makes use of the same instructions used by (deterministic) algorithms, plus the special GO TO BOTH X,Y command. This instruction simultaneously transfers the execution flow of a computer program to two instructions labeled X and Y. This very powerful statement behaves as if it creates two parallel threads of the algorithm currently under execution, one continuing from the instruction labeled X and the other from that labeled Y. The repeated application of this command can create very strong algorithms with an unlimited level of parallelism. As an example, we consider the nondeterministic algorithm ND-01KSP$(n,a,b,c,B)$ in Figure 2.10 that solves KNAPSACK, i.e., the decision version of the knapsack problem.

The first part of the algorithm consists of the loop from line 1 to 7, which is used to create $2^n$ parallel execution threads. Line 1 is used to implement a loop exploring all variables $x_1, \ldots, x_n$, starting from $x_1$. Line 2 duplicates each currently active thread in the execution flow of the algorithm: line 3 sets $x_i$ to 0 in the first thread, while line 5 sets $x_i$ to 1 in the second. At the end of the execution of the loop in lines 1 to 7, there are $2^n$ threads of the algorithm, all of them running in parallel. Every variable $x_i$, for $i = 1, \ldots, n$, is set to 0 in $2^{n-1}$ threads and to 1 in the other $2^{n-1}$.

In the second part of the algorithm, each parallel thread verifies in line 8 if the solution $x_1, \ldots, x_n$ that it contains is feasible and if its total utility is greater than or equal to $B$. If so, then the algorithm returns "yes" and stops.

We observe that the first part of the algorithm is equivalent to the creation of $2^n$ concise certificates for KNAPSACK, each of them running in a different thread and corresponding to a different assignment of 0-1 values to variables $x_1, \ldots, x_n$. The second part of the algorithm running in each thread checks the certificate it stores and answers "yes" if it is valid. Since all possible certificates are explored in parallel, there will be always at least one thread that will answer "yes" for any "yes" instance. Once again, we observe that nothing is required for the "no" instances.

We say that a nondeterministic algorithm runs in polynomial time if the first thread to reach the "yes" answer runs in time polynomial in the size of the instance. Although the construction of parallel computers with an arbitrarily large number of processors (i.e., with unlimited parallelism) is unlikely at least in the near future, nondeterministic algorithms provide a very useful and powerful tool. In particular, they can be used to provide an alternative to Definition 2.6 of the class *NP*:

**Definition 2.7 (Class *NP*).** A decision problem $\mathscr{P}$ belongs to the class *NP* if and only if any of its "yes" instances can be solved in polynomial time by a nondeterministic algorithm.

In addition to containing all problems in $P$, the class *NP* also contains the decision versions of many optimization problems and arises very naturally in the study of the complexity of combinatorial optimization problems.

### 2.2.5 Polynomial transformations and NP-complete problems

Solving a computational problem often becomes easy as soon as we assume the existence of an efficient algorithm for solving a second problem, which is equivalent to the first.

**Definition 2.8 (Polynomial-time reduction).** Let $\mathscr{P}_1$ and $\mathscr{P}_2$ be two decision problems. We say that there is a *polynomial-time reduction* from problem $\mathscr{P}_1$ to $\mathscr{P}_2$ if and only if the first can be solved by an algorithm $\mathscr{A}_1$ that amounts to a polynomial number of calls to an algorithm $\mathscr{A}_2$ for solving problem $\mathscr{P}_2$.

As a consequence, if problem $\mathscr{P}_1$ polynomially reduces to $\mathscr{P}_2$ and there is a polynomial-time algorithm for $\mathscr{P}_2$, then there is also a polynomial-time algorithm for problem $\mathscr{P}_1$. A polynomial-time transformation is a special case of a polynomial-time reduction that is particularly relevant in the context of complexity theory.

**Definition 2.9 (Polynomial-time transformation).** Let $\mathscr{P}_1$ and $\mathscr{P}_2$ be two decision problems. We say that there is a *polynomial-time transformation* from problem $\mathscr{P}_1$ to problem $\mathscr{P}_2$ if an instance $\mathscr{J}_2$ of $\mathscr{P}_2$ can be constructed in polynomial time from any instance $\mathscr{J}_1$ of $\mathscr{P}_1$, such that $\mathscr{J}_1$ is a "yes" instance of $\mathscr{P}_1$ if and only if $\mathscr{J}_2$ is a "yes" instance of $\mathscr{P}_2$.

We give two examples of polynomial-time transformations between problems in *NP*. We first show that CLIQUE polynomially transforms to INDEPENDENT SET. Let an instance $\mathscr{J}_1$ of CLIQUE be defined by a graph $G = (V,U)$ and an integer $B$. Let $\bar{G} = (V,\bar{U})$ be the complement of $G$: for every pair of nodes $i,j \in V$, there is an edge $(i,j) \in \bar{U}$ if and only if the pair $i,j$ does not constitute an edge in $U$. Therefore, an instance $\mathscr{J}_2$ of INDEPENDENT SET defined by the complement of $G$ and the same integer $B$ can be constructed in time $O(|V|^2)$ such that $\mathscr{J}_1$ is a "yes" instance of CLIQUE if and only if $\mathscr{J}_2$ is a "yes" instance of INDEPENDENT SET (see Figure 2.11 for the illustration of the transformation from CLIQUE to INDEPENDENT SET).

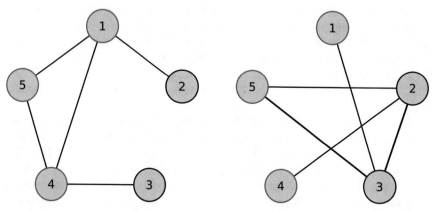

(a) Maximum clique in the original graph $G$      (b) Maximum independent set in $\bar{G}$

**Fig. 2.11** Polynomial transformation from CLIQUE to INDEPENDENT SET: Nodes 1, 4, and 5 form a maximum clique of the original graph $G$ in (a), while the same nodes correspond to a maximum independent set of the complement $\bar{G}$ of $G$ in (b). The instances defined by $G$ and $\bar{G}$ are "yes" instances for any $B \le 3$ and "no" instances for any $B > 3$.

As a second example, we show that HAMILTONIAN CYCLE polynomially transforms to TSP. Let an instance $\mathscr{J}_1$ of HAMILTONIAN CYCLE be defined by a directed graph $G = (V,A)$. First, associate a city of the TSP instance with every node in $V$. For every pair of cities $i, j \in V$, we set the distance $d_{ij} = 1$ if $(i, j) \in A$, $d_{ij} = 2$ otherwise. Next, set $B = |V|$. Therefore, an instance $\mathscr{I}_2$ of TSP can be constructed in time $O(|V|^2)$ such that $\mathscr{I}_1$ is a "yes" instance of HAMILTONIAN CYCLE if and only if $\mathscr{I}_2$ is a "yes" instance of TSP. In fact, there is a Hamiltonian cycle in $G$ if and only if there exists a tour of length $B = |V|$ visiting all cities corresponding to the nodes in $V$.

A polynomial-time transformation can be seen as a polynomial-time reduction that makes a single call to algorithm $\mathscr{A}_2$, exactly at the end of algorithm $\mathscr{A}_1$, and spends the rest of the time constructing the instance $\mathscr{I}_2$ of problem $\mathscr{P}_2$. With this definition, we can introduce a very important subclass of the problems in $NP$.

**Definition 2.10 (*NP*-complete problems).** A decision problem $\mathscr{P} \in NP$ is said to be *NP-complete* if every other problem in $NP$ can be transformed to it in polynomial time.

$NP$-complete problems have therefore a very important property: if there is a polynomial-time algorithm for any one of them, then there are also polynomial-time algorithms for all other problems in $NP$.

The proof that a problem is $NP$-complete involves two main steps: (1) proving that it is in $NP$ and (2) showing that all other problems in $NP$ can be transformed to it in polynomial time. The second part is often the hardest and is usually proved by showing that another problem already proved to be $NP$-complete is polynomially transformable to the problem on hand. SAT was the first problem to be explicitly proved to be $NP$-complete. Other $NP$-completeness results followed by polynomial transformations originating with SAT, showing that 3-SAT (a particular case of SAT, in which every clause has exactly three variables or their complements), KNAPSACK, CLIQUE, INDEPENDENT SET, TSP, STEINER TREE IN GRAPHS, INTEGER PROGRAMMING, HAMILTONIAN CYCLE, GRAPH COLORING, and GRAPH PLANARITY are also $NP$-complete, among many other problems.

We notice that special cases of $NP$-complete problems do not necessarily need to be $NP$-complete and hard to solve. As an example, we recall that CLIQUE is $NP$-complete. We now consider the complexity of the decision problem PLANAR CLIQUE, which corresponds to a restriction of CLIQUE to planar graphs. We know from graph theory that a planar graph cannot have a clique with five or more nodes. Therefore, a maximum clique of a planar graph $G = (V, U)$ can have at most four nodes and can be found by exhaustive search in time $O(|V|^4)$ or even faster by more specialized algorithms. Therefore, PLANAR CLIQUE is indeed a polynomially solvable special case of CLIQUE and, consequently, belongs to $P$.

As a second example, we consider KNAPSACK, the decision version of the knapsack problem that can be solved in nonpolynomial time $O(n \cdot b)$ by a straightforward dynamic programming algorithm. Suppose now that we are only interested in a restricted set of KNAPSACK instances, for which the maximum total weight is limited to $b \leq n$. In this case, the dynamic programming algorithm runs in $O(n^2)$ time. Therefore, although KNAPSACK is in $NP$, we conclude that KNAPSACK restricted to $b \leq n$ is in $P$.

### 2.2.6 NP-hard problems

We say that a problem $\mathscr{P}$ is *NP-hard* if all problems in *NP* are polynomially transformable to $\mathscr{P}$, but its membership to *NP* cannot be established. We notice that although $\mathscr{P}$ is certainly as hard as any problem in *NP*, in this case it does not qualify to be called *NP*-complete.

Besides its use to describe decision problems that are not proved to be in *NP*, the term *NP*-hard is also used to refer to optimization problems (which are certainly not in *NP*, since they are not decision problems) whose decision versions are *NP*-complete.

For example, we can say that the maximum clique problem, the knapsack problem, and the traveling salesman problem introduced as combinatorial optimization problems in Section 1.2 are all *NP*-hard, since the decision problems CLIQUE, KNAPSACK, and TSP are *NP*-complete, respectively.

### 2.2.7 The class co-NP

A problem $\bar{\mathscr{P}}$ is said to be the *complement* of problem $\mathscr{P}$ if every "yes" instance of $\bar{\mathscr{P}}$ is a "no" instance of $\mathscr{P}$ and vice-versa. We recall the definition of the CLIQUE decision problem:

CLIQUE: Given a graph $G = (V,U)$ and an integer $B$, is there a clique in $G$ with at least $B$ nodes?

It is easy to show that CLIQUE belongs to *NP*, since every "yes" instance has a concise certificate, defined by a list with at least $B$ nodes of $G$. We now consider the following complementary version of the same problem:

CLIQUE COMPLEMENT: Given a graph $G = (V,U)$ and an integer $B$, is it true that there is no clique in $G$ with at least $B$ nodes?

It is clear that every "yes" instance of CLIQUE is a "no" instance of CLIQUE COMPLEMENT and vice-versa. However, there is no proof to date that CLIQUE COMPLEMENT belongs to *NP*, since the only known strategy for proving that there is no clique with $B$ or more nodes consists in listing all cliques in $G$, counting the number of nodes in each of them, and verifying that any of them has fewer than $B$ nodes. This clique list is indeed a certificate, but it is not concise since its has exponential length.

We now consider the case of a problem that belongs to $P$:

SPANNING TREE: Given a graph $G = (V,U)$, weights $d_{ij}$ associated with every edge $(i,j) \in U$, and an integer $B$, is there a spanning tree of $G$ whose length is less than or equal to $B$?

The same algorithm that solves any "yes" or "no" instance of SPANNING TREE can also be used to solve SPANNING TREE COMPLEMENT:

> SPANNING TREE COMPLEMENT: Given a graph $G = (V, U)$, weights $d_{ij}$ associated with every edge $(i, j) \in U$, and an integer $B$, is it true that there is no spanning tree of $G$ whose length is less than or equal to $B$?

Since a polynomial algorithm that solves any problem in $P$ can also be used to solve its complement in polynomial time by simply replacing any "yes" answer to the former by a "no" answer to the latter and vice-versa, we can conclude that the complement of any problem in $P$ is also in $P$.

This same argument cannot be used to prove that the complement of a problem in $NP$ is also in $NP$. This leads to the definition of a new complexity class:

**Definition 2.11 (Class co-NP).** A decision problem $\mathscr{P}$ belongs to the class co-NP if its complement is in $NP$.

### 2.2.8 Pseudo-polynomial algorithms and strong NP-completeness

**Definition 2.12 (Pseudo-polynomial algorithms).** An algorithm $\mathscr{A}$ for a problem $\mathscr{P}$ is said to be *pseudo-polynomial* if there is a polynomial function $p$ such that $\mathscr{A}$ solves any instance of $\mathscr{P}$ in time bounded by $p(L, M)$, where $L$ and $M$ are, respectively, the length of a reasonable encoding of this instance and the largest integer appearing in this instance.

We observe that whenever the largest integer $M$ appearing in any instance of a problem solvable by a pseudo-polynomial algorithm is bounded by a polynomial function in the size of the instance, then algorithm $\mathscr{A}$ becomes polynomial. As an example, we recall that KNAPSACK can be solved in pseudo-polynomial time $O(n \cdot b)$ by a dynamic programming algorithm. This gives rise to a very efficient algorithm for the case where the maximum weight capacity $b$ is bounded and small. For instance, this dynamic programming algorithm runs in time $O(n^2)$ whenever $b \leq n$.

**Definition 2.13 (Strongly NP-complete problems).** A problem $\mathscr{P}$ is said to be *strongly NP-complete* if it remains $NP$-complete even if there is a polynomial function $p$ such that the largest integer $M$ appearing in each of the instances of $\mathscr{P}$ is bounded by $p(L)$, where $L$ is the length of a reasonable encoding of the instance.

TSP, CLIQUE, and HAMILTONIAN CYCLE are examples of strongly $NP$-complete problems. On the other hand, KNAPSACK is a typical example of a problem that is not strongly $NP$-complete, since it can be solved in polynomial time whenever the maximum capacity $b$ is bounded by a polynomial function on the number $n$ of variables.

## 2.2.9 PSPACE and the polynomial hierarchy

We have discussed up to this point the complexity of decision problems exclusively in terms of the computational time or the number of operations needed for their solution.

However, we can also consider the computational requirements in terms of the amount of space or memory that is needed for solving a decision problem. With this idea in mind, we present the definition of a new class of decision problems:

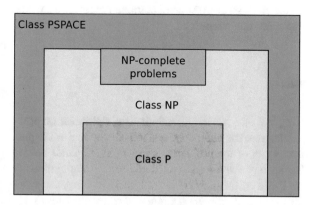

**Fig. 2.12** Complexity classes and the polynomial hierarchy.

**Definition 2.14 (Class *PSPACE*).** A decision problem $\mathscr{P}$ belongs to the class *PSPACE* if there exists an algorithm $\mathscr{A}$ that solves any of its instances using an amount of space (or memory) that is bounded by a polynomial in the length of its input.

Any algorithm that takes a polynomial amount of time to be solved cannot consume more than a polynomial amount of space, since it cannot write more than a fixed number of symbols (or words) at any of its operations. Therefore, it is clear that $P \subseteq PSPACE$.

However, what can we say about the relationship between the classes *NP* and *PSPACE*? In fact, we can prove that $NP \subseteq PSPACE$ also holds, although this might seem unexpected at first sight. We first notice that any nondeterministic algorithm for solving a decision problem $\mathscr{P} \in NP$ can be simulated by a deterministic algorithm that generates all possible concise certificates for any of its "yes" instances, one after the other. Furthermore, since $\mathscr{P} \in NP$, this deterministic algorithm can check each of these certificates in polynomial time, using an amount of space that is bounded by a polynomial in the length of the input of $\mathscr{P}$. Although this deterministic algorithm will run in exponential time, the total amount of space will be polynomial, since each certificate can be erased after it is checked and the space it occupied can be freed and reused for checking the next certificate. Therefore, $NP \subseteq PSPACE$. Using a similar argument, we can also prove that $co\text{-}NP \subseteq PSPACE$.

We conclude this section with Figure 2.12, which depicts the basic polynomial hierarchy and complexity classes. The preceding discussion has shown that any decision problem that can be solved in polynomial time by either a deterministic or a nondeterministic algorithm (or, alternatively, sequentially or in parallel), can also be solved using a polynomial amount of space. Therefore, even problems that take an exponential amount of time can be solved in polynomial space. Since polynomiality is considered as a limitation for any scarce resource such as time or space, we can say that time requirements become critical (i.e., superpolynomial) before space does. This observation supports the consideration of time as the main and critical scarce resource considered in the analysis and design of computer algorithms, which in practice very rarely involve space considerations.

## 2.3 Solution approaches

There are many more *NP*-complete and *NP*-hard combinatorial optimization problems than those presented to illustrate the main concepts introduced in this chapter. In fact, we can say that the majority of the problems of practical relevance belong to these classes. The fact that such problems are considered to be computationally intractable does not preclude the need for their solution. In addition to general superpolynomial exact methods to solve them, a great amount of research is devoted to identifying special cases or situations that can be solved exactly in reasonable time, or to developing approximate algorithms that are able to efficiently find high-quality solutions. Some of these approaches are quickly discussed below:

1. *Superpolynomial-time exact algorithms:* Theoretical developments in polyhedral theory, combined with efficient algorithm design and data structures and advances in computer hardware, have made it possible to solve even very large instances of *NP*-complete and *NP*-hard problems. Methods such as *branch-and-bound* and *branch-and-cut* are routinely applied to exactly solve large instances in affordable computation time. Such strategies cannot be discarded, in particular in the case of real-life instances whose sizes are often limited in practice.

2. *Pseudo-polynomial algorithms:* These algorithms form a subclass of superpolynomial-time algorithms. Pseudo-polynomial algorithms can be very efficient in practice whenever the maximum integer appearing in any instance of a given problem is small. We have already noticed that since KNAPSACK can be solved in pseudo-polynomial time $O(n \cdot b)$ by dynamic programming, this algorithm becomes very efficient for the case where the maximum weight capacity $b$ is bounded and small. Pseudo-polynomial algorithms can therefore become very practical and attractive in some situations, in spite of the fact that they are, essentially, superpolynomial-time algorithms.

3. *Polynomially solvable special cases:* It is often the case that although the general formulation of some specific problem is *NP*-complete or *NP*-hard, interesting or practical instances can be solved exactly in polynomial time. If one is interested

in solving exclusively these special instances, the fact that the general problem is intractable is less relevant and exact approaches can be used. Some examples follow:

- We have seen that CLIQUE is *NP*-complete. However, if one considers only planar graphs $G = (V,U)$, the special, restricted case of CLIQUE in planar graphs (or PLANAR CLIQUE) can be exactly solved by exhaustive enumeration in polynomial time $O(|V|^4)$ or even faster by direct application of Kuratowski's theorem.
- Although SAT is *NP*-complete (as is 3-SAT), its special case 2-SAT in which each clause has exactly two literals (a literal is a Boolean variable or its complement) can be solved exactly in polynomial time.

4. *Approximation algorithms:* These are algorithms that build feasible solutions that are not necessarily optimal, but whose objective function value can be shown to be within a guaranteed difference from the exact optimal value. Although in some cases this gap can be reasonable, for most problems it can be quite large.

5. *Heuristics:* A heuristic is essentially any algorithm that provides a feasible solution for a given problem, without necessarily providing a guarantee of performance in terms of solution quality or computation time. Heuristic methods can be classified into three main groups:

- *Constructive* heuristics are those that build a feasible solution from scratch. Greedy and semi-greedy algorithms, to be introduced in Chapter 3, are examples of constructive heuristics.
- *Local search* or *improvement* procedures start from a feasible solution and improve it by successive small modifications until a locally optimal solution is found. Although they provide high-quality solutions close to the optimum in many cases, in some situations they can become prematurely stuck in low-quality locally optimal solutions. Local search heuristics and their variants will be explored in Chapter 4.
- *Metaheuristics* are general high-level procedures that coordinate simple heuristics and rules to find good-quality solutions to computationally difficult optimization problems. Among them, we find simulated annealing, tabu search, greedy randomized adaptive search procedures, genetic algorithms, scatter search, variable neighborhood search, ant colony optimization, and others. Metaheuristics are based on distinct paradigms and offer different mechanisms to escape from locally optimal solutions (as opposed to greedy algorithms or local search methods). They are among the most effective solution strategies for solving combinatorial optimization problems in practice and very often produce much better solutions than those obtained by the simple heuristics and rules they coordinate. Metaheuristics have been applied to a wide array of academic and real-world problems. The customization (or instantiation) of a metaheuristic to a given problem yields a heuristic for this problem.

## 2.4 Bibliographical notes

Fundamental references for the shortest path problem, the minimum spanning tree problem, the maximum clique problem, the knapsack problem, the traveling salesman problem, and the Steiner problem in graphs that were revisited and formulated in Section 1.2 have been already presented in Section 1.7.

The foundations of the theory of computational complexity appeared in Cobham (1964) and Edmonds (1965; 1975), where informal references to $P$, $NP$, and related concepts are made. The landmark reference on the theory of $NP$-completeness is the seminal paper of Cook (1971), in which the author proved that SAT and 3-SAT are $NP$-complete. This work was closely followed by that of Karp (1972), in which its consequences were discussed and explored, leading to results establishing the $NP$-completeness of several other problems. A tutorial on the theory of $NP$-completeness was presented by Karp (1975). A discussion about strong $NP$-completeness first appeared in Garey and Johnson (1978).

Garey and Johnson (1979) is the most influential textbook on computational complexity theory. It introduced the theory of $NP$-completeness and computer intractability. The exposition and the basic notions of computational complexity presented in Section 2.2 follow closely the textbook by Papadimitriou and Steiglitz (1982). These ideas were further developed in Papadimitriou (1994) and Yannakakis (2007).

Accounts of integer programming methods that were cited in Section 2.3, such as branch-and-bound and branch-and-cut, can be found in textbooks by Schrijver (1986), Nemhauser and Wolsey (1988), Wolsey (1998), and Bertsimas and Weismantel (2005), among others. The pseudo-polynomial dynamic programming algorithm for the knapsack problem appeared in many references, in particular in the textbook of Martello and Toth (1990). It is well known that the stable set problem, the maximum clique problem, the chromatic number problem, and the clique cover problem are $NP$-complete for general graphs (Garey and Johnson, 1979). However, Grötschel et al. (1984) showed that the weighted versions of these problems can be solved in polynomial time for perfect graphs. Kuratowski's theorem was originally published in Kuratowski (1930). Krom (1967) described the first polynomial-time algorithm for 2-SAT. Early discussions about approximation algorithms appeared in Johnson (1974) and Garey and Johnson (1976). The reader is also referred to the textbooks of Vazirani (2001) and Williamson and Shmoys (2011). The first fit decreasing algorithm for bin packing is a classical example of an approximation algorithm, guaranteeing that no packing it generates will use more than 11/9 times the optimal number of bins (Johnson, 1973).

As noticed in the previous chapter, the textbooks by Nilsson (1971; 1982) and Pearl (1985) are fundamental references on the origins, principles, and applications of A* and other heuristic search methods. Cormen et al. (2009) presented a good coverage of greedy algorithms. Hoos and Stützle (2005) report in detail the foundations and applications of stochastic local search, while Michelis et al. (2007) discuss theoretical aspects of local search.

Glover and Kochenberger (2003) and Gendreau and Potvin (2010) collected thorough and complete accounts of metaheuristics, with a large coverage of the subject and detailed chapters about each of them. Other tutorials can also be found in Reeves (1993) and Burke and Kendall (2005; 2014). Some books provide detailed accounts of individual metaheuristics, see, e.g., van Laarhoven and Aarts (1987) and Aarts and Korst (1989) for simulated annealing, Glover and Laguna (1997) for tabu search, and Michalewicz (1996) and Goldberg (1989) for genetic algorithms. Previous surveys and tutorials about greedy randomized adaptive search procedures and their extensions and applications were authored by Feo and Resende (1995), Festa and Resende (2002; 2009a;b), Ribeiro (2002), Pitsoulis and Resende (2002), and Resende and Ribeiro (2003b; 2005a;b; 2010).

# Chapter 3
# Solution construction and greedy algorithms

This chapter addresses the construction of feasible solutions. We begin by considering greedy algorithms and show their relationship with matroids. We then consider adaptive greedy algorithms, a generalization of greedy algorithms. Next, we present semi-greedy algorithms, obtained by randomizing adaptive greedy algorithms. The chapter concludes with a discussion of solution repair procedures.

## 3.1 Greedy algorithms

As we saw in Chapter 2, a feasible solution $S$ of a combinatorial optimization problem is a subset of the elements of the ground set $E = \{1, \ldots, n\}$. Since certain subsets of ground set elements can lead to infeasibilities, by definition a feasible solution cannot contain any such subset. If $c_i$ denotes the contribution to the objective function value of ground set element $i \in E$, then we assume in this discussion that the objective function value of a solution $S$ is $f(S) = \sum_{i \in S} c_i$.

Many algorithms for combinatorial optimization problems build a solution incrementally from scratch, where at each step, a single ground set element is added to the partial solution under construction. A ground set element to be added at each step cannot be such that its combination with one or more previously added elements leads to an infeasibility. We call such an element *feasible* and denote by $\mathscr{F}$ the set of all feasible elements at the time a given step is performed. Since the set of candidate elements $\mathscr{F}$ may contain more than one element, an algorithm designed to build a feasible solution for some problem must have a mechanism to select the next feasible ground set element from $\mathscr{F}$ to be added to the partially built solution under construction.

From among all yet unselected feasible elements, a *greedy algorithm* for minimization always chooses one of least cost. Figure 3.1 shows the pseudo-code of a greedy algorithm. The solution $S$ to be constructed and its cost $f(S)$ are initialized to $\varnothing$ and 0, respectively, in lines 1 and 2. In line 3, the set $\mathscr{F}$ of candidate elements is initialized with all feasible ground set elements. The construction of the solution is

© Springer Science+Business Media New York 2016
M.G.C. Resende, C.C. Ribeiro, *Optimization by GRASP*,
DOI 10.1007/978-1-4939-6530-4_3

```
begin GREEDY;
1   S ← ∅;
2   f(S) ← 0;
3   𝓕 ← {i ∈ E : S∪{i} is not infeasible};
4   while 𝓕 ≠ ∅ do
5       i* ← argmin{c_i : i ∈ 𝓕};
6       S ← S∪{i*};
7       f(S) ← f(S) + c_{i*};
8       𝓕 ← {i ∈ 𝓕\{i*} : S∪{i} is not infeasible};
9   end-while;
10  return S, f(S);
end GREEDY.
```

**Fig. 3.1** Pseudo-code of a greedy algorithm for a minimization problem.

done in the while loop in lines 4 to 9, ending when $\mathcal{F}$ becomes empty. In line 5, the feasible ground set element $i^*$ having least cost is selected. Then, in lines 6 and 7, respectively, the solution under construction and its cost are updated to account for the inclusion of $i^*$ in the solution under construction. In line 8, the set $\mathcal{F}$ is updated, taking into account that $i^*$ is now part of solution $S$. Solution $S$ and its cost are returned in line 10.

The algorithm shown in Figure 3.1 is devised for minimization problems. For the case of maximization, the argmin operator in line 5 of the pseudo-code is simply replaced by argmax, which selects a candidate element of maximum cost. We next show examples of greedy algorithms for some combinatorial optimization problems.

**Minimum spanning tree problem - Greedy algorithm**

Recall from Chapter 2 that in the minimum spanning tree problem, the ground set is the set of edges $U$ and $d_{ij}$ is the length of edge $(i, j) \in U$. A greedy algorithm for the minimum spanning tree problem is shown in Figure 3.2. The solution $S$ to be constructed and its cost $f(S)$ are initialized to $\varnothing$ and 0, respectively, in lines 1 and 2. The set of feasible ground set elements is initialized in line 3 with all the edges in $U$. A feasible edge of least length is selected in line 5 and added to the spanning tree in line 6, with the length of the partial tree being updated in line 7. All edges whose inclusion in the current solution would create a cycle (i.e., an infeasibility) are removed from the set $\mathcal{F}$ of feasible candidate elements in line 8. The solution $S$ and its cost are returned in line 10.

This greedy algorithm for the minimum spanning tree problem is known as Kruskal's algorithm. As we shall see later in this chapter, this algorithm will always produce an optimal solution for this problem.                                    ∎

```
begin GREEDY-MST;
1   S ← ∅;
2   f(S) ← 0;
3   ℱ ← U;
4   while ℱ ≠ ∅ do
5       (i*, j*) ← argmin{d_{ij} : (i, j) ∈ ℱ};
6       S ← S ∪ {(i*, j*)};
7       f(S) ← f(S) + d_{i*j*};
8       ℱ ← {(i, j) ∈ ℱ \ {(i*, j*)} : S ∪ {(i, j)} does not have a cycle};
9   end-while;
10  return S, f(S);
end GREEDY-MST.
```

**Fig. 3.2** Pseudo-code of Kruskal's greedy algorithm for the minimum spanning tree problem.

## Knapsack problem - Greedy algorithm

As we saw in Chapter 2, the ground set for the knapsack problem consists of the set $I$ of items to be packed. Each item $i \in I$ has weight $a_i$ and utility $c_i$. The knapsack can accommodate a maximum weight of $b$. We assume, without loss of generality, that $a_i \leq b$, for all $i \in I$. A greedy algorithm for the knapsack problem is shown in Figure 3.3. The solution $S$ to be constructed and its cost $f(S)$ are initialized to $\emptyset$ and 0, respectively, in lines 1 and 2. The set of feasible ground set elements is initialized in line 3 with all items in $I$. A feasible item of greatest utility per unit weight is selected in line 5 and added to the knapsack in line 6, with the total utility of the partial solution being updated in line 7. All items whose inclusion in the current solution would overflow the knapsack (i.e., create an infeasibility) are removed from the set $\mathscr{F}$ of feasible ground set elements in line 8. The solution $S$ and its cost are returned in line 10.

As opposed to the greedy algorithm for the minimum spanning tree problem, this greedy algorithm for the knapsack problem will not always find an optimal solution. Consider the following counter-example with three items, where the item weight vector $a = (3, 2, 2)$, the item utility vector $c = (12, 7, 6)$, and the knapsack weight capacity $b = 4$. The utility per unit weight of each item is $c_1/a_1 = 12/3 = 4, c_2/a_2 = 7/2 = 3.5$, and $c_3/a_3 = 6/2 = 3$. Consequently, the greedy algorithm considers the items in the order 1, 2, 3. Since the weight of item 1 is 3, then it is included in the solution. Since items 2 and 3 both have weight 2, neither can be included in the solution together with item 1. Therefore, the total utility of the greedy solution is 12. However, note that a solution consisting of items 2 and 3, but not item 1, is feasible (since its total weight is 4, which equals the capacity of the knapsack) and its utility is 13, which is greater than 12, the utility of the greedy solution. ∎

```
begin GREEDY-01KSP;
1   S ← ∅;
2   f(S) ← 0;
3   ℱ ← I;
4   while ℱ ≠ ∅ do
5       i* ← argmax{cᵢ/aᵢ : i ∈ ℱ};
6       S ← S∪{i*};
7       f(S) ← f(S)+cᵢ*;
8       ℱ ← {i ∈ ℱ\{i*} : ∑ⱼ∈S∪{i} aⱼ ≤ b};
9   end-while;
10  return S, f(S);
end GREEDY-01KSP.
```

**Fig. 3.3** Pseudo-code of a greedy algorithm for the knapsack problem.

### Steiner tree problem in graphs - Greedy algorithm

We describe here the distance network heuristic for the Steiner tree problem in graphs. Recall that, in this problem, we are given a graph $G = (V,U)$, where the node set is $V = \{1,2,\ldots,n\}$ and the edge set $U$ is formed by unordered pairs of points $i,j \in V$, with $i \neq j$. There is a length $d_{ij}$ associated with each edge $(i,j) \in U$ and a subset $T \subseteq V$ of terminal nodes that have to be connected by a minimum length subtree of $G$.

The distance network heuristic is based on the construction of the complete graph of distances $D(G)$, whose node set is formed exclusively by the terminal nodes $T$ of the original graph $G$. The length of the edge $(i,j)$ of graph $D(G)$ is equal to the length of the shortest path between $i$ and $j$ in graph $G$, for every $i,j \in T$, with $i \neq j$. Therefore, each edge of the complete graph $D(G)$ may be seen as a "super-edge" representing the shortest path between its extremities in $G$.

Once the graph of distances $D(G)$ has been built, the distance network heuristic consists basically in the computation of a minimum spanning tree of $D(G)$. This is followed by the replacement of each super-edge of the minimum spanning tree by the edges in the shortest path between its extremities in the original graph. We observe that the shortest paths in $G$, associated with the super-edges of the graph of distances, are not necessarily disjoint. Therefore, since the same edge of the original graph may appear in more than one shortest path, it is possible that the length of the Steiner tree in the original graph is smaller than that of the minimum spanning tree of $D(G)$.

The pseudo-code for the main building blocks of the distance network heuristic for the Steiner tree problem in graphs is given in Figure 3.4. Algorithm GREEDY-SPG is indeed a greedy algorithm whenever Kruskal's greedy algorithm is used to compute the minimum spanning tree in line 5.

Figure 3.5(a) displays an instance of the Steiner tree problem in graphs, whose optimal solution appears in Figure 3.5(b). Figure 3.6 shows the application of the GREEDY-SPG distance network heuristic to this instance. The shortest path from terminal 1 to 2 corresponds to the node sequence 1 - 5 - 2 and has length 2. The selected alternative path from terminal 1 to 3 is given by the node sequence 1 - 6 - 3

---

**begin** GREEDY-SPG;
1    **forall** $i, j \in T : i \neq j$ **do**
2        Compute the shortest path $P_{ij}(G)$ between $i$ and $j$ in $G = (V, U)$ and
             let $d(P_{ij}(G))$ be its length;
3    **end-forall**;
4    Build the graph of distances $D(G)$, with edge lengths given by $d(P_{ij}(G))$
         for every $i, j \in T$, with $i \neq j$;
5    Compute a minimum spanning tree of $D(G)$;
6    **return** a minimum Steiner tree of $G = (V, U)$ recovered from the minimum
         spanning tree by replacing each edge $(i, j)$ of the latter by the corresponding
         and previously computed shortest path $P_{ij}(G)$ between $i$ and $j$ in $G$;
**end** GREEDY-SPG.

---

**Fig. 3.4** Pseudo-code of the greedy distance network heuristic for the Steiner tree problem in graphs.

and has length 4. The shortest path from terminal 1 to 4 is formed by nodes 1 - 5 - 7 - 9 - 4 and has also length 4. Similarly, the shortest path from node 2 to 3 is formed by nodes 2 - 5 - 7 - 9 - 3 and has the same length 4. The selected alternative path from terminal 2 to 4 is given by the node sequence 2 - 8 - 4 and has length 4. The shortest path from terminal 3 to 4 corresponds to the node sequence 3 - 9 - 4 and has length 2. The graph of distances $D(G)$ appears in Figure 3.6(a). Figure 3.6(b) depicts a minimum spanning tree of $D(G)$. Finally, Figure 3.6(c) shows the Steiner tree for the original graph recovered from the minimum spanning tree computed for $D(G)$. We observe that the solution found by the heuristic has length 8. Therefore, it is not optimal.                                                                    ∎

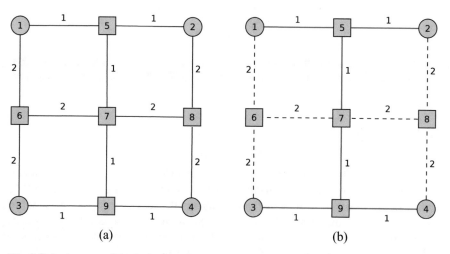

(a)                                                   (b)

**Fig. 3.5** An instance of the Steiner problem in graphs (a) and a minimum Steiner tree with total length 6 (b).

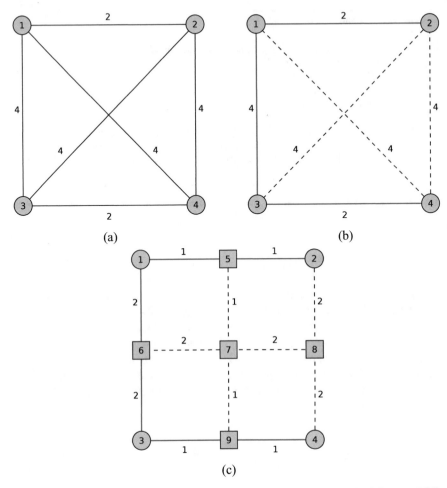

**Fig. 3.6** Application of the GREEDY-SPG distance network heuristic: (a) graph of distances $D(G)$, (b) minimum spanning tree of $D(G)$ with length 8, and (c) Steiner tree with length 8, which is not optimal.

## 3.2 Matroids

We stated earlier in this chapter that the greedy algorithm for the minimum spanning tree problem always produces an optimal solution. In this section, we make use of the theory of matroids to show that this is indeed the case. This theory is useful in showing many situations in which a greedy algorithm does find an optimal solution.

A *matroid* $\mathcal{M} = (\mathcal{S}, \mathcal{I})$ is a combinatorial structure defined by a finite nonempty set $\mathcal{S}$ of elements and a nonempty family $\mathcal{I}$ of *independent subsets* of $\mathcal{S}$, such that:

*Property 3.1.* $\mathcal{I}$ is *hereditary*, i.e., $\varnothing \in \mathcal{I}$ and all proper subsets of a set $I \in \mathcal{I}$ are also in $\mathcal{I}$.

*Property 3.2.* If $I'$ and $I''$ are sets in $\mathscr{I}$ and $|I'| < |I''|$, then there exists an element $e \in I'' \setminus I'$ such that $I' \cup \{e\} \in \mathscr{I}$.

As an example of a matroid, consider the *graphic matroid* $\mathscr{M}_G = (\mathscr{S}_G, \mathscr{I}_G)$ defined on the graph $G = (V, U)$, where $V$ is the set of nodes and $U$ the set of edges of $G$. In this matroid, $\mathscr{S}_G$ is defined to be the set $U$ of edges and $\mathscr{I}_G$ is such that if $U' \subseteq U$, then $U' \in \mathscr{I}_G$ if and only if the graph induced in $G$ by $U'$ is acyclic.

To see that the graphic matroid is indeed a matroid, we must verify properties 3.1 and 3.2 above. Property 3.1 clearly holds, since a graph with an empty set of edges is acyclic and all subgraphs of an acyclic graph are also acyclic. Suppose now that $I'$ and $I''$ are the edge sets of two forests in $G$ such that $|I'| < |I''|$. To verify property 3.2, we must show that there is some edge $e \in I'' \setminus I'$ such that the graph induced in $G$ by the edge set $I' \cup \{e\}$ is acyclic. Since the number of trees in any forest induced in $G$ by the edge set $I$ can be proved by induction to be equal to $|V| - |I|$, then the forest induced by $I'$ has more trees than the one induced by $I''$. Consequently, there must exist some edge $e \in I''$ having its extremities in two disjoint trees in the forest induced by $I'$. Therefore, $e \in I'' \setminus I'$. Since adding an edge between two disjoint trees does not create a cycle, then $I' \cup \{e\} \in \mathscr{I}$.

An important property of a matroid is that all its maximal independent subsets are of the same size. In the graphic matroid, a maximal independent subset is the largest edge set that induces an acyclic graph in $G$, i.e., the $|V| - 1$ edges of any spanning tree of $G$.

A *weighted matroid* is a matroid $\mathscr{M} = (\mathscr{S}, \mathscr{I})$ with a weight function $w : \mathscr{S} \to \mathbb{R}^+$ that assigns a positive weight $w(x)$ to each element $x \in \mathscr{S}$. This weight function is also defined for any subset $U' \subseteq \mathscr{S}$ as $w(U') = \sum_{x \in U'} w(x)$. In a graphic matroid, if $w(e)$ denotes the weight of edge $e \in U$ and $U' \subseteq U$ is a subset of the edges of $G$, then $w(U')$ denotes the total weight of the edges in set $U'$. A natural optimization problem on a weighted matroid is to find a maximum-weight independent set. Since $w(x) > 0$ for all $x \in \mathscr{S}$, then any maximum-weight independent set is maximal. A minimum-weight spanning tree of the graph $G = (V, U)$ is simply a maximum-weight independent set on the weighted graphic matroid $\mathscr{M}_G = (\mathscr{S}_G, \mathscr{I}_G)$ with weights $w'(e) = \bar{w} - w(e)$, where $\bar{w} = 1 + \max\{w(e) : e \in U\}$ and $w(e)$ is the weight of edge $e \in U$.

A greedy algorithm for finding a maximum-weight independent set of any weighted matroid $\mathscr{M} = (\mathscr{S}, \mathscr{I})$ is shown in Figure 3.7. At each iteration, the algorithm adds to the solution the element of maximum weight that maintains independence within the solution. In case no element can be added to the initial solution while maintaining independence, the algorithm returns $\varnothing$, which by definition is independent. Otherwise, line 8 guarantees that only elements that maintain feasibility are considered for inclusion in the solution under construction. Therefore, this algorithm returns in line 10 an independent subset $S$ of $\mathscr{S}$.

We now state some properties of weighted matroids and of the greedy algorithm for finding a maximum-weight independent set of a weighted matroid $\mathscr{M} = (\mathscr{S}, \mathscr{I})$ that show the correctness of the algorithm.

*Property 3.3.* Let $i^* \in \mathscr{S}$ be the first element selected in line 5 of the greedy algorithm in Figure 3.7. Then, there exists an optimal subset $U' \subseteq \mathscr{S}$ such that $i^* \in U'$.

```
begin GREEDY-MATROID;
1   S ← ∅;
2   f(S) ← 0;
3   𝓕 ← {i ∈ 𝓢 : S∪{i} ∈ 𝓘};
4   while 𝓕 ≠ ∅ do
5       i* ← argmax{w(i) : i ∈ 𝓕};
6       S ← S∪{i*};
7       f(S) ← f(S)+w(i*);
8       𝓕 ← {i ∈ 𝓕\{i*} : S∪{i} ∈ 𝓘};
9   end-while;
10  return S, f(S);
end GREEDY-MATROID.
```

**Fig. 3.7** Pseudo-code of a greedy algorithm for finding a maximum-weight independent set of a weighted matroid.

*Property 3.4.* If $i \in \mathscr{S}$ and $\varnothing \cup \{i\} \notin \mathscr{I}$, then $U' \cup \{i\} \notin \mathscr{I}$, for any independent subset $U' \subseteq \mathscr{S}$.

*Property 3.5.* Let $i^* \in \mathscr{S}$ be the first element selected in line 5 of the greedy algorithm in Figure 3.7. The remaining problem reduces to finding a maximum-weight independent set of the weighted matroid $\mathscr{M}' = (\mathscr{S}', \mathscr{I}')$, where $\mathscr{S}' = \{i \in \mathscr{S} : \{i^*, i\} \in \mathscr{I}\}$, $\mathscr{I}' = \{U' \subseteq \mathscr{S} \setminus \{i^*\} : U' \cup \{i^*\} \in \mathscr{I}\}$, and the weight function for $\mathscr{M}'$ is the same for $\mathscr{M}$, although restricted to $\mathscr{S}'$.

Property 3.3 says that once an element is added to $S$, it will never be removed from $S$ later. This is so because there is some optimal subset $U' \subseteq \mathscr{S}$ that contains that element. Property 3.4 tells us that if an element is initially disregarded from $S$, then we can definitely discard it since it will never be added to solution $S$ later. Finally, Property 3.5 states that once an element is added to $S$, the problem reduces itself to the same problem on a reduced weighted matroid.

Summarizing, Properties 3.1 to 3.5 ensure that the greedy algorithm will always find an optimal solution whenever it is applied to a combinatorial optimization problem defined over a weighted matroid. For this reason, a greedy algorithm for the minimum spanning tree problem will always produce an optimal solution, while a greedy algorithm applied to the knapsack problem will not necessarily find an optimum.

## 3.3 Adaptive greedy algorithms

The greedy algorithm of Figure 3.1, as well as the other greedy algorithms described in the previous section, selects an element $i^*$ of the set of feasible candidate elements $\mathscr{F}$ as

$$i^* \leftarrow \operatorname{argmin}\{c_i : i \in \mathscr{F}\},$$

where $c_i$ is the cost associated with the inclusion of element $i \in \mathscr{F}$ in the solution. In all of these algorithms, this cost is constant. Therefore, the elements can be sorted in the increasing order of their costs in a preprocessing step. The while loop in lines 4 to 9 of the algorithm in Figure 3.1 will then scan the elements in this order. Upon considering the $k$-th element, if its inclusion in the solution causes an infeasibility, then it is discarded for good and the next element is considered. Otherwise, it is added to the solution and the next element is considered.

Although the greedy algorithm described in Figure 3.1 is applicable in many situations, such as to the minimum spanning tree problem and to the knapsack problem (in which case it will not always find an optimal solution), there are other situations where the cost of the contribution of an element is affected by the previous choices of elements made by the algorithm. We shall call these *adaptive greedy* algorithms.

---

**begin** ADAPTIVE-GREEDY;
1   $S \leftarrow \varnothing$;
2   $f(S) \leftarrow 0$;
3   $\mathscr{F} \leftarrow \{i \in E : S \cup \{i\} \text{ is not infeasible}\}$;
4   Compute the greedy choice function $g(i)$ for all $i \in \mathscr{F}$;
5   **while** $\mathscr{F} \neq \varnothing$ **do**
6      $i^* \leftarrow \mathrm{argmin}\{g(i) : i \in \mathscr{F}\}$;
7      $S \leftarrow S \cup \{i^*\}$;
8      $f(S) \leftarrow f(S) + c_{i^*}$;
9      $\mathscr{F} \leftarrow \{i \in \mathscr{F} \setminus \{i^*\} : S \cup \{i\} \text{ is not infeasible}\}$;
10     Update the greedy choice function $g(i)$ for all $i \in \mathscr{F}$;
11 **end-while**;
12 **return** $S, f(S)$;
**end** ADAPTIVE-GREEDY.

**Fig. 3.8** Pseudo-code of a generic adaptive greedy algorithm for a minimization problem.

---

Figure 3.8 shows the pseudo-code of an adaptive greedy algorithm for a minimization problem. As before, a solution $S$ is constructed, one element at a time. This solution and its cost are initialized in lines 1 and 2, respectively. The initial set of feasible elements from the ground set $E$ is determined in line 3. The *greedy choice function* $g(i)$ measures the suitability of element $i$ to be included in the partial solution $S$, for all $i \in \mathscr{F}$. The values of the greedy choice function are initialized in line 4. The while loop in lines 5 to 11 constructs the solution. In line 6, a candidate element with minimum greedy choice function value is selected. This element is included in the solution in line 7. The value of the cost function is updated in line 8. The set of feasible candidate elements is updated in line 9 and the values of the greedy choice function are updated in line 10, for all remaining feasible candidate elements. Construction ends when there are no more feasible candidate elements to be added, i.e., when $\mathscr{F} = \varnothing$. Solution $S$ and its cost are returned in line 12.

We next give examples of adaptive greedy algorithms for four combinatorial optimization problem: the minimum spanning tree problem, the set covering problem, the maximum clique problem, and the traveling salesman problem.

**Minimum spanning tree problem – Adaptive greedy algorithm**

In Section 3.1 we saw a greedy algorithm for the minimum spanning tree problem. The first example of an adaptive greedy algorithm is one for the same problem. As before, we are given a graph $G = (V,U)$, where $V$ is the set of nodes and $U$ is the set of weighted edges. Let $d_{ij}$ be the weight of edge $(i,j) \in U$.

An adaptive greedy approach for this problem is to grow the set of spanned nodes of the tree, at each step adding a new edge with the least weight among all edges with only one endpoint in the set of already spanned nodes. The other endpoint of this edge is then added to the set of spanned nodes. This is repeated until all nodes are spanned. Figure 3.9 shows the pseudo-code of this adaptive greedy algorithm for the minimum spanning tree problem. In lines 1 and 2, respectively, the set $S$ of edges in the spanning tree and its weight $f(S)$ are initialized. The set $\mathcal{V}$ of nodes yet to be spanned is initialized in line 3. The loop in lines 4 to 7 is used to initialize the greedy choice function $g(i)$ in lines 5 and the pointer $\pi(i)$ to the other endpoint of the minimum weight edge connecting node $i$ to the current set $V \setminus \mathcal{V}$ of spanned nodes in line 6, for every $i \in \mathcal{V}$. In line 8, a node $j \in \mathcal{V}$ is chosen to be placed in the set of spanned nodes and, in line 9, its greedy choice function is set to 0. The main loop of the algorithm, in lines 10 to 24, is repeated until the set $\mathcal{V}$ of nodes yet to be spanned becomes empty. In line 11, node $j$ is added to the set of spanned nodes (in fact, it is removed from the set of nodes yet to be spanned). Node $i$ is set in line 12 to be the closest node to $j$ yet to be spanned. In the first iteration of the loop, $i$ is 0 and, consequently, lines 14 and 15 are not computed. For all other iterations, edge $(i,j)$ is added to the spanning tree $S$ in line 14 and the partial cost $f(S)$ of the spanning tree is updated in line 15. Lines 17 to 22 update the greedy choice function $g(k)$ and the pointer $\pi(k)$ for each node $k$ adjacent to $j$ that has not yet been spanned. In line 23, a yet unspanned node $j$ with minimum greedy choice function value is chosen to become spanned. The minimum-weight spanning tree $S$ and its weight $f(S)$ are returned in line 25.

This adaptive greedy algorithm for the minimum spanning tree problem is known as Prim's algorithm. ∎

**Minimum cardinality set covering problem – Adaptive greedy algorithm**

Given a set $I = \{1,\ldots,m\}$ of objects, let $\{P_1,\ldots,P_n\}$ be a collection of finite subsets of $I$ such that $\cup_{j=1}^{n} P_j = I$, with a non-negative cost $c_j$ associated with each subset $P_j$, for $j = 1,\ldots,n$. A subset $\hat{J} \subseteq J = \{1,\ldots,n\}$ is a *cover* of $I$ if $\cup_{j \in \hat{J}} P_j = I$. The cost of a cover $\hat{J}$ is given by $\sum_{j \in \hat{J}} c_j$. The *set covering problem* consists in finding a minimum cost cover. In the *minimum cardinality set covering problem*, we seek

```
begin ADAPTIVE-GREEDY-MST;
1   S ← ∅;
2   f(S) ← 0;
3   𝒱 ← V;
4   for i ∈ 𝒱 do
5       g(i) ← ∞;
6       π(i) ← 0;
7   end-for;
8   Let j be any node in 𝒱;
9   g(j) ← 0;
10  while 𝒱 ≠ ∅ do
11      𝒱 ← 𝒱 \ {j};
12      i ← π(j);
13      if i ≠ 0 then
14          S ← S ∪ {(i, j)};
15          f(S) ← f(S) + d_{ij};
16      end-if;
17      for k ∈ 𝒱 : (j, k) ∈ U do
18          if d_{jk} < g(k) then
19              g(k) ← d_{jk};
20              π(k) ← j;
21          end-if;
22      end-for;
23      j ← argmin{g(k) : k ∈ 𝒱};
24  end-while;
25  return S, f(S);
end ADAPTIVE-GREEDY-MST.
```

**Fig. 3.9** Pseudo-code of Prim's adaptive greedy algorithm for the minimum spanning tree problem.

a cover of minimum cardinality, which is equivalent to setting $c_j = 1$ in the set covering problem, for $j = 1, \ldots, n$. Let the $m \times n$ binary matrix $A = \{a_{ij}\}$ be such that for all $i \in I$ and for all $j \in J$, $a_{ij} = 1$ if and only if $i \in P_j$; $a_{ij} = 0$, otherwise. A solution $\hat{J}$ of the minimum cardinality set covering problem can be represented by a binary $n$-vector $x$, where $x_j = 1$ if and only if $j \in \hat{J}$; $x_j = 0$ otherwise, for $j = 1, \ldots, n$. An integer programming formulation for the minimum cardinality set covering problem is then

$$\min \left\{ \sum_{j \in J} x_j : \sum_{j \in J} a_{ij} \cdot x_j \geq 1 \ \forall i \in I, \ x_j \in \{0, 1\} \ \forall j \in J \right\}.$$

We say that column $j$ of matrix $A$ covers row $i$ if $a_{ij} = 1$. A greedy approach to this problem is to select columns of matrix $A$, one at a time, such that each selected column covers the maximum number of yet-uncovered rows of $A$. Let $g(j)$ be the greedy choice function which measures the number of yet-uncovered rows of $A$ that would become covered if the still unused column $j$ were to be added to the cover under construction. Initially, we set $g(j) = \sum_{i \in I} a_{ij}$, for all $j = 1, \ldots, n$. We denote by $j^*$ the first column selected by the greedy algorithm, which is the one that maximizes $g(j)$, for $j = 1, \ldots, n$: $g(j^*) = \max_{j \in J} g(j)$. Once column $j^*$ is placed in

```
begin ADAPTIVE-GREEDY-MIN-CARDINALITY-SET-COVERING;
1   S ← ∅;
2   f(S) ← 0;
3   ℱ ← {1,...,n};
4   𝒞 ← ∅;
5   g(j) ← ∑ᵢ₌₁ᵐ aᵢⱼ for all j ∈ ℱ;
6   while 𝒞 ≠ I do
7       j* ← argmax{g(j) : j ∈ ℱ};
8       S ← S ∪ {j*};
9       f(S) ← f(S) + 1;
10      ℱ ← ℱ \ {j*};
11      for i ∈ I \ 𝒞 : aᵢ,ⱼ* = 1 do
12          𝒞 ← 𝒞 ∪ {i};
13          for j ∈ ℱ : aᵢⱼ = 1 do
14              g(j) ← g(j) − 1;
15          end-for;
16      end-for;
17  end-while;
18  return S, f(S);
end ADAPTIVE-GREEDY-MIN-CARDINALITY-SET-COVERING.
```

**Fig. 3.10** Pseudo-code of an adaptive greedy algorithm for minimum cardinality set covering.

the partial solution under construction, every unselected column that covers a row newly covered by column $j^*$ must have its $g(j)$ value updated, since $g(j)$ measures the number of yet-uncovered rows that will be covered with the inclusion of column $j$ in the solution. The adaptive greedy algorithm repeats this column selection and greedy choice function update process until all rows of $A$ are covered.

Figure 3.10 shows the pseudo-code of an adaptive greedy algorithm for the minimum cardinality set covering problem. Lines 1 to 5 initialize the cover $S$, its cost $f(S)$, the set of potential cover elements $\mathscr{F}$, the set of covered row indices $\mathscr{C}$, and the greedy choice function $g(j)$ for each potential cover element $j \in \mathscr{F}$. The cover is constructed in the while loop in lines 6 to 17. In line 7, a column $j^*$ that maximizes the greedy choice function is selected. This column is included in the cover in line 8, while the cover's cost $f(S)$ is updated in line 9. Element $j^*$ is removed from the set of potential cover elements in line 10. The greedy choice function is updated in the for loop in lines 11 to 16, which scans all uncovered rows that have just became covered by $j^*$. The index of each such row is inserted in the set $\mathscr{C}$ of covered rows in line 12. The value of the greedy choice function $g(j)$ is updated in line 14 for each column $j$ other than $j^*$ that also covers that row. The algorithm terminates when a cover is produced. A cover $S$ and its cost $f(S)$ are returned in line 18.                                                                                   ∎

### Maximum clique problem – Adaptive greedy algorithm

We now give an adaptive greedy algorithm for the maximum clique problem. Given an undirected graph $G = (V, U)$, we recall that a clique is any subset of nodes of

```
begin ADAPTIVE-GREEDY-MAX-CLIQUE;
1    S ← ∅;
2    f(S) ← 0;
3    𝓕 ← V;
4    g(v) ← deg_𝓕(v) for all v ∈ 𝓕;
5    while 𝓕 ≠ ∅ do
6        v' ← argmax{g(v) : v ∈ 𝓕};
7        S ← S ∪ {v'};
8        f(S) ← f(S) + 1;
9        𝓕 ← {v ∈ 𝓕 \ {v'} : v is adjacent to all nodes in S};
10       g(v) ← deg_𝓕(v) for all v ∈ 𝓕;
11   end-while;
12   return S, f(S);
end ADAPTIVE-GREEDY-MAX-CLIQUE.
```

**Fig. 3.11** Pseudo-code of an adaptive greedy algorithm for maximum clique.

$G$ that are mutually adjacent. In the maximum clique problem, we want to find a largest cardinality clique in $G$. An adaptive greedy algorithm for the maximum clique problem builds a clique, one node at a time. Initially, all nodes are candidates to be included in the clique. We shall call the candidate set $\mathcal{F}$. A natural measure of suitability for a node $v \in V$ to be the first node included in the clique is its degree, which is equal to the number of nodes adjacent to it. Let us denote this greedy choice function by $g(v)$, for all $v \in \mathcal{F}$. Once the node with maximum degree is placed in the clique, all nodes that are not adjacent to it can no longer be considered for placement in the clique. Let us redefine $\mathcal{F}$ as the set of remaining nodes that can be added to the current clique. The greedy choice function $g(v)$ for all nodes $v \in \mathcal{F}$ must be updated to account for the fact that the clique now consists of the first selected node. The suitability of a node $v \in \mathcal{F}$ to be the next node to be included in the clique is related with the number of nodes adjacent to it in $\mathcal{F}$. The adaptive greedy algorithm repeats this node selection and greedy choice function update process until the candidate list becomes empty, i.e., $\mathcal{F} = \varnothing$.

Figure 3.11 shows the pseudo-code for an adaptive greedy algorithm for the maximum clique problem. Lines 1 to 4 initialize the clique $S$, its cost $f(S)$, the set $\mathcal{F}$ of yet unselected potential clique elements, and, for each potential clique node $v \in \mathcal{F}$, its initial greedy choice function value $g(v)$ that is set to $\deg_{\mathcal{F}}(v)$, which represents the number of nodes that are adjacent to $v$ in $G$ and belong to $\mathcal{F}$ (or, alternatively, the degree of node $v$ with respect to the nodes in $\mathcal{F}$). The clique is constructed in the while loop in lines 5 to 11. In line 6, a node $v'$ maximizing the greedy choice function is selected. This node is included in the clique $S$ in line 7, while the clique's size is updated in line 8. The set of potential clique nodes is updated in line 9 and consists of all yet unselected nodes that are adjacent to all nodes in the current clique $S$. The greedy choice function is updated in line 10. The algorithm terminates when a maximal clique is produced, i.e., when $\mathcal{F} = \varnothing$. The largest clique found and its cardinality are returned in line 12.                                        ∎

```
begin ADAPTIVE-GREEDY-TSP;
1   S ← ∅;
2   f(S) ← 0;
3   Let i be any node in V and set i₀ ← i;
4   F ← V \ {i₀};
5   while F ≠ ∅ do
6       H ← {j ∈ F : (i,j) ∈ U};
7       g(j) ← d_{ij} for all j ∈ H;
8       j' ← argmin{g(j) : j ∈ H};
9       S ← S ∪ {(i,j')};
10      f(S) ← f(S) + d_{i,j'};
11      F ← F \ {j'};
12      i ← j';
13  end-while;
14  S ← S ∪ {(i,i₀)};
15  f(S) ← f(S) + d_{i,i₀};
16  return S, f(S);
end ADAPTIVE-GREEDY-TSP.
```

**Fig. 3.12** Pseudo-code of the nearest neighbor adaptive greedy algorithm for the traveling salesman problem.

### Traveling salesman problem – Adaptive greedy algorithm

The next example of an adaptive greedy algorithm is known as the *nearest neighbor heuristic* for the traveling salesman problem. We are given a graph $G = (V,U)$, where $V$ is the set of nodes and $U$ is the set of weighted edges. Let $d_{ij}$ be the length (or weight) of edge $(i,j) \in U$.

An adaptive greedy approach for this problem is to grow the set of visited nodes of the tour, starting from any initial node. Denote by $v$ the last visited node of the partial tour under construction. At each step we add to the tour a nearest unvisited node adjacent to $v$. This is repeated until the tour visits all nodes.

Figure 3.12 shows the pseudo-code of this algorithm. In lines 1 and 2, the set $S$ of edges in the tour and its total length $f(S)$ are initialized. In line 3, we select any initial node $i \in V$ to start the tour and save it as $i_0$ to be used later. The set of unvisited nodes $F$ is initialized in line 4. The main loop of the algorithm, in lines 5 to 13, is repeated until the set $F$ of unvisited nodes becomes empty. In line 6, we build the set $H$ of candidate nodes that can be added to the tour following the last added node $i$. For each candidate node, the greedy choice function is set in line 7. Node $j'$ is set to be the nearest unvisited node adjacent to $i$ in line 8. Edge $(i,j')$ is added to the Hamiltonian cycle under construction in line 9 and the length $f(S)$ of the tour is updated in line 10. The set $F$ of candidateunvisited nodes is updated in line 11 and $j'$ is made the last visited node in line 12. The while loop terminates when all nodes have been visited, i.e., when $F$ becomes empty. At this point, we add a return edge connecting the last visited node $i$ with the initial node $i_0$ in line 14.

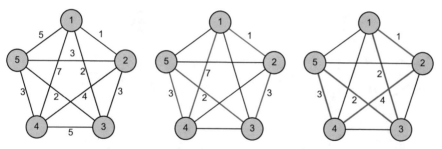

**Fig. 3.13** Examples of a TSP instance solved with two adaptive greedy algorithms. The leftmost graph shows all edge lengths. The one in the middle shows a tour of length 16 produced by the nearest neighbor adaptive greedy algorithm that grows the path from one end of the partial path. The rightmost graph shows the tour of length 12 produced by a variant of this adaptive greedy algorithm that grows the partial path from both of its extremities.

The length of the tour is updated in line 15 and the solution and its total length are returned in line 16.

We remark that if the graph $G = (V, U)$ is not complete, then it is possible that at some iteration in line 6 there is no edge connecting node $i$ with an unvisited node. Note also that if the graph is not complete then in line 14 the return edge $(i, i_0)$ may not exist. Therefore, a sufficient condition for this algorithm to find a feasible solution is that the graph be complete.

An example of the application of the nearest neighbor adaptive greedy algorithm of Figure 3.12 to the leftmost graph in Figure 3.13 is described in the following. The algorithm starts by selecting some node to be the start of the tour. Suppose node 1 is selected as the starting node. From node 1, the distances to nodes 2, 3, 4, and 5 are, respectively, 1, 2, 7, and 5. Since $d_{12} = 1$ is the smallest of the distances, node 2 is selected to be the next node in the tour and the partial tour becomes $1 \to 2$. From node 2, the distances to nodes 3, 4, and 5 (nodes not yet in the tour) are, respectively, 3, 4, and 3. Since $d_{23} = d_{25} = 3$ is the smallest distance from node 2 to a yet unselected node, either node 3 or node 5 could be selected as the next node in the tour. Suppose node 3 is chosen. The partial tour is now formed by $1 \to 2 \to 3$. From node 3, the distances to nodes 4 and 5 are, respectively, 5 and 2. Since $d_{35} = 2$ is the smallest distance from node 3 to any of the yet unselected nodes, then node 5 is chosen next to be in the partial tour, which becomes $1 \to 2 \to 3 \to 5$. The only yet unselected node is node 4 and it is then selected to be the next node on the tour. Consequently, the full tour is $1 \to 2 \to 3 \to 5 \to 4 \to 1$. The length of the tour is $d_{12} + d_{23} + d_{35} + d_{54} + d_{41} = 1 + 3 + 2 + 3 + 7 = 16$. It is shown in the middle graph of Figure 3.13.

The adaptive greedy algorithm of Figure 3.12 always extends the path from the last node to be added, i.e., the path is grown out from only one side of the partial path. If instead of only considering one side of the partial path, we considerboth sides, we get a new modified adaptive greedy algorithm for the TSP. Again consider node 1 as the initial node of the path. As before, from node 1, the distances to nodes 2, 4, and 5 are, respectively, 1, 2, 7, and 5. Since $d_{12} = 1$ is the smallest of the

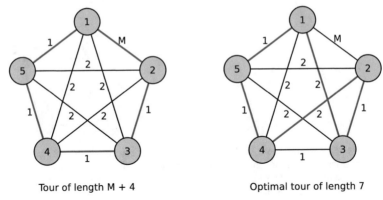

Fig. 3.14 Example of an instance of the traveling salesman problem on which, assuming $M > 3$, the nearest neighbor adaptive greedy algorithm always produces a tour of length $M + 4$ that may be arbitrarily bad as $M$ grows, since there exists an optimal tour of length 7 shown in red on the graph on the right of the figure.

distances, node 2 is selected to be the next node in the tour and the partial tour becomes $1 \to 2$. The two extremities of the path are nodes 1 and 2. As before, from node 2, the distances to nodes 3, 4, and 5 (nodes not yet in the tour) are, respectively, 3, 4, and 3. From node 1, the distances to nodes 3, 4, and 5, are, respectively, 2, 7, and 5. Since $d_{13} = 2$ is the smallest of the lengths, node 3 is selected to be the next node in the tour. It is connected to node 1 and the partial tour becomes $3 \to 1 \to 2$. The two extremities of the path are now nodes 3 and 2. From node 2, the distances to nodes 4 and 5 are, respectively, 4 and 3, while from node 3 the distances to nodes 4 and 5 are, respectively, 5 and 2. Since $d_{35} = 2$ is the smallest of the lengths, node 5 is selected to be the next node in the tour. It is connected to node 3 and the partial tour becomes $5 \to 3 \to 1 \to 2$. Node 2 can now only connect to node 4, which in turns connects to node 5. The final tour becomes $4 \to 5 \to 3 \to 1 \to 2$ with a corresponding length of $d_{45} + d_{53} + d_{31} + d_{12} + d_{24} = 3 + 2 + 2 + 1 + 4 = 12 < 16$. This solution improves the previous one and appears as the rightmost graph of Figure 3.13. ∎

We observe that even if the graph is complete, the nearest neighbor adaptive greedy algorithm may still find a very bad solution for some instances. Consider the graph on the left side of Figure 3.14, where $M > 3$ is an arbitrarily large number. If the nearest neighbor adaptive greedy algorithm starts from node 1, it produces a tour of length $M + 4$ containing all the edges of length 1 and edge $(1, 2)$ of length $M$. This tour is shown in red on the graph on the left of the figure. It can be arbitrarily longer than the optimal tour of length 7, which is shown on the right of the figure.

### Steiner tree problem in graphs – Adaptive greedy algorithm

Once again, recall that we are given a graph $G = (V, U)$, a length $d_{ij}$ associated with each edge $(i, j) \in U$, and a subset $T \subseteq V$ of terminal nodes that have to be connected by a minimum length subtree of $G$. The adaptive greedy heuristic for the Steiner tree problem in graphs may be seen as an extension of Prim's algorithm for the minimum spanning tree problem presented earlier in this section. At each iteration, the closest yet unconnected terminal node is connected to the current partial tree by a minimum shortest path.

A pseudo-code for this heuristic is given in Figure 3.15. The algorithm starts in line 1 from any randomly selected terminal node $s \in T$, which is used to initialize the Steiner tree in line 2. The set of terminal nodes already connected by the Steiner tree is initialized with terminal $s$ in line 3. Next, in line 4, the algorithm computes the shortest path $SP(i,S)$ from each yet unconnected terminal $i \in T \setminus M$ to node $s$. The loop in lines 5 to 10 connects one new terminal in each iteration, until a Steiner tree connecting all terminals is built. The closest terminal $s$ to the current partial tree is selected in line 6. The set of connected terminal nodes is expanded by node $s$ in line 7 and all nodes and edges in the path $SP(s,S)$ are added to the Steiner tree $S$ in line 8. In line 9, the shortest path $SP(i,S)$ from each yet unconnected terminal $i \in T \setminus M$ to the updated tree $S$ is recomputed. The Steiner tree $S$ is returned in line 11.

```
begin ADAPTIVE-GREEDY-SPG;
1   Randomly select any terminal node s ∈ T;
2   Initialize the Steiner tree S with the terminal node s;
3   M ← {s};
4   Compute the shortest path SP(i,S) from every yet unconnected terminal i ∈ T \ M
        to the closest node in the partial tree S;
5   while M ≠ T do
6       s ← argmin{d(SP(i,S)) : i ∈ T \ M};
7       M ← M ∪ {s};
8       Add all nodes and edges of SP(s,S) to the Steiner tree S;
9       Update the shortest path SP(i,S) from every yet unconnected terminal i ∈ T \ M
            to the closest node in the partial tree S;
10  end-while;
11  return S;
end ADAPTIVE-GREEDY-SPG.
```

**Fig. 3.15** Pseudo-code of the adaptive greedy heuristic for the Steiner tree problem in graphs.

Figure 3.16 illustrates the application of the ADAPTIVE-GREEDY-SPG heuristic to the same instance in Figure 3.5. The Steiner tree is initialized in Figure 3.16(a) with terminal node 1 that has been randomly selected from the set of terminals. The length of the shortest path from terminal 2 to node 1 is two, while that from terminal 3 is four and that from terminal 4 is also four. Since terminal 2 is the closest to node 1, it is the next to be connected and the path 1 - 5 - 2 is added to the partial Steiner tree in Figure 3.16(b). The shortest paths from terminals 3 and 4 to the partial Steiner tree are updated. The new lengths of the shortest paths from either terminal 3 or 4 to the partial Steiner tree become equal to three and any of these terminals may be selected at the next iteration. Suppose that terminal 4 is selected to be added to the set of connected terminals. The path 4 - 9 - 7 - 5 is incorporated into the tree in Figure 3.16(c). Terminal 3 is the last to be connected to the previously selected terminal nodes and the path 3 - 9 is added to complete a Steiner tree in Figure 3.16(d). ∎

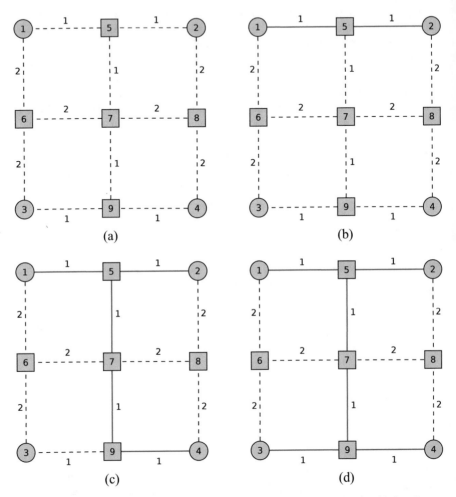

**Fig. 3.16** Application of the ADAPTIVE-GREEDY-SPG: (a) terminal node 1 is added to the tree that is initially empty, (b) terminal 2 is the next to be connected, since it is the closest to terminal node 1, (c) terminal 4 is connected to the tree, and (d) terminal node 3 is the last to be connected forming a Steiner tree with length 6, which is optimal.

## 3.4 Semi-greedy algorithms

Consider the graph shown in Figure 3.17 and suppose we wish to find a shortest
Hamiltonian cycle in this graph applying the nearest neighbor adaptive greedy algorithm presented in Figure 3.12. The algorithm starts from any node and repeatedly
moves from the current node to its nearest unvisited node. Suppose the algorithm
were to start from node 1, in which case it should move next to either node 2 or 3. If
it moves to node 2, then it must necessarily move next to node 3 and then to node 4.
Since there is no edge connecting node 4 to node 1, the algorithm will fail to find
a tour. By symmetry, the same situation occurs if it were to start from node 4. Now
suppose the algorithm starts from node 2. Node 3 is the nearest to node 2 and from
node 3 it can move either to node 1 or node 4, failing in either case to find a tour.
Again, by symmetry, the same situation occurs if one were to start from node 3.
Therefore, this adaptive greedy algorithm fails to find a tour, no matter which node
it starts from.

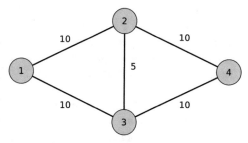

**Fig. 3.17** Example of a traveling salesman problem instance for which the nearest neighbor adaptive greedy algorithm fails to find an optimal solution, while a semi-greedy algorithm succeeds.

  Now, consider the following randomized version of the same adaptive greedy algorithm. This randomized variant starts from any node and repeatedly moves, with
equal probability, to one of its two nearest unvisited nodes. Starting from node 1,
it then moves to either node 2 or node 3 with equal probability. Suppose it were to
move to node 2. Now, again with equal probability, it moves to either node 3 or node
4. On the one hand, if it were to move to node 3, it would fail to find a tour. On the
contrary, by moving to node 4, it would then go to node 3, and then back to node 1,
thus finding a tour of length 40. Therefore, there is a 50% probability that the algorithm will find a tour if it starts from node 1. By applying this algorithm repeatedly,
the probability that it will eventually find the optimal cycle quickly approaches one.
For example, after only ten attempts, the probability that this algorithm finds the
optimal solution is over 99.9%.
  Algorithms like the one above, which add randomization to a greedy or adaptive greedy algorithm, are called *semi-greedy* or *randomized-greedy* algorithms.
Figure 3.18 shows the pseudo-code of a generic semi-greedy algorithm for a minimization problem. This pseudo-code is similar to that of the greedy algorithm in

```
begin SEMI-GREEDY;
1   S ← ∅;
2   f(S) ← 0;
3   𝓕 ← {i ∈ E : S ∪ {i} is not infeasible};
4   while 𝓕 ≠ ∅ do
5       Let RCL be a subset of lowest-cost elements of 𝓕;
6       Let i* be a randomly chosen element from RCL;
7       S ← S ∪ {i*};
8       f(S) ← f(S) + c_{i*};
9       𝓕 ← {i ∈ 𝓕 \ {i*} : S ∪ {i} is not infeasible};
10  end-while;
11  return S, f(S);
end SEMI-GREEDY.
```

**Fig. 3.18** Pseudo-code of a semi-greedy algorithm for a minimization problem.

Figure 3.1, differing only in how the ground set element is chosen from the set $\mathscr{F}$ of feasible candidate ground set elements (lines 5 and 6). In line 5, a subset of lowest-cost elements of set $\mathscr{F}$ is placed in a *restricted candidate list* (RCL). In line 6, a ground set element is selected at random from the RCL to be incorporated into the solution in line 7. Although random selection usually assigns equal probabilities to each RCL element, probabilities proportional to the quality of each element can also be used.

Two simple schemes to define a restricted candidate list are *cardinality-based* and *quality-based*. In the former, the $k$ least-costly feasible candidate ground set elements of set $\mathscr{F}$ are placed in the RCL. In the latter, let $c_{\min} = \min\{c_i : i \in \mathscr{F}\}$ and $c_{\max} = \max\{c_i : i \in \mathscr{F}\}$. Furthermore, let $\alpha$ be such that $0 \le \alpha \le 1$. The RCL is formed by all ground set elements $i \in \mathscr{F}$ satisfying $c_{\min} \le c_i \le c_{\min} + \alpha(c_{\max} - c_{\min})$. We observe that setting $\alpha = 0$ corresponds to an implementation of a pure greedy algorithm, since a lowest-cost element will always be selected at any iteration. On the other hand, setting $\alpha = 1$ leads to a completely random algorithm, since any new element may be added with equal probability at any iteration. Later in this book, we present other variants and applications of semi-greedy algorithms.

## 3.5 Repair procedures

Suppose that in the greedy algorithm of Figure 3.1 (or in the adaptive greedy algorithm of Figure 3.8 or, still, in the semi-greedy algorithm of Figure 3.18) we reach a situation in which $\mathscr{F} = \emptyset$ but $S$ is not yet a feasible solution. We illustrate this situation with two problems: the knapsack problem with an equality constraint and the traveling salesman problem.

### Knapsack problem with an equality constraint – Infeasible construction

In the knapsack problem, we seek a solution maximizing the total utility and whose total weight is at most equal to the maximum weight capacity of the knapsack $b$. In case of the *knapsack problem with an equality constraint*, we also seek a solution maximizing the total utility, but whose total weight is exactly equal to the maximum capacity of the knapsack $b$.

A naive adaptation of the greedy algorithm for the knapsack problem, whose pseudo-code is shown in Figure 3.3, will not always find a feasible solution. Consider the following counter-example with three items, where the item weight vector is $a = (3,2,2)$, the item utility vector is $c = (3,1,1)$, and the knapsack capacity is $b = 4$. The utility per unit weight of each item is $c_1/a_1 = 3/3 = 1$ and $c_2/a_2 = c_3/a_3 = 1/2 = 0.5$. Consequently, the greedy algorithm considers the items in the order $1,2,3$. Since the weight of item 1 is 3, then it would be included in the solution. Since items 2 and 3 both have weight equal to 2, neither can be included in the solution together with item 1 because otherwise the equality constraint would be violated. Note, however, that if the items were to be packed in the reverse order, then items 2 and 3 would be packed and a feasible solution produced. ■

### Traveling salesman problem – Infeasible construction

Consider again the graph shown in Figure 3.17, on which we wish to find a tour with minimum length by applying the nearest neighbor greedy algorithm described in Figure 3.12. If the tour were to start at node 1, then we could add either arc $(1,2)$ or $(1,3)$ without causing any infeasibility. If we chose arc $(1,2)$, then from node 2 we could add either arc $(2,3)$ or $(2,4)$. Since the greedy choice is to add arc $(2,3)$, we do so. From node 3, all ground set elements lead to infeasibility: if we were to add arc $(3,1)$, we would get a sub-tour; if we add arc $(3,4)$ we get a path that cannot be extended to form a tour. ■

We saw in Section 3.4 that one way to try to produce a feasible solution is to add randomization to the greedy algorithm, thus repeatedly applying the resulting semi-greedy algorithm until a feasible solution is produced.

Another way is through a *repair procedure*. A repair procedure undoes erroneous selections made by the construction procedure and attempts to correct them so that a feasible solution can be found.

A possible strategy for implementing a repair procedure consists in removing the last element added to the solution and attempting to add another feasible (but not necessarily greedy) element. In the above example for the traveling salesman problem, arc $(2,3)$ would be removed from the solution. The only remaining feasible element that could be added from node 2 is arc $(2,4)$. By doing so, we easily construct a feasible solution by then adding arcs $(4,3)$ and $(3,1)$ to the tour. An extension of this strategy consists in backtracking, if the removal of the last added element is not sufficient to recover feasibility.

A generalization of the previous strategy, which was based on backtracking and replacing the last added element, is to repeatedly apply destructive modifications to the solution, followed by constructive steps that attempt to recover feasibility. However, all these strategies are problem-specific and are difficult to generalize. Examples will be presented later in this book.

## 3.6 Bibliographical notes

Kruskal (1956) proposed the greedy algorithm for the minimum weighted spanning tree problem described in Section 3.1. Greedy algorithms for knapsack problems were discussed by Martello and Toth (1990). A more efficient implementation of the distance network heuristic for the Steiner tree problem in graphs was developed by Melhorn (1988), based on Voronoi diagrams.

Edmonds (1971) established the connection between weighted matroids and greedy algorithms. Chapter 7 of Lawler (1976) and Chapter 16 of Cormen et al. (2009) cover greedy algorithms and an introduction to matroid theory. Many of the properties of matroids listed in Section 3.2 are proved there. Matroid theory was introduced by Whitney (1935) and was independently discovered by Takeo Nakasawa (see Nishimura and Kuroda (2009) for a historical note and English translation of his original work). Pitsoulis (2014) offered an in-depth coverage of matroids.

Prim (1957) developed the adaptive greedy algorithm for the minimum spanning tree problem described in Section 3.3. Prim's algorithm was originally proposed by Jarník (1930). The adaptive greedy algorithm for the set covering problem was first described by Johnson (1974) and studied by Chvátal (1979). The adaptive greedy algorithm for the maximum clique problem was based on an adaptive greedy algorithm for finding maximum independent sets proposed by Feo et al. (1994). Heuristics for the traveling salesman problem were discussed by Lawler et al. (1985), Gutin and Punnen (2002), and Applegate et al. (2006). The shortest path adaptive greedy heuristic for the Steiner tree problem in graphs was developed by Takahashi and Matsuyama (1980).

Semi-greedy algorithms presented in Section 3.4 were first introduced by Hart and Shogan (1987) and independently developed by Feo and Resende (1989). Bang-Jensen et al. (2004) characterized cases where the greedy algorithm fails and applied their results to the traveling salesman problem and to the minimum bisection problem.

Examples of the repair procedures described in Section 3.5 were reported, e.g., by Duarte et al. (2007a), Duarte et al. (2007b), and Mateus et al. (2011).

# Chapter 4
# Local search

Local search methods start from any feasible solution and visit other (feasible or infeasible) solutions, until a feasible solution that cannot be further improved is found. Local improvements are evaluated with respect to neighboring solutions that can be obtained by slight modifications applied to a solution being visited. We introduce in this chapter the concept of solution representation, which is instrumental in the design and implementation of local search methods. We also define neighborhoods of combinatorial optimization problems and moves between neighboring solutions. We illustrate the definition of a neighborhood by a number of examples for different problems. Local search methods are introduced and different implementation issues are discussed, such as neighborhood search strategies, quick cost updates, and candidate list strategies.

## 4.1 Solution representation

We consider that any solution $S$ for a combinatorial optimization problem is defined by a subset of the elements of the ground set $E$. A feasible solution is one that satisfies all constraints of the problem. We denote by $F$ the set of feasible solutions for this problem and by $\hat{F}$ the set formed by all subsets of ground set elements, which includes all feasible and infeasible solutions. We assume in the following that the objective function value of any (feasible or infeasible) solution $S$ is given by $f(S) = \sum_{i \in S} c_i$, where $c_i$ denotes the contribution to the objective function value of the ground set element $i \in E$.

**Maximum clique problem – Solution representation**

Let $G = (V, U)$ be a graph, in which we seek a maximum cardinality clique. In the case of the maximum clique problem, the ground set $E$ corresponds to the set of nodes $V = \{1, \ldots, n\}$. Every solution $S$ can be represented by a binary vector

© Springer Science+Business Media New York 2016
M.G.C. Resende, C.C. Ribeiro, *Optimization by GRASP*,
DOI 10.1007/978-1-4939-6530-4_4

$(x_1, \ldots, x_n)$, in which $x_i = 1$ if node $i$ belongs to $S$, $x_i = 0$ otherwise, for every $i = 1, \ldots, n$. The set of feasible solutions $F \subseteq \hat{F} = 2^E$ is formed by all subsets of $V$ in which all nodes are pairwise adjacent. ∎

### Knapsack problem – Solution representation

In the case of the knapsack problem, one has a set $I = \{1, \ldots, n\}$ of items to be placed in a knapsack. Integer numbers $a_i$ and $c_i$ represent, respectively, the weight and the utility of each item $i \in I$. We assume that each item fits in the knapsack by itself and denote by $b$ its maximum total weight. As for the previous problem, every solution $S$ can be represented by a binary vector $(x_1, \ldots, x_n)$, in which $x_i = 1$ if item $i$ is selected, $x_i = 0$ otherwise, for every $i = 1, \ldots, n$. A solution $S = (x_1, \ldots, x_n)$ belongs to the feasible set $F$ if $\sum_{i \in I} a_i \cdot x_i \leq b$. ∎

### Steiner tree problem in graphs – Solution representation

Let $G = (V, U)$ be a graph, where the node set is $V = \{1, \ldots, n\}$ and the edge set is $U$. We recall that, in the Steiner tree problem in graphs, we are also given a subset $T \subseteq V$ of terminal nodes that have to be connected. A Steiner tree $S = (V', U')$ of $G$ is a subtree of $G$ that connects all nodes in $T$, i.e., $T \subseteq V' \subseteq V$.

Given any subset $V'$ of nodes such that $T \subseteq V' \subseteq V$, we note that any spanning tree of the graph induced in $G$ by $V'$ is also a Steiner tree of $G$ connecting all terminal nodes in $T$. Therefore, any Steiner tree of $G$ connecting the terminal nodes may be obtained by selecting a subset of optional nodes $W \subseteq V \setminus T$ and computing a minimum spanning tree of the graph $G(W \cup T)$ induced in $G$ by $V' = W \cup T$.

As a consequence, every solution $S$ of the Steiner tree problem can be represented by a binary vector $(x_1, \ldots, x_n)$, in which $x_i = 1$ if node $i \in V' = W \cup T$; $x_j = 0$ otherwise, for every $i = 1, \ldots, n$. We notice that this solution representation is very similar to those adopted for the maximum clique and knapsack problems.

Figure 4.1 illustrates these ideas for an instance of the Steiner tree problem in graphs. This instance is depicted in Figure 4.1(a) and has 15 nodes. The terminal nodes 1, 2, 3, and 4 are represented by circles, while the optional nodes 5 to 15 correspond to squares. The graph induced by the terminal nodes 1, 2, 3, and 4 and the optional nodes 6, 7, 10, 13, and 14 is shown in Figure 4.1(b), with the edges of the corresponding minimum spanning tree marked in red. This minimum spanning tree of length 72 contains edge (6,10), which is not needed to form a Steiner tree of the original graph. Node 10 is removed from the set of optional nodes and the graph induced by the terminal nodes 1, 2, 3, and 4 and the optional nodes 6, 7, 13, and 14 is shown in Figure 4.1(c). The new minimum spanning tree has a smaller length 64 and its edges are marked in red. To conclude this example, we show in Figure 4.2 a still better Steiner tree with length 62 for the same instance, formed by the red edges and containing the optional nodes 9, 11, 12, and 13. ∎

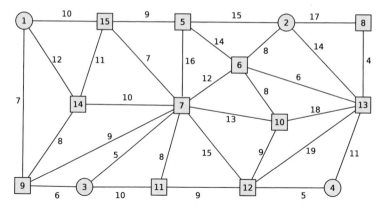

(a) Instance of the Steiner tree problem in a graph with 15 nodes and four terminals (nodes 1, 2, 3, and 4).

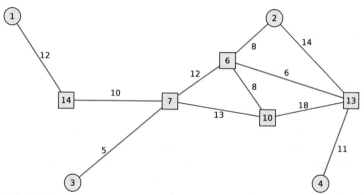

(b) The Steiner tree defined by the optional nodes 6, 7, 10, 13, and 14 is formed by the red edges and has length 72.

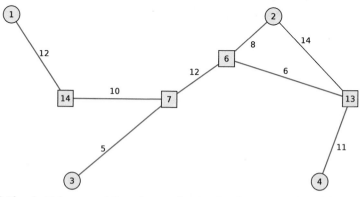

(c) If node 10 is removed from the set of optional nodes, then the new Steiner tree defined by the optional nodes 6, 7, 13, and 14 has a smaller length 64.

**Fig. 4.1** Solution representation for the Steiner tree problem in graphs.

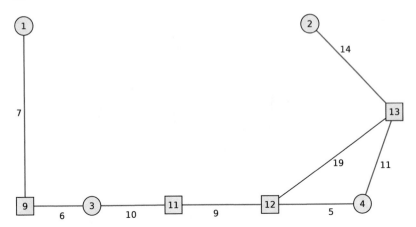

**Fig. 4.2** Smaller Steiner tree with length 62 for the instance in Figure 4.1(a).

## Traveling salesman problem – Solution representation

Let $V = \{1,\ldots,n\}$ be the set of cities a traveling salesman has to visit, with non-negative lengths $d_{ij}$ associated with each pair of cities $i, j \in V$.

Any tour visiting each of the $n$ cities exactly once corresponds to a feasible solution. Every feasible solution $S$ can be represented by a binary vector $(x_1,\ldots,x_m)$, where $m = n(n-1)/2$ and $x_k = 1$ if the edge indexed by $k$ belongs to the corresponding tour, $x_k = 0$ otherwise, for every $k = 1,\ldots,m$. However, this representation applies to any edge subset, regardless if it corresponds to a tour or not. Therefore, the edge subset $\{k = 1,\ldots,m : x_k = 1\}$ must define a tour for this solution to be feasible.

Figure 4.3 illustrates a complete graph with four nodes. Numbers on the six edges represent their indices. Every solution can be represented by a binary vector $(x_1,x_2,x_3,x_4,x_5,x_6)$. There are three different tours, corresponding to the incidence vectors $(1,1,1,1,0,0)$, $(1,0,1,0,1,1)$, and $(0,1,0,1,1,1)$.

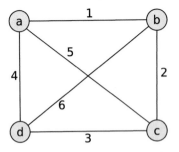

**Fig. 4.3** Hamiltonian cycles on a complete graph with four nodes.

We notice that any solution to the traveling salesman problem can alternatively be represented by a circular permutation $(\pi_1,\ldots,\pi_n)$ of the $n$ cities, with $\pi_i \in V$

for every $i = 1, \ldots, n$ and $\pi_i \neq \pi_j$ for every $i, j = 1, \ldots, n : i \neq j$. This permutation is associated with the tour defined by the edges $(\pi_1, \pi_2)$, $(\pi_2, \pi_3)$, $\ldots$, $(\pi_{n-1}, \pi_n)$, and $(\pi_n, \pi_1)$. Referring to Figure 4.3, the three tours represented by the incidence vectors $(1, 1, 1, 1, 0, 0)$, $(1, 0, 1, 0, 1, 1)$, and $(0, 1, 0, 1, 1, 1)$ correspond, respectively, to the circular permutations $(a, b, c, d)$, $(a, b, d, c)$, and $(a, c, d, b)$. ■

This discussion illustrates the fact that alternative representations can exist for a combinatorial optimization problem. The choice of one over another can lead to easier implementations or faster algorithms. In some occasions, it can also be helpful to work simultaneously with two different representations for every solution, since each of them can be more effective than the other for the implementation of some specific operations. We find the following schemes among the most frequently used solution representation techniques:

- *0-1 incidence vector*: This representation is typically used whenever the ground set is partitioned into two subsets, one of them corresponding to the elements that belong to the solution, while the others do not. This representation was applied in the three examples described above.
- *Generalized incidence vector*: This representation is often used whenever the ground set has to be partitioned into a number of subsets, each of them with a different interpretation. We can cite as examples the graph coloring problem and its variants, in which different colors have to be assigned to adjacent nodes of a graph. Another example is that of the vehicle routing and scheduling problem, in which a number of vehicles have to be assigned to clients. Still another example is that of the bin packing problem, in which a number of items have to be accommodated in different bins with the same size.
- *Permutation*: This representation typically applies to scheduling problems in which one is interested in establishing an optimal order for the execution of a number of tasks. It was illustrated above in the context of the traveling salesman problem.

## 4.2 Neighborhoods and search space graph

A *neighborhood* of a solution $S \in F$ can be defined by any subset of $F$. More formally, a neighborhood is a mapping that associates each feasible solution $S \in F$ with a subset $N(S) = \{S_1, \ldots, S_p\}$ of feasible solutions also in $F$.

Each solution $S' \in N(S)$ can be reached from $S$ by an operator called *move*. Normally, two neighboring solutions $S$ and $S' \in N(S)$ differ only by a few elements and a move from a solution $S$ consists simply in changing one or more elements in $S$. Usually, $S \in N(S')$ whenever $S' \in N(S)$.

The *search space graph* $\mathcal{G} = (F, M)$ has a node set that corresponds to the set $F$ of feasible solutions. The edge set $M$ of the search space graph is such that there is an edge $(S, S') \in M$ between two solutions $S, S' \in F$ if and only if $S' \in N(S)$ and $S \in N(S')$. An extended search space graph may be similarly defined, encompassing not only the set of feasible solutions $F$ but, instead, the whole set $\hat{F} = 2^E$ formed by all subsets of elements of the ground set $E$.

Figure 4.4 displays an example of an instance of a combinatorial problem in which the set $F$ is formed by 16 feasible solutions depicted in a square grid and represented by $S(i, j)$, for $i, j = 1, \ldots, 4$.

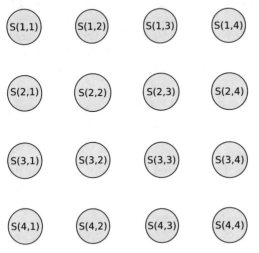

**Fig. 4.4** Set of 16 feasible solutions of a combinatorial optimization problem.

A search space graph associated with the above problem instance can be created by imposing a neighborhood definition on the node set $F$. Neighborhood $N_1$ is defined such that any solution $S(i, j)$ has neighbors $S(i+1, j)$, $S(i-1, j)$, $S(i, j+1)$, and $S(i, j-1)$, whenever they exist. Figure 4.5 displays the corresponding search space graph.

However, other different neighborhoods can be defined and imposed on the same set of feasible solutions. Another neighborhood $N_2$ can be defined, such that any solution $S(i, j)$ has neighbors $S(i+1, j+1)$, $S(i+1, j-1)$, $S(i-1, j+1)$, and $S(i-1, j-1)$, whenever they exist. Figure 4.6 displays the search space graph defined by this neighborhood.

We observe that some pairs of solutions are closer within one neighborhood or another. For instance, six moves are necessary to traverse the search space graph defined by neighborhood $N_1$ from $S(1,1)$ to $S(4,4)$, although only three are necessary if neighborhood $N_2$ is applied. On the other hand, every feasible solution is reachable from any other one if neighborhood $N_1$ is used. Contrarily, if the search starts from a solution $S(i, j)$ where $i + j$ is odd and takes moves defined by neighborhood $N_2$, only half of the solutions in the search space graph are reachable. However, if $i + j$ is even, then only the other half of the solutions can be visited. Therefore, the search space graph is not connected in this case. This can lead to implementation difficulties and can even make it impossible for the search procedure to find good solutions located at the part of the graph that cannot be reached from the initial solution.

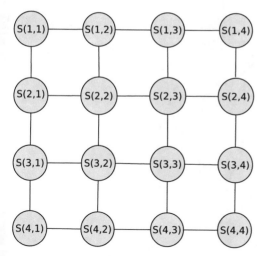

**Fig. 4.5** Search space graph with 16 feasible solutions and neighborhood $N_1$.

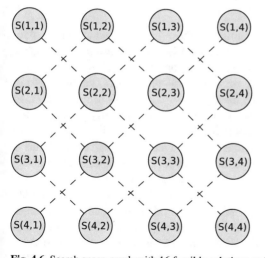

**Fig. 4.6** Search space graph with 16 feasible solutions and neighborhood $N_2$.

Since each of the neighborhoods $N_1$ and $N_2$ leads to different search paths through the set $F$ of feasible solutions, a natural idea is to combine them into a single neighborhood. Therefore, neighborhood $N_3$ can be defined as the union of $N_1$ and $N_2$: within this new neighborhood, any feasible solution $S(i, j)$ has up to eight neighbors $S(i+1, j), S(i-1, j), S(i, j+1), S(i, j-1), S(i+1, j+1), S(i+1, j-1), S(i-1, j+1)$, and $S(i-1, j-1)$, whenever they exist. Figure 4.7 displays the search space graph defined by this enlarged neighborhood.

At this point, it is very insightful to present some analogies between these three neighborhoods on the $4 \times 4$ grid of feasible solutions of our illustrative combinatorial optimization problem and the way chess pieces move on a chess board.

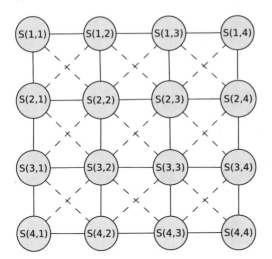

**Fig. 4.7** Search space graph with 16 feasible solutions and neighborhood $N_3$.

Moves along neighborhood $N_1$ are similar to those of rooks, which move along rows and columns of the chess board. Moves following neighborhood $N_2$ are equivalent to those of bishops, which traverse the diagonals of the chess board. While the white bishop can visit only white squares of the chess board, the black bishop can visit only the black squares. Each bishop can visit only half of the chess board squares, in the same way as moves within neighborhood $N_2$ starting from any feasible solution in Figure 4.4 can visit only half of the set of feasible solutions. The queen is stronger than both a rook and a bishop, since it can perform all kinds of moves, along rows, columns, and diagonals. Analogously, neighborhood $N_3$ entails all moves that can be performed within neighborhoods $N_1$ and $N_2$.

The above discussion illustrates the notion that different neighborhoods can be defined and used in the implementation of a local search method. The larger the neighborhood, the denser will be the search space graph and the shorter will be the paths connecting any two solutions. However, the use of large neighborhoods requires the evaluation of more neighboring solutions, leading to larger computation times during the investigation of the current solution.

It is important to note that the search space does not need to be formed exclusively by feasible solutions in $F$, but can also contain any subset of the set $\hat{F} = 2^E$ formed by all solutions, either feasible or infeasible. The definitions presented in this section remain the same, with the search space graph being now defined over $\hat{F}$ and not only over $F$. In this situation, the search can visit feasible and infeasible solutions but, in any case, it must terminate at a feasible solution. Working with more complex search space graphs, which include infeasible solutions, can be essential in some cases to ensure connectivity between any pair of feasible solutions.

In conclusion, finding an appropriate neighborhood and the best way to explore it is a crucial step towards the implementation of effective and efficient local search methods.

### Knapsack problem – Neighborhood and search space graph

We have already seen that any solution of the knapsack problem can be represented by a binary vector $(x_1, \ldots, x_n)$, in which $x_i = 1$ if item $i$ is selected, $x_i = 0$ otherwise, for every $i = 1, \ldots, n$. A solution $S = (x_1, \ldots, x_n)$ belongs to the feasible set $F$ if $\sum_{i=1}^{n} a_i \cdot x_i \leq b$. In this context, a move from any solution amounts to complementing the value of any single variable among $x_1, \ldots, x_n$, while keeping the others fixed. Each solution has exactly $n$ neighbors and the full search space graph defined over the set $\hat{F}$ formed by all feasible and infeasible solutions is an $n$-dimensional hypercube, as depicted in Figure 4.8. ∎

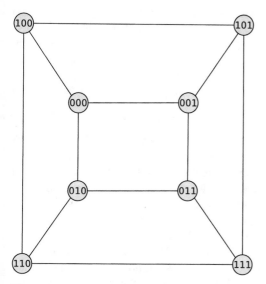

**Fig. 4.8** Search space graph for a knapsack problem with three items.

### Traveling salesman problem – Neighborhood and search space graph

We consider the case of a traveling salesman problem defined over a set $V = \{1, \ldots, n\}$ of cities that have to be visited exactly once. We have already seen that a feasible solution is a tour visiting each of the cities in $V$ exactly once. Any feasible solution of the traveling salesman problem can be represented by a circular permutation $(\pi_1, \pi_2, \ldots, \pi_{n-1}, \pi_n)$ of the $n$ cities, with $\pi_i \in V$ for every $i = 1, \ldots, n$ and $\pi_i \neq \pi_j$ for every $i, j = 1, \ldots, n : i \neq j$. This circular permutation is equivalent to any of the $n$ linear permutations $(\pi_1, \pi_2, \ldots, \pi_{n-1}, \pi_n)$, $(\pi_2, \pi_3, \ldots, \pi_n, \pi_1)$, $\ldots$, and $(\pi_n, \pi_1, \ldots, \pi_{n-2}, \pi_{n-1})$, each of them originating at a different city. All of them correspond to the same tour $(\pi_1, \pi_2)$, $(\pi_2, \pi_3)$, $\ldots$, $(\pi_{n-1}, \pi_n)$, $(\pi_n, \pi_1)$. The three tours in Figure 4.3 correspond to the circular permutations $(a, b, c, d)$, $(a, b, d, c)$, and $(a, c, d, b)$.

The search space graph has exactly $n!$ nodes, each of them corresponding to a permutation of the $n$ cities to be visited. Several neighborhood definitions are possible in this context. Neighborhood $N_1$ is defined as that formed by all permutations that can be obtained by exchanging the positions of two consecutive cities of the current permutation. Any solution $(\pi_1, \ldots, \pi_{i-1}, \pi_i, \ldots, \pi_n)$ has exactly $n-1$ neighbors within neighborhood $N_1$, each of them defined by a different permutation $(\pi_1, \ldots, \pi_i, \pi_{i-1}, \ldots, \pi_n)$ characterized by the swap of cities $\pi_{i-1}$ and $\pi_i$, for $i = 2, \ldots, n$. Figure 4.9 illustrates the search space graph corresponding to this neighborhood for a symmetric traveling salesman problem with four cities. Every solution has exactly three neighbors. As an example, the neighbors of solution $(1, 2, 3, 4)$ are $(2, 1, 3, 4)$, $(1, 3, 2, 4)$, and $(1, 2, 4, 3)$.

Neighborhood $N_2$ is defined by associating a solution $(\pi_1, \ldots, \pi_i, \ldots, \pi_j, \ldots, \pi_n)$ with all $n(n-1)/2$ neighbors $(\pi_1, \ldots, \pi_j, \ldots, \pi_i, \ldots, \pi_n)$ that can be obtained by exchanging the positions of any two cities $\pi_i$ and $\pi_j$, for $i, j = 1, \ldots, n : i \neq j$. Considering the same example of a symmetric traveling salesman problem with four cities, every solution has exactly six neighbors. In particular, the same solution $(1, 2, 3, 4)$ has now $(2, 1, 3, 4)$, $(1, 3, 2, 4)$, $(1, 2, 4, 3)$, $(3, 2, 1, 4)$, $(1, 4, 3, 2)$, and $(4, 2, 3, 1)$ as its neighbors.

We can also define a third neighborhood $N_3$ for the same problem by associating a solution $(\pi_1, \ldots, \pi_{i-1}, \pi_i, \pi_{i+1}, \ldots, \pi_j, \ldots, \pi_n)$ with all $n(n-1)/2$ neighbors $(\pi_1, \ldots, \pi_{i-1}, \pi_{i+1}, \ldots, \pi_i, \pi_j, \ldots, \pi_n)$ that can be obtained by moving city $\pi_i$ to position $j$, with $1 \leq i < j \leq n$, and shifting by one position to the left all cities between positions $i+1$ and $j$. Considering once again the same example of a traveling salesman problem with four cities as above, every solution has also exactly six neighbors. In particular, solution $(1, 2, 3, 4)$ has now $(2, 1, 3, 4)$, $(2, 3, 1, 4)$, $(2, 3, 4, 1)$, $(1, 3, 2, 4)$, $(1, 3, 4, 2)$, and $(1, 2, 4, 3)$ as its neighbors.

Each solution has three neighbors in neighborhood $N_1$ and six neighbors in neighborhoods $N_2$ and $N_3$, for this example. Neighbors in neighborhoods $N_2$ and $N_3$ are not the same. Therefore, even if the feasible solution set is the same in the three examples, the search space graphs defined by the three neighborhoods have different edge sets and, consequently, are different. Figure 4.10 superimposes neighborhoods $N_1$, $N_2$, and $N_3$, illustrating the neighbors of solution $(1, 2, 3, 4)$ within each of the three neighborhoods.

We recall that each tour $(\pi_1, \pi_2)$, $(\pi_2, \pi_3)$, $\ldots$, $(\pi_{n-1}, \pi_n)$, $(\pi_n, \pi_1)$ corresponds to exactly $n$ linear permutations of the cities to be visited. Considering the same example of a traveling salesman problem with four cities, the tour that starts at city 1, visits cities 2, 3, and 4 in this order, and then returns to city 1, can be represented by any one of the following four linear permutations: $(1, 2, 3, 4)$, $(2, 3, 4, 1)$, $(3, 4, 1, 2)$, and $(4, 1, 2, 3)$. Furthermore, if the problem is symmetric, then each tour can be traversed in two opposite directions with the same total traveled distance. Therefore, permutations $(4, 3, 2, 1)$, $(1, 4, 3, 2)$, $(2, 1, 4, 3)$, and $(3, 2, 1, 4)$ also correspond to the

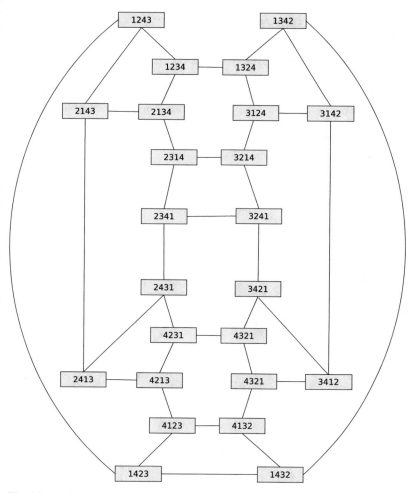

**Fig. 4.9** Search space graph defined by neighborhood $N_1$ for a traveling salesman problem with four cities.

same tour, but traversed in reverse order. In the general case, each tour corresponds to a circular permutation of the $n$ cities. Since each circular permutation can be indistinctly traversed in two different orders, the number of feasible solutions in $F$ can be reduced from $n!$ to $(n-1)!/2$, leading to a more compact representation of the problem. However, this reduction is not sufficient to make the solution of large problems easier, since the size of the search space graph remains superpolynomial in the number of cities. ∎

## 4.3 Implementation strategies

We have shown that the search space can be seen as a graph whose vertices correspond to feasible solutions that are connected by edges associated with neighboring solutions. A path in the search space graph consists of a sequence of feasible solutions, in which any two consecutive solutions are neighbors of each other.

This definition of the search space graph can be enlarged to allow also for vertices that correspond to infeasible solutions. In this case, paths in the search space can visit feasible as well as infeasible solutions.

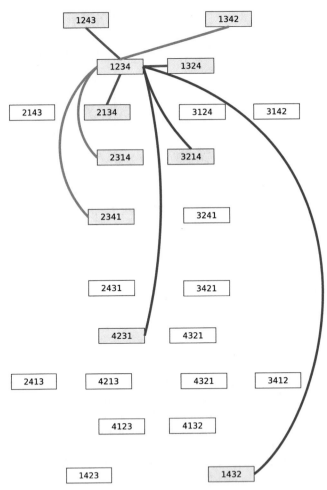

**Fig. 4.10** Neighborhoods $N_1$, $N_2$, and $N_3$ of solution $(1,2,3,4)$ in the search space graph of a traveling salesman problem with four cities. Nodes connected by red edges belong to the three neighborhoods. Nodes connected by blue edges are those within neighborhood $N_2$, while those connected by green edges belong to neighborhood $N_3$.

Given any instance of a minimization problem defined by a finite ground set $E = \{1,\ldots,n\}$, a set of feasible solutions $F \subseteq 2^E$, and an objective function $f : 2^E \to \mathbb{R}$, we have noted in Section 2.1 that we seek a globally optimal solution $S^* \in F$ such that $f(S^*) \leq f(S)$, $\forall S \in F$.

We have also seen in Section 4.2 that a neighborhood of a feasible solution $S \in F$ is defined by a mapping that associates $S$ with a subset $N(S)$ of feasible solutions also in $F$. A solution $S^+$ is said to be a *local optimum* for a minimization problem with respect to neighborhood $N$ if and only if $f(S^+) \leq f(S)$, $\forall S \in N(S^+)$. We notice that a global optimum is also locally optimal with respect to any neighborhood, while a local optimum is not necessarily a global optimum.

Local search methods can be viewed as a traversal of the search space graph starting from any given solution and stopping whenever some optimality condition is met. In most cases, a local search procedure is made to stop after a locally optimal solution is encountered. Metaheuristics such as tabu search extend the search beyond the first local optimum found, offering different escape mechanisms. The effectiveness and efficiency of a local search method depend on several factors, such as the starting solution, the neighborhood structure, and the objective function being optimized. The main components or phases of a local search method are

**Phase 4.1** – *Start: Construction of the initial solution, from where the search starts. Methods that may be applied to build an initial solution have already been discussed in Chapter 3 and will be further considered in Chapter 7.*

**Phase 4.2** – *Neighborhood search: Application of a subordinate heuristic or search strategy to find an improving solution in the neighborhood of the current solution. Neighborhood search strategies will be discussed along this section.*

**Phase 4.3** – *Stop: Interruption of the search by a stopping criterion, which in most cases consists in the identification that a locally optimal solution has been found. Stopping criteria for the neighborhood search will be considered at different points of this chapter.*

## 4.3.1 Neighborhood search

We consider in the following first-improving, best-improving, and other variants and strategies for the implementation of the neighborhood search.

At any iteration of an *iterative improvement* or *first-improving* neighborhood search strategy, the algorithm moves from the current solution to any neighbor with a better value for the objective function. In general, the new solution is the first-improving solution identified along the neighborhood search. The pseudo-code in Figure 4.11 describes a local search procedure based on a first-improving strategy for a minimization problem. The search starts from a given initial solution $S$. A flag

```
begin FIRST-IMPROVING(S);
1  improvement ← .TRUE.;
2  while improvement = .TRUE. do
3     improvement ← .FALSE.;
4     forall S' ∈ N(S) while improvement = .FALSE. do
5        if f(S') < f(S) then
6           S ← S';
7           improvement ← .TRUE.;
8        end-if;
9     end-forall;
10 end-while;
11 return S;
end FIRST-IMPROVING.
```

**Fig. 4.11** Pseudo-code of a first-improving local search procedure for a minimization problem.

indicating whether or not an improving solution was found is set in line 1. The loop in lines 2 to 10 is performed until it becomes impossible to replace the current solution with a better neighbor. The flag is reset to .FALSE. in line 3 at the beginning of a new iteration. The loop in lines 4 to 9 visits every neighbor $S' \in N(S)$ of the current solution $S$ until an improving solution is found. If the test in line 5 detects that $S'$ is better than the current solution $S$, then the latter is replaced by the former in line 6 and the flag is reset to .TRUE. in line 7, indicating that a better solution was found. The algorithm returns the local optimum $S$ in line 11.

```
begin BEST-IMPROVING(S);
1  improvement ← .TRUE.;
2  while improvement = .TRUE. do
3     improvement ← .FALSE.;
4     f_best ← ∞;
5     forall S' ∈ N(S) do
6        if f(S') < f_best then
7           S_best ← S';
8           f_best ← f(S');
9        end-if;
10    end-forall;
11    if f_best < f(S) then
12       S ← S_best;
13       improvement ← .TRUE.;
14    end-if;
15 end-while;
16 return S;
end BEST-IMPROVING.
```

**Fig. 4.12** Pseudo-code of a best-improving local search procedure for a minimization problem.

At any iteration of a *best-improving* local search strategy, the algorithm moves from the current solution to the best of its neighbors, whenever this neighbor improves upon the current solution. The pseudo-code in Figure 4.12 describes a local search procedure based on a best-improving strategy for a minimization problem. As in the previous algorithm, the search starts from any given initial solution $S$. A flag indicating whether or not an improving solution was found is set to .TRUE. in line 1. The loop in lines 2 to 15 is performed until it becomes impossible to replace the current solution with a better neighbor. The flag is reset to .FALSE. in line 3 at the beginning of a new iteration. The variable $f_{best}$ that stores the best objective function value over all neighbors of the current solution $S$ is set to a large value in line 4. The loop in lines 5 to 10 visits every neighbor $S' \in N(S)$ of the current solution $S$. If the test in line 6 detects that a new neighbor $S'$ is better than the current best neighbor, then the current best is replaced by the improved neighbor in line 7 and the best objective function value $f_{best}$ in the neighborhood is updated in line 8. In line 11, we compare the current solution $S$ with its best neighbor $S_{best}$. If $f_{best}$ is less than $f(S)$, then the current solution is updated in line 12 and the flag is reset to .TRUE. in line 13, indicating that a better solution was found. The algorithm returns the local optimum $S$ in line 16.

To illustrate the main ideas discussed in this section, we consider the traveling salesman problem instances with four cities whose search space graph was represented in Figure 4.9. We assume that the edge lengths are such that the values of the objective function for each solution are those depicted in Figure 4.13. For this minimization problem, there is only one local minimum, which is, necessarily, also a global optimum. The node of the search space graph corresponding to this solution is colored red. We observe that independently of the starting solution and of the neighborhood search strategy, the local search always stops at the global optimum.

We now suppose that the edge lengths are modified and that the solution costs are now as depicted in Figure 4.14. Considering this new situation, there are six nodes of the search space graph corresponding to locally optimal solutions: two nodes colored green have their objective function values equal to 49, two colored blue have their objective function values equal to 48, and two colored red have their objective function values equal to 46. Among those, we observe that only the red nodes correspond to globally optimal solutions. Furthermore, we notice that the solution obtained by local search varies, depending on both the starting solution and the neighborhood search strategy.

### 4.3.2 Cost function update

The complexity of each neighborhood search iteration depends not only on the number of neighbors of each visited solution, but also on the efficiency of the computation of the cost function value for each neighbor.

Efficient implementations of neighborhood search usually compute the cost of each neighbor $S'$ by updating the cost of the current solution $S$, instead of calculating it from scratch, avoiding repetitive and unnecessary calculations, as illustrated in the two examples that follow.

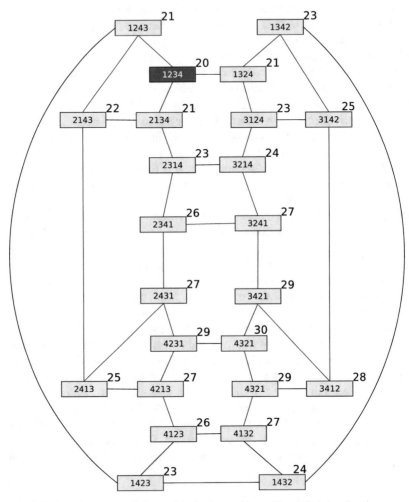

**Fig. 4.13** Search space graph for a minimization problem with a unique local optimum.

### Knapsack problem – Cost function update

Consider a solution $S$ for the knapsack problem, represented by a binary vector $(x_1, \ldots, x_n)$, in which $x_i = 1$ if item $i$ is selected, $x_i = 0$ otherwise, for every item $i = 1, \ldots, n$. The cost of solution $S$ is given by $f(S)$. Consider now a neighbor solution $S' \in N(S)$ that differs from $S$ by a single element, i.e., $S' = (x'_1, \ldots, x'_n)$, with

$x'_j = 1 - x_j$ for some $j \in \{1, \ldots, n\}$ and $x'_i = x_i$ for every $i = 1, \ldots, n : i \neq j$. The cost $f(S')$ can be computed from scratch in time $O(n)$ by adding up the costs of all items for which $x'_j = 1$. However, this value can be computed much faster in time $O(1)$ as $f(S') = f(S) + c_j$ if $x_j = 0$ and $x'_j = 1$, or as $f(S') = f(S) - c_j$ if $x_j = 1$ and $x'_j = 0$. ∎

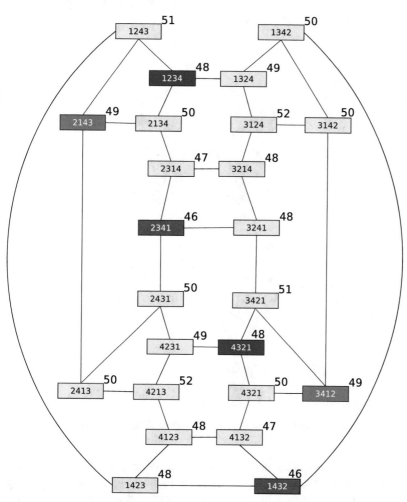

**Fig. 4.14** Search space graph for a minimization problem with six local optima.

## Traveling salesman problem – Cost function update

In Section 4.2 we examined three different neighborhoods for the traveling salesman problem. Here, we discuss the implementation of a local search procedure for the traveling salesman problem based on a different neighborhood definition.

Recall that every solution $S$ for an instance of the traveling salesman problem can be represented by an incidence binary vector $(x_1, \ldots, x_m)$, where $m = n(n-1)/2$ and $x_j = 1$ if the edge indexed by $j$ belongs to the corresponding tour, $x_j = 0$ otherwise, for $j = 1, \ldots, m$. Figure 4.15 (a) depicts an example involving a 5-vertex weighted graph, whose edges are numbered as indicated in Figure 4.15 (b). Figure 4.15 (c) shows the initial solution $S$ corresponding to the incidence vector $(1,1,1,1,1,0,0,0,0,0)$, whose cost is 17.

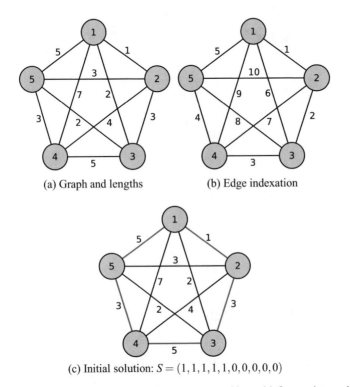

(a) Graph and lengths        (b) Edge indexation

(c) Initial solution: $S = (1,1,1,1,1,0,0,0,0,0)$

**Fig. 4.15** Instance of the traveling salesman problem with five vertices and its initial solution.

The *2-opt neighborhood* for the traveling salesman problem is defined by replacing any pair of nonadjacent edges of solution $S$ by the unique pair of edges that recreates a Hamiltonian cycle. Figure 4.16 displays the five neighbors of the initial solution $S = (1,1,1,1,1,0,0,0,0,0)$ together with their costs, indicating for each of them the eliminated edges by dashed lines. Suppose that the neighbors are generated and examined from left to right. In that case, a first-improving neighborhood search strategy would return the second generated neighbor (with cost 16) as the improving solution. However, if a best-improving strategy were applied, then it would return

the fourth neighbor as the best one (with cost 14). Assuming that this fourth neighbor is selected and becomes the new current solution $S = (1,0,1,1,0,1,0,0,0,1)$,
Figure 4.17 displays its five neighbors. The best of those is the second from left to
right (with cost 12). Since this solution cannot be improved by any of its neighbors,
then it is a local optimum and the search is interrupted.

Each solution has exactly $n(n-1)/2 - n$ neighbors. As for the case of the knapsack problem, the cost of each neighbor $S'$ can be recomputed in time $O(1)$ from the
cost of solution $S$, by simply taking the cost $f(S)$, subtracting the lengths of the two
removed edges, and adding those of the two that replaced them.

The *3-opt neighborhood* for the traveling salesman can be defined by taking three
nonadjacent edges of the current solution and replacing them with any of the only
four possible combinations of three edges that recreate a tour, as illustrated in Figure 4.18. In that case, the number of neighbors increases to $O(n^3)$ and the search
becomes slower, although more solutions can be investigated andbetter neighbors

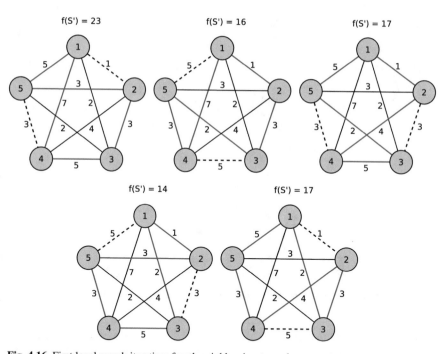

**Fig. 4.16** First local search iteration: fourth neighbor is returned.

might be possibly found. This neighborhood can be generalized to any $k \le m$: neighborhood $k$-opt is formed by all solutions that can be obtained by replacing $k$ edges
from the current solution by $k$ others that do not belong to it, so as to recreate a new
tour.                                                                                        ∎

### 4.3.3 Candidate lists

Candidate list strategies correspond to different techniques that make it possible
to implement local search methods in the most efficient ways by dealing faster or
with fewer neighbors instead of the full neighborhood. Basically, these candidate
list strategies provide a number of techniques to speed up the local search either by
restraining the number of neighbors investigated (for instance, when the neighborhood is very large) or by avoiding repetitive computations that can be saved from
one iteration to the next (typically, whenever the computation of the objective function is expensive).

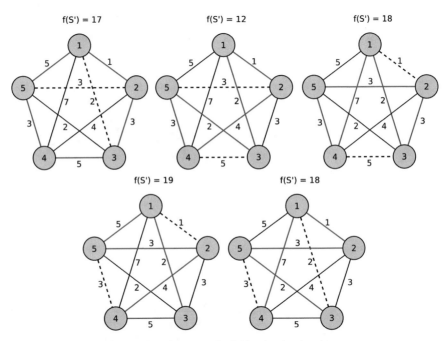

**Fig. 4.17** Second local search iteration: second neighbor is a local optimum.

Instead of describing the types of candidate list strategies that are useful for the
efficient implementation of local search, we illustrate with an example of one of the
most simple and effective variants often used.

We assume that a best-improving neighborhood search strategy is applied as part
of a local search method to solve a minimization problem. We also assume that the
current solution $S$ with cost $f(S)$ at a given local search iteration has $p$ neighboring
solutions, each of them associated with a move indexed by $j = 1, \ldots, p$. Each neighbor is obtained by applying one of the $p$ moves to the current solution $S$. Each move
$j = 1, \ldots, p$ may correspond, for instance, to flipping a 0-1 variable indexed by $j$,
or to interchanging the values of the $j$-th pair of variables, or, still, to removing the

$j$-th pair of edges in a 2-opt neighborhood for the traveling salesman problem and replacing them by the unique pair of edges that makes it possible to recover a tour. We denote by $S \oplus \{j\}$ the solution obtained by applying the move indexed by $j$ to the current solution $S$. The incremental cost associated with the move indexed by $j$ is computed and stored in $\Delta(j)$, for $j = 1, \ldots, p$. Therefore, the cost of each neighbor is given by $f(S \oplus \{j\}) = f(S) + \Delta(j)$, for $j = 1, \ldots, p$. In particular, suppose that $p = 10$ and $\Delta(1) = -10, \Delta(2) = 1, \Delta(3) = -8, \Delta(4) = -12, \Delta(5) = 4, \Delta(6) = -3,$ $\Delta(7) = -5, \Delta(8) = 6, \Delta(9) = -1,$ and $\Delta(10) = 5$. Since we are dealing with a minimization problem and a best-improving strategy is being used, the best-improving move is that indexed by $j^* = \text{argmin}\{\Delta(j) : \Delta(j) < 0, j = 1, \ldots, p\}$. Consequently, $j^* = 4$ in this example and the search moves to solution $S \oplus \{4\}$, whose cost is $f(S \oplus \{4\}) = f(S) + \Delta(4) = f(S) - 12$.

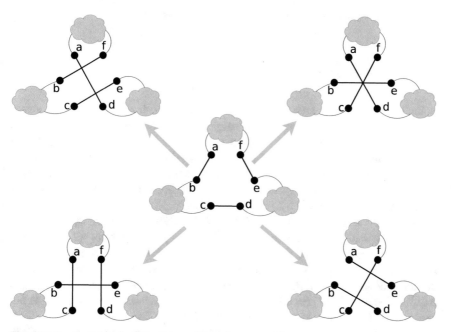

**Fig. 4.18** 3-opt neighborhood for the traveling salesman problem.

At this time, the best-improving neighborhood search strategy would investigate the full neighborhood of solution $S \oplus \{4\}$. Instead, we consider a candidate list formed by all yet unselected negative cost moves $j = 1, \ldots, p : j \neq j^*$ and $\Delta(j) < 0$. Therefore, all other possible moves from $S \oplus \{4\}$ are discarded and the candidate list is formed exclusively by the yet unselected negative cost moves indexed by 1, 3, 6, 7, and 9. In addition to reducing the number of neighbors investigated, we use the already-available values of $\Delta(j)$ as estimates of the new incremental costs associated with each move from $S \oplus \{4\}$. The best-improving strategy selects $j^* = 1$ as the best candidate move and only now the true value of the incremental cost $\Delta(1)$

associated with the application of the move indexed by 1 to solution $S \oplus \{4\}$ will be recomputed. If this move remains an improving move (i.e., if the recently updated value $\Delta(1)$ is negative), then it is applied to $S \oplus \{4\}$ and the search moves to solution $S \oplus \{4\} \oplus \{1\}$ with cost $f(S \oplus \{4\} \oplus \{1\}) = f(S) + \Delta(4) + \Delta(1)$. Otherwise, if $\Delta(1) \geq 0$, then this move became nonimproving. Therefore, it can be discarded from the candidate list and the procedure resumes from the reduced candidate list formed by the still remaining moves, which are indexed by 3, 6, 7, and 9.

This procedure continues until the candidate list is exhausted and becomes empty. The incremental costs of all possible moves from the current solution are fully reevaluated and the search continues from this updated candidate list, until a local optimum is found.

Different variants of candidate lists strategies have been proposed in the literature and successfully applied to a number of problems. Additional references are given in the bibliographical notes presented at the end of this chapter.

### 4.3.4 Circular search

In the case of a local search procedure following a first-improving strategy, a very effective strategy for exploring a full neighborhood or a candidate list consists in performing a circular search. A circular search amounts to using a circular list of candidate moves. As before, consider an example in which local search is applied to a minimization problem. Again, we assume that the current solution $S$ with cost $f(S)$ at a given local search iteration has $p$ neighboring solutions, each of them associated with a move indexed by $j = 1, \ldots, p$. Each neighbor is obtained by applying one of the $p$ moves to the current solution $S$. As before, we denote by $S \oplus \{j\}$ the solution obtained by applying the move indexed by $j$ to the current solution $S$. The incremental cost associated with the move indexed by $j$ is computed and stored in $\Delta(j)$, for $j = 1, \ldots, p$. Suppose that the moves are investigated in ascending order of their indices $j = 1, \ldots, p$, until the first-improving neighbor $j' \geq 2$ is found, i.e., $\Delta(j') < 0$ and $\Delta(j) \geq 0$ for all $j = 1, \ldots, j' - 1$. At this point, the new solution becomes $S \oplus \{j'\}$ and the search would resume from the first move, i.e., from $j = 1$. However, since in the last iteration this was not an improving move since $\Delta(1) \geq 0$, it most likely will still be a nonimproving move. The same applies to all moves $j = 2, \ldots, j' - 1$. Therefore, we profess that a more effective strategy resumes the search from $j = j' + 1$, instead of from $j = 1$. In this context, the move defined by $j = p$ is followed by that indexed by $j = 1$, as if they were organized as a circular list. The first-improving local search strategy stops at a local optimum as soon as a complete tour of this circular list is performed without any improvement in the current solution.

The use of a candidate list strategy using a circular search to implement a local search method based on a first-improving strategy can speed up the search by several orders of magnitude, without any loss in terms of solution quality.

## 4.4 Ejection chains and perturbations

Simple moves as those described in Section 4.2 can be extended to define broader neighborhoods associated with compound moves. These can be seen as sequences of simple moves that introduce structural changes in the current solution. Algorithms that incorporate compound moves are called variable depth methods, since the number of components of a compound move may vary from step to step.

*Ejection chains* are variable depth methods that make use of a sequence of interrelated, consecutive simpler moves to create more complex moves. They are designed to induce successive changes following an initial move that entailed infeasibilities in the neighbor solution, until feasibility is recovered. We say that an ejection chain has a variable depth because the number of simpler moves that are needed to recover feasibility may vary from one iteration to another. The length of the sequence of simple consecutive moves that lead to a compound move may be long and the evaluation of a compound move defining an ejection chain can be very costly in terms of computational effort. Therefore, the full exploration of neighborhoods defined by ejection chains can hardly be done. However, ejections chains may be employed as random or biased perturbations to destroy the structure of a local optimum, followed by the generation of a new solution that still shares part of the original local optimum that originated it.

Many metaheuristics make use of diversification strategies to drive the search towards unexplored regions of the search space. In particular, iterated local search (ILS) explores perturbation moves to escape from locally optimal solutions, obtaining new, different initial solutions from where the search restarts. Ejection chains are a very attractive alternative to generate perturbation moves in the context of diversification steps in tabu search or iterated local search.

We illustrate the use of ejection chains with an application to the traveling tournament problem. In this problem, one seeks to schedule the games of a compact round robin tournament involving $n$ teams. Each team plays exactly once with every other team. The tournament takes $n - 1$ rounds to be completed and every team plays exactly one game in each round. The tournament can be represented by the complete graph $K_n$, in which each node represents one team and each edge corresponds to the game between the two teams represented by its two extremities. Each round can be seen as a 1-factor of $K_n$, containing exactly $n/2$ edges. The goal is to minimize the total distance traveled by the teams, subject to constraints on the number of consecutive games each team plays at home and away. The discussion that follows regards exclusively feasibility issues.

Simple and easy moves to be applied within a local search method for the traveling tournament problem consist, for example, of team swaps (the opponents of any two teams are interchanged over all rounds) or round swaps (the games played in any two rounds are interchanged), as illustrated in Figure 4.19.

However, moves such as round swaps and team swaps are not sufficient to make
the search space graph connected and some solutions may be unreachable from the
initial solution. The game rotation neighborhood consists in enforcing any given
game to be played in any particular round. Only this neighborhood is capable of
breaking the inner structure of the 1-factorization corresponding to the initial solution and making the search space graph connected. Nevertheless, whenever a game
is removed from the round where it is currently being played and enforced to be
played in a different round, infeasibilities are created in both rounds. Appropriate
modifications to avoid clashes of teams playing more than once in the same round
should be applied. To eliminate these infeasibilities, new edges have to be succes-
sively removed from one round and reassigned to another, until a feasible solution is
recovered. The sequence of reassignments of games to new rounds gives rise to an
ejection chain, as illustrated in Figures 4.20 to 4.22 for an instance of the traveling
tournament problem with six teams.

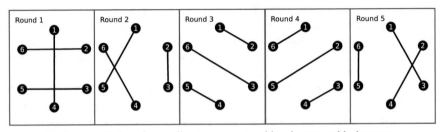

(a) Current solution of a traveling tournament problem instance with six teams.

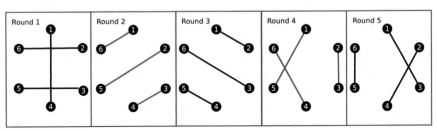

(b) Example of a round swap move applied to the solution in (a): rounds 2 and 4 (edges in red)
are interchanged and all games played in one of them are reassigned to the other.

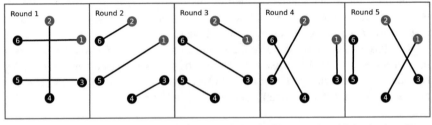

(c) Example of a team swap move applied to the solution in (b): teams 1 and 2 (nodes in red) are
interchanged and all their opponents are swapped over all rounds.

**Fig. 4.19** Round swap and team swap moves for the traveling tournament problem.

In Figure 4.20(a), game $(1,3)$ is selected to be reassigned from round 5 to round 2. In the seven steps in Figures 4.20(b) to 4.22(b) one game is reassigned to a new round and another game is selected for reassignment. Finally, in Figure 4.22(c) game $(2,3)$ is assigned to round 5 and a feasible solution is reached.

To conclude, we may say that ejection chains are then based on the principle of generating compound sequences of moves by linked steps in which changes in selected elements cause other elements to be ejected from their current state, position, or value assignment. Although they are often costly to be used as regular moves in a local search method in terms of computation time, they are very effective to generate perturbations and for search diversification.

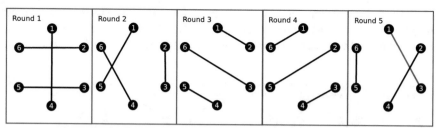

(a) Game $(1,3)$ will be removed from round 5 to be played in round 2.

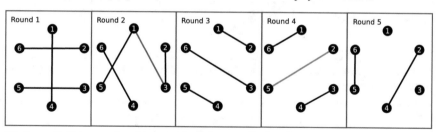

(b) Game $(1,3)$ is now played in round 2, while game $(2,5)$ will be removed from round 4 to be played in round 2.

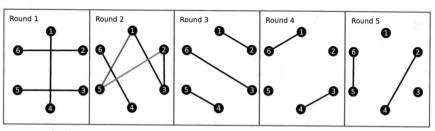

(c) Game $(2,5)$ is now played in round 2, while game $(1,5)$ will be removed from round 2 to be played in round 4.

**Fig. 4.20** Ejection chain moving game $(1,3)$ from round 5 to round 2: first three moves.

## 4.5 Going beyond the first local optimum

We have shown that local search methods always stop at the first local optimum, from which they are unable to escape. In the following, we illustrate two extended variants of local search that may overcome this limitation.

### 4.5.1 Tabu search and short-term memory

Tabu search is a memory-based metaheuristic whose philosophy is to derive and exploit a collection of principles of intelligent problem solving. The method is based on procedures designed to cross boundaries of feasibility or local optimality, which are often treated as barriers. It guides a local search procedure to explore the solution

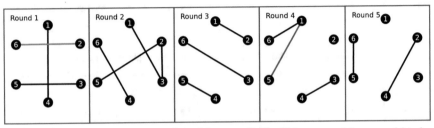

(a) Game $(1,5)$ is now played in round 4, while game $(2,6)$ will be removed from round 1 to be played in round 4.

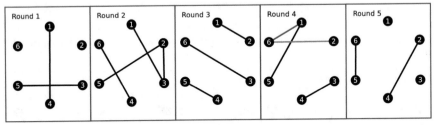

(b) Game $(2,6)$ is now played in round 4, while game $(1,6)$ will be removed from round 4 to be played in round 1.

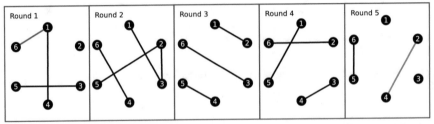

(c) Game $(1,6)$ is now played in round 1, while game $(2,4)$ will be removed from round 5 to be played in round 1.

**Fig. 4.21** Ejection chain moving game $(1,3)$ from round 5 to round 2: intermediate three moves.

space beyond local optimality. In its simplest version, which makes exclusive use of short-term memory to avoid cycling, tabu search may be seen as a powerful extension of a local search procedure that accepts nonimproving moves to escape from locally optimal solutions, until some alternative stopping criterion is satisfied.

In this context, a reasonable strategy to extend a pure local search method beyond optimality consists in accepting only a small, limited number of nonimproving moves in order to give the search a chance of escaping from the first local optimum found, without either being trapped or increasing too much the computation time. A short-term memory has to be used to keep track of recent nonimproving moves whose reversal should be forbidden to avoid cycling, i.e., visiting a solution that has been already visited in a previous iteration.

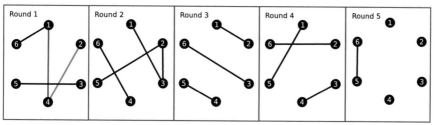

(a) Game $(2,4)$ is now played in round 1, while game $(1,4)$ will be removed from round 1 to be played in round 5.

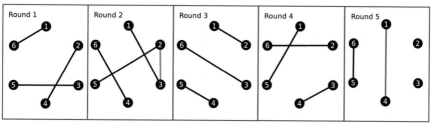

(b) Game $(1,4)$ is now played in round 5, while game $(2,3)$ will be removed from round 2 to be played in round 5.

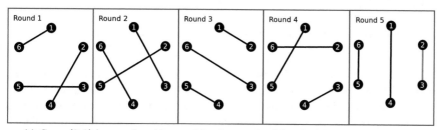

(c) Game $(2,3)$ is now played in round 5 and a new feasible schedule is finally obtained.

**Fig. 4.22** Ejection chain moving game $(1,3)$ from round 5 to round 2: final three moves.

## 4.5.2 *Variable neighborhood descent*

A local optimum with respect to some neighborhood is not necessarily optimal with respect to another neighborhood. For example, a locally optimal solution for the traveling salesman problem under the 2-opt neighborhood may not be optimal within the 3-opt neighborhood. Changes of neighborhoods can be successfully performed within a local search procedure. They are crucial and instrumental in some cases. Exploring different neighborhoods in an appropriate order can save a significant amount of computation time.

In this context, small neighborhoods or those whose elements can be quickly evaluated may be explored first, with the search moving to progressively larger or more complex neighborhoods as locally optimal solutions are found within the lower order neighborhoods. This local search strategy is called *variable neighborhood descent* (VND) and its main steps are presented in the pseudo-code of Figure 4.23 for a minimization problem, in which $\mathcal{N}^k$ denotes the $k$-th neighborhood to be explored, for $k = 1, \ldots, k_{max}$.

```
begin VND(S);
1    improvement ← .TRUE.;
2    while improvement = .TRUE. do
3        improvement ← .FALSE.;
4        k ← 1;
5        while k ≤ kmax do
6            S' ← BEST-IMPROVING(S, N^k);
7            if f(S') < fS then
8                S ← S';
9                k ← 1;
10               improvement ← .TRUE.;
11           else k ← k + 1;
12           end-if;
13       end-while;
14   end-while;
15   return S;
end VND.
```

**Fig. 4.23** Pseudo-code of a VND local search procedure for a minimization problem.

At any iteration of a VND local search strategy, the algorithm moves from the current solution to the best of its neighbors within the current neighborhood, whenever the best neighbor improves upon the current solution. The search starts from a given initial solution $S$. A flag indicating whether or not an improving solution was found is set to .TRUE. in line 1. The loop in lines 2 to 14 is performed until it becomes impossible to replace the current solution by a better neighbor. The flag is reset to .FALSE. in line 3 at the beginning of a new iteration. The index of the current neighborhood is set initially to 1 in line 4. The loop in lines 5 to 13 consecutively investigates all neighborhoods. Line 6 returns the best neighbor solution $S'$

of the current solution $S$ within neighborhood $\mathcal{N}^k$. If the test in line 7 detects that $S'$ is better than the current solution $S$, then $S$ is replaced by $S'$ in line 8, the current neighborhood is reset to 1 in line 9, and the flag is reset to .TRUE. in line 10, indicating that a better solution was found. Otherwise, the neighborhood counter is increased by one, indicating that a higher-order neighborhood will be investigated next. The algorithm returns the local optimum $S$ with respect to all neighborhoods in line 15.

## 4.6 Final remarks

We have seen that local search methods start from an initial solution and iteratively improve it until a local optimum is found. They are memoryless search methods that are very sensitive to the initial solution and stop at the first local optimum, being unable to escape from it. They are also sensitive to the neighborhood structure and to the search strategy applied to explore each neighborhood.

Metaheuristics are general high-level procedures that coordinate simple heuristics and rules to find good (often optimal) approximate solutions to computationally difficult combinatorial optimization problems. Among them, we find simulated annealing, tabu search, GRASP, VNS, genetic algorithms, scatter search, ant colonies, and others. They are based on distinct paradigms and offer different mechanisms to escape from locally optimal solutions, as opposed to greedy algorithms or local search methods. Metaheuristics are among the most effective solution strategies for solving combinatorial optimization problems in practice and they have been applied to a large variety of areas and situations. The customization (or instantiation) of some metaheuristic to a given problem yields a heuristic to the problem.

In the next chapter we give an introductory presentation to the fundamentals of the GRASP (which stands for *greedy randomized adaptive search procedures*) metaheuristic and is one of the most effective approximate solution methods for hard combinatorial optimization problems. GRASP is one of the alternatives to overcome some of the main limitations of basic local search methods, such as the sensitivity to the initial solution and stopping at the first local optimum.

## 4.7 Bibliographical notes

Local search methods are a common component of a number of metaheuristics. Hoos and Stützle (2005) defined stochastic local search algorithms to be methods based on local search that make use of randomization to generate or select candidate solutions for combinatorial optimization problems. Yagiura and Ibaraki (2002) traced the history of local search since the work of Croes (1958), where a local search algorithm for the traveling salesman problem was proposed. Kernighan and Lin (1970) and Lin and Kernighan (1973) were early proponents of local search for, respectively, graph partitioning and the traveling salesman problem. Michelis et al. (2007) discussed theoretical aspects of local search. The simplex algorithm developed by Dantzig (1953) can be seen as a local search algorithm for solving linear programming problems.

The principles discussed in Sections 4.1 to 4.3 of this chapter are usually described in books, chapters, and articles devoted to metaheuristics, such as simulated annealing, tabu search, GRASP, variable neighborhood search, iterated local search, genetic algorithms, and ant colony optimization, which all make use of local search. In particular, we refer the reader to the aforementioned book of Hoos and Stützle (2005), to the textbook on tabu search by Glover and Laguna (1997), to different chapters in the handbooks edited by Glover and Kochenberger (2003), Gendreau and Potvin (2010), and Burke and Kendall (2005; 2014), and to the book chapters by Rego and Glover (2002) and Yagiura and Ibaraki (2002).

We showed in Section 4.4 that simple moves can be extended to define broader neighborhoods associated with compound moves that are called ejection chains. They were defined by Rego and Glover (2002) as variable depth methods that generate a sequence of interrelated simple moves to create more complex moves. We refer the reader to Cao and Glover (1997), Glover (1996a), Glover and Punnen (1997), Pesch and Glover (1997), Glover (1991), and Rego (1998) for a comprehensive description of ejection chains and their applications to the traveling salesman problem and other optimization problems in graphs. Dorndorf and Pesch (1994), Laguna et al. (1995), Yagiura et al. (2004), and Cavique et al. (1999) applied ejection chains to problems in other domains. The traveling tournament problem was introduced in the seminal paper by Easton et al. (2001). Ribeiro and Urrutia (2007) presented a successful application of ejection chains to generate perturbations in the context of a hybridization of GRASP with iterated local search to solve the mirrored variant of the traveling tournament problem. We also refer the reader to Kendall et al. (2010) and Ribeiro (2012) for surveys of optimization problems in sports. Accounts of the iterated local search metaheuristic (ILS) were presented by Lourenço et al. (2003), Martin and Otto (1996), and Martin et al. (1991).

We showed in Section 4.5 that tabu search and variable neighborhood descent represent two extended variants of local search that make it possible to go beyond the first local optimum encountered. Tabu search is a memory-based metaheuristic proposed and developed by Glover (1989; 1990) in two seminal papers, see also Glover and Laguna (1997). A similar idea was independently developed by Hansen (1986) in what was called the steepest-ascent mildest-descent method. The use of a tabu search procedure based on a small short-term memory to replace a pure local search method for going beyond the first local optimum was successfully used by Souza et al. (2004) in the context of a GRASP heuristic developed for the capacitated minimum spanning tree problem. The idea of using nested neighborhoods to improve a local search procedure was around for a long time, see, e.g., the aforementioned applications to the traveling salesman problem (Lin, 1965; Lin and Kernighan, 1973) and to graph partitioning (Kernighan and Lin, 1970).

Variable neighborhood descent was completely described and its VND acronym coined by Mladenović and Hansen (1997), see also Hansen and Mladenović (2002; 2003). VND was used to implement the local search phase of GRASP heuristics for the Steiner problem in graphs (Martins et al., 2000) and for the phylogeny problem (Andreatta and Ribeiro, 2002), among other applications.

# Chapter 5
# GRASP: The basic heuristic

This chapter presents the basic structure of a greedy randomized adaptive search procedure (or, more simply, GRASP). We first introduce random and semi-greedy multistart procedures and show how solutions produced by both procedures differ. The hybridization of a semi-greedy procedure with a local search method constitutes a GRASP heuristic. The chapter concludes with some implementation details, including stopping criteria.

## 5.1 Random multistart

A *multistart* procedure is an algorithm which repeatedly applies a solution construction procedure and outputs the best solution found over all trials. Each trial, or iteration, of a multistart procedure is applied under different conditions. An example of a random multistart procedure for minimization is shown in Figure 5.1. The algorithm repeatedly generates random solutions. Similar to the GREEDY algorithm presented in Figure 3.1 of Chapter 3, a new random solution is generated in line 3 by adding to the partial solution (initially empty) a new feasible ground set element,

---

**begin** RANDOM-MULTISTART;
1   $f^* \leftarrow \infty$;
2   **while** stopping criterion not satisfied **do**
3       $S \leftarrow$ RandomSolution;
4       **if** $f(S) < f^*$ **then**
5           $S^* \leftarrow S$;
6           $f^* \leftarrow f(S)$;
7       **end-if**;
8   **end-while**;
9   **return** $S^*$;
**end** RANDOM-MULTISTART.

---

**Fig. 5.1** Pseudo-code of a random multistart procedure for a minimization problem.

© Springer Science+Business Media New York 2016
M.G.C. Resende, C.C. Ribeiro, *Optimization by GRASP*,
DOI 10.1007/978-1-4939-6530-4_5

one element at a time. Unlike in the greedy algorithm, each ground set element is chosen at random from the set of candidate ground set elements. If the solution has a better objective function value than the incumbent solution, then it is saved and made the incumbent in lines 5 and 6. The best solution over all iterations is returned as the solution of the random multistart algorithm.

## 5.2 Semi-greedy multistart

In Chapter 3 we introduced the adaptive greedy algorithm whose pseudo-code was given in Figure 3.8. In line 5 of the pseudo-code, the index of the next ground set element to be added to the partial solution under construction is chosen. Since there may exist more than one index $i \in \mathscr{F}$ that minimizes the greedy $g(i)$, the algorithm needs to have a tie-breaking rule. If ties are not broken at random, then embedding a greedy algorithm in a multistart strategy would be useless since the same solution would be produced at each iteration of the multistart procedure.

We also introduced in Figure 3.18 of Chapter 3 the semi-greedy construction procedure which adds randomization to the greedy algorithm. The semi-greedy algorithm can also be embedded in a multistart framework as shown in Figure 5.2. This algorithm is almost identical to the random multistart method, except that solutions are generated with a semi-greedy procedure instead of at random. Note that each invocation of procedure SemiGreedy($\cdot$) in line 3 of Figure 5.2 is independent of the others, therefore producing independent solutions.

```
begin SEMI-GREEDY-MULTISTART;
1   f* ← ∞;
2   while stopping criterion not satisfied do
3       S ← SEMI-GREEDY;
4       if f(S) < f* then
5           S* ← S;
6           f* ← f(S);
7       end-if;
8   end-while;
9   return S*;
end SEMI-GREEDY-MULTISTART.
```

**Fig. 5.2** Pseudo-code of a semi-greedy multistart procedure for a minimization problem.

The semi-greedy procedure can use either a quality-based or a cardinality-based restricted candidate list (RCL), as described in Section 3.4. In the former case, a quality-enforcing parameter $\alpha$ regulates how random or how greedy the construction will be. In a minimization problem, the value $\alpha = 0$ leads to a purely greedy construction, since it places in the RCL all ground set elements $i \in \mathscr{F}$ for which the cost associated with its inclusion in the solution is minimum. On the other hand, the value $\alpha = 1$ leads to a random construction, since it now places all feasible ground set elements in the RCL. Other values of $\alpha$, i.e., $0 < \alpha < 1$, mix greediness and randomness in the construction. The same outcomes can be achieved with a cardinality-based restricted candidate list. A cardinality-enforcing parameter value $k = 1$ places in the RCL only one candidate element, with minimum cost associated with its inclusion in the solution and therefore corresponds to a greedy construction. On the other hand, a parameter value $k = |\mathscr{F}|$ places all candidate elements in the RCL, and therefore corresponds to a random construction. Other values of $k$, i.e., $2 \leq k \leq |\mathscr{F}| - 1$, mix greediness and randomness in the construction.

Conversely, in the case of a maximization problem, a value $\alpha = 1$ would lead to a greedy construction, since it would place in the RCL all ground set elements $i \in \mathscr{F}$ for which the cost associated with the inclusion of the element in the solution is maximum. A value $\alpha = 0$ would lead to a random construction, since it would place all feasible ground elements in the RCL.

Figure 5.3 shows the distribution of solution values produced by a random multistart procedure and by a semi-greedy multistart algorithm with the RCL parameter $\alpha = 0.85$ on an instance of the *maximum covering problem*. In this problem, we are given a set of $m$ demand points, each with an associated weight, a set of $n$ potential facility locations, each of which can provide service (or cover) to a given subset of the demand points, and are asked to find $p \leq n$ facility locations such that the sum of the weights of the demand points covered by the $p$ facility points is maximized. The figure compares the two distributions with the greedy solution value and the best known solution value for this problem instance. It illustrates four important points:

1. Semi-greedy solutions are on average much better than random solutions.
2. There is more variance in the solution values produced by a random multistart method than by a semi-greedy multistart algorithm.

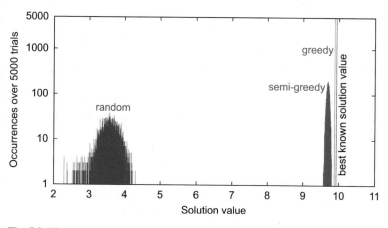

**Fig. 5.3** Distributions of 5000 random solution values and 5000 semi-greedy solution values for an instance of the maximum covering problem. The figure also shows the values of the greedy solution and of the best known solution for this problem instance.

3. The greedy solution is on average better than both the random and the semi-greedy solutions but, even if ties are broken at random, it has less variance than the random or semi-greedy solutions.
4. Random, semi-greedy, and greedy solutions are usually sub-optimal.

Figure 5.4 further illustrates the three first points above. It shows the distribution of 1000 independent semi-greedy solutions with RCL parameters $\alpha = 0$ (random), 0.2, 0.4, 0.6, 0.8, and 1 (greedy) on an instance of the *maximum weighted satisfiability problem*. In this problem, we are given $m$ disjunctive clauses $\mathscr{C}_1, \ldots, \mathscr{C}_m$, with corresponding real-valued weights $w_1, w_2, \ldots, w_m$, involving the Boolean variables $x_1, \ldots, x_n$ and their complements. The problem is to find a truth assignment of 0 (false) and 1 (true) values to these variables such that the sum of the weights of the satisfiable clauses (i.e., clauses that evaluate to true) is maximized.

## 5.3 GRASP

Local search was introduced in Chapter 4 as a solution improvement procedure that is given a starting solution and iteratively explores the neighborhood of the current solution, looking for one with a better objective function value. If such solution is found, it is made the current solution and the algorithm proceeds with a new iteration. Local search ends when no solution in the neighborhood of the current solution has a better objective function value than the current solution. In this case, the current solution is called a local optimum.

We observed that in designing a local search algorithm, one is usually given an objective function but has the flexibility to choose one or more neighborhood structures. We also observed that given an objective function and a neighborhood structure, the success of a local search algorithm to find a global optimum solution will depend on the starting solution it uses, among other factors. Regardless of the neighborhood search strategy used, some starting solutions always lead to a global optimum, while others always lead to a local optimum that is not globally optimal. Others, yet, can lead to either a global optimum or to a local optimum that is not globally optimal, depending on the search strategy used. Having the capability of producing different starting solutions for the local search, one would like to increase the likelihood of producing at least one starting solution that leads to a global optimum with the application of a local search procedure.

A *greedy randomized adaptive search procedure* (GRASP) is the hybridization of a semi-greedy algorithm with a local search method embedded in a multistart framework. The method consists of multiple applications of local search, each starting from a solution generated with a semi-greedy construction procedure. The best local optimum, over all GRASP iterations, is returned as the solution provided by the algorithm.

Figure 5.5 illustrates a basic GRASP heuristic for minimization. After initializing the value of the incumbent in line 1, the GRASP iterations are carried out in the while loop in lines 2 to 12. A solution $S$ is constructed with a semi-greedy algorithm

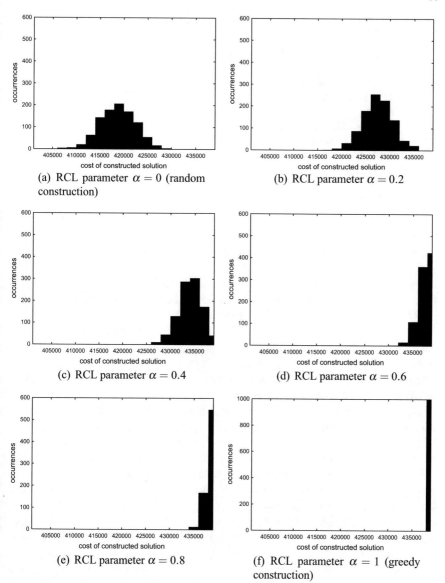

**Fig. 5.4** Distribution of semi-greedy solution values as a function of the quality-based RCL parameter $\alpha$ (1000 repetitions were recorded for each value of $\alpha$) on an instance of the maximum weighted satisfiability problem.

in line 3. As shown in Chapter 3, a semi-greedy algorithm may not always generate a feasible solution. When this happens, a repair procedure must be invoked in line 5 to make changes in $S$ so that it becomes feasible (alternatively, solution $S$ may be simply discarded and followed by a new run of the semi-greedy algorithm, until a feasible solution is built). In line 7, local search is applied starting from a feasible solution provided by the semi-greedy algorithm or, if necessary, by the repair

```
begin GRASP;
1   f* ← ∞;
2   while stopping criterion not satisfied do
3       S ← SEMI-GREEDY;
4       if S is not feasible then
5           S ← Repair(S);
6       end-if;
7       S ← LOCAL-SEARCH(S);
8       if f(S) < f* then
9           S* ← S;
10          f* ← f(S);
11      end-if;
12  end-while;
13  return S*;
end GRASP.
```

**Fig. 5.5** Pseudo-code of a basic GRASP heuristic for minimization.

procedure. We use LOCAL-SEARCH to denote any variant of the local search methods considered in Sections 4.3.1 and 4.5.2, such as FIRST-IMPROVEMENT, BEST-IMPROVEMENT, or VND. If the objective function value $f(S)$ of the local minimum produced in line 7 is better than the objective function value $f*$ of the incumbent, then the local minimum is made the incumbent in line 9 and its objective function value is recorded as $f*$ in line 10. The while loop is repeated until some stopping criterion is satisfied. The best solution found over all GRASP iterations is returned in line 13 as the GRASP solution.

Figure 5.6 shows the distribution of the solution values obtained after local search is applied to 1000 solutions built by the semi-greedy algorithm as a function of the RCL quality-enforcing parameter $\alpha$ of the semi-greedy construction procedure for an instance of the maximum weighted satisfiability problem. This figure is similar to Figure 5.4, with the difference being that here local search is applied to the semi-greedy solution. For each value of $\alpha$, 1000 GRASP iterations were carried out and a histogram was produced showing the frequency of solution values in different cost ranges. The distributions show that the variance of the solution values decreases as $\alpha$ increases, as already observed in Figure 5.4 for the distributions of the solution values obtained by semi-greedy construction. As occurs with semi-greedy construction, GRASP solutions improve on average as we move from a totally random construction to a greedy construction. However, they differ in one important way from those of Figure 5.4. The best solution found, over all 1000 runs, improves as we move from random to semi-greedy construction (until some value of parameter $\alpha$), and then deteriorates as $\alpha$ approaches 1. This is illustrated better in Figure 5.7, where we superimpose two plots: one for the average solution value and the other for the best solution value, both displayed as a function of $\alpha$.

Figures 5.8 and 5.9 show, respectively, the objective function values for solutions of an instance of the maximum covering problem constructed with random and semi-greedy algorithms, each followed by local search in a multistart procedure. For each iteration, the plots show in red the value of the constructed solution and in blue the value of the local maximum solution. The iterations are sorted in increasing

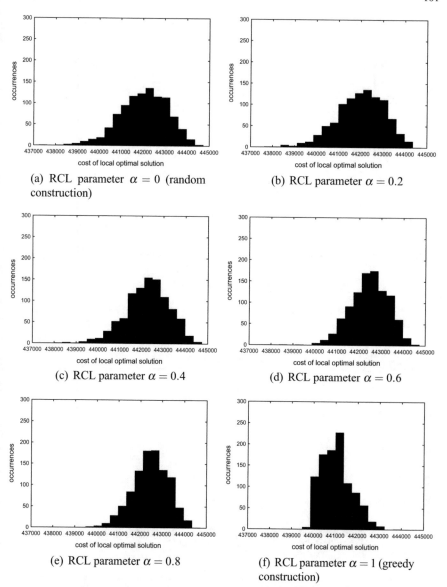

**Fig. 5.6** Distribution of the solution values obtained after local search as a function of the quality-based parameter $\alpha$ of the semi-greedy construction procedure (1000 repetitions for each value of $\alpha$) on an instance of the maximum weighted satisfiability problem.

order of the values of their local maxima. The figures illustrate further why the iterations of a random multistart method with local search are longer than those of a GRASP. While random construction can be slightly faster than semi-greedy construction, this does not compensate for the poor quality of the randomly constructed solutions when compared to the semi-greedy solutions. Note that the average value of solutions constructed with the random approach is 3.55, while the average value

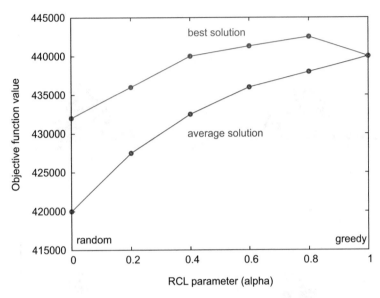

**Fig. 5.7** Best and average solution values for GRASP as a function of the RCL parameter $\alpha$ for 1000 GRASP iterations on an instance of the maximum weighted satisfiability problem.

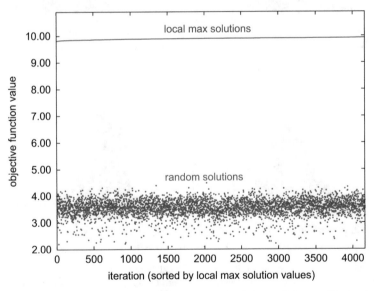

**Fig. 5.8** Random construction and local maximum solution values, sorted by local maximum values, for an instance of the maximum covering problem.

of solutions constructed with the semi-greedy algorithm is 9.70. This indicates that the path taken by local search to a local optimum from the semi-greedy solution is much shorter than the path taken from the random solution.

Figure 5.10 shows, for an instance of the maximum weighted satisfiability problem, the effect of different RCL parameter values on the Hamming distance between

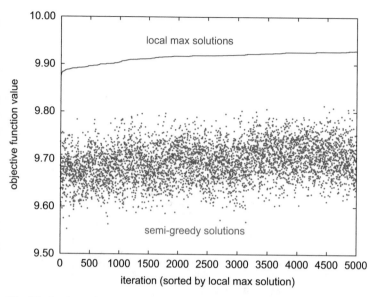

**Fig. 5.9** Semi-greedy construction and local maximum solution values, sorted by local maximum values, for an instance of the maximum covering problem.

the constructed solutions and the corresponding local maxima, the number of moves made by local search, and the local search and total running times. The figure shows a strong correlation between Hamming distance, number of moves taken by local search, and local search running time. Figure 5.11 displays, again for the same instance of the maximum covering problem considered in Figures 5.8 and 5.9, the best objective function solution value as a function of running time for GRASP (with $\alpha = 0.85$), random multistart (GRASP with $\alpha = 0$) with local search, and greedy multistart (GRASP with $\alpha = 1$) with local search. While greedy multistart with local search fails to find the best known solution of value 9.92926, GRASP finds it after only 126 seconds, while random multistart with local search takes 152,664 seconds to reach that solution, i.e., over one thousand times longer.

## 5.4 Accelerating GRASP

The local search phase takes considerably longer than the construction phase in most GRASP applications. Many times, a single execution of the local search algorithm may be more time-consuming than the overall time spent in constructing all starting solutions along the main GRASP loop. In addition to quick cost updates, candidate lists, and circular search strategies already explored in Section 4.3, a number of filtering strategies have been used to speedup the time spent with local search in GRASP.

*Filtering* consists basically in not applying local search at every GRASP iteration, but only to the most promising solutions built by the construction phase. One strategy is to apply local search only if the semi-greedy solution is better than a

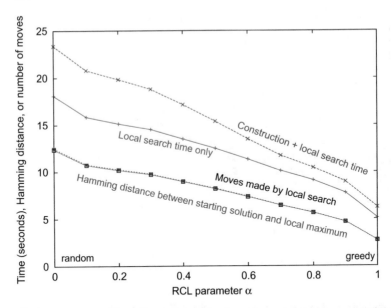

**Fig. 5.10** Total GRASP running time, total local search running time, average Hamming distance between constructed solution and local maximum, and average number of local search moves as a function of the RCL parameter $\alpha$ on 1000 GRASP iterations on an instance of the maximum weighted satisfiability problem. Note that since the local search traverses a 1-flip neighborhood, the curve for the number of moves made by local search coincides with the curve for the Hamming distance between the starting solution and the local maximum.

given threshold. Another strategy often applied consists in considering an integer parameter $\eta$ and applying local search only to the best solution built along the $\eta$ previous applications of the construction phase. Since the local search phase is usually much more time-consuming, this strategy may lead to a reduction by a factor of up to $\eta$ in the total time needed to perform a number of iterations.

Another filtering strategy consists in calculating, observing, and making use of performance statistics computed along successive applications of local search. We describe a typical, simple use of this idea. Considering a minimization problem, one can keep track of the maximum relative reduction $\rho$, resulted from applying local search, in the value of the starting solution $S$ created by the construction algorithm in the same iteration. Although the value of $\rho$ may increase as the number of iterations grows, it is likely to quickly stabilize. Since it becomes progressively more and more unlikely that local search will cause a reduction in the objective function $f(S)$ that is greater than $\rho$, we may discard the application of local search whenever $f(S) > \rho \cdot f^*$, where $f^*$ is the value of the incumbent solution, with low risk of missing good starting solutions.

Other variations of these strategies exist and have been applied. Although all of them contribute to significantly accelerate the main GRASP loop, there is always the possibility that some good initial solutions may be discarded by filtering, leading to a small deterioration in the quality of the best solution obtained at the end of the algorithm.

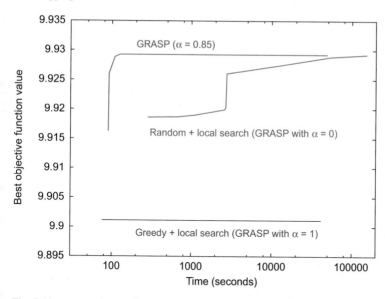

**Fig. 5.11** Best solution value for GRASP, random multistart with local search, and greedy multistart with local search as a function of time (in seconds) for an instance of the maximum covering problem. GRASP reaches the best solution in less than one thousandths of the time taken by random multistart with local search.

## 5.5 Stopping GRASP

The main drawback of most metaheuristics is the absence of effective stopping criteria. Most implementations of these algorithms stop after performing a given maximum number of iterations or a given maximum number of consecutive iterations without improvement in the best known solution value, or after the stabilization of the set of elite solutions found along the search. In some cases, the algorithm can perform an exaggerated and unnecessary number of iterations, when the best solution is found quickly (as often happens in GRASP implementations). In other situations, the algorithm can stop just before the iteration in which an optimal solution would be found. Dual bounds can be used to implement quality-based stopping rules, but they are often hard to compute or far from the optimal value, which make them unusable in both situations.

Randomization plays a very important role in the design of metaheuristics. Effective probabilistic stopping rules can be applied to randomized metaheuristics.

### 5.5.1 Probabilistic stopping rule

Let $X$ be a random variable that denotes the objective function value of the local minima obtained at each iteration of a GRASP heuristic for a minimization problem. Let the probability density function and cumulative probability distribution of

$X$ be denoted, respectively, by $f_X(\cdot)$ and $F_X(\cdot)$. Let $f_k$ be the solution value obtained at iteration $k$ of a GRASP heuristic and $f_1, \ldots, f_k$ be a sample formed by the solution values obtained along the first $k$ iterations. Let $f_X^k(\cdot)$ and $F_X^k(\cdot)$ be, respectively, estimates of the probability density function and of the cumulative probability distribution of the random variable $X$, obtained after the first $k$ GRASP iterations.

Let $UB^k$ be the value of the best solution found along the first $k$ GRASP iterations. Therefore, the probability of finding a solution value smaller than or equal to $UB^k$ in the next iteration can be estimated by

$$F_X^k(UB^k) = \int_{-\infty}^{UB^k} f_X^k(\tau)d\tau.$$

For sake of computational efficiency, the value of $F_X^k(UB^k)$ can be recomputed periodically or whenever the value of the incumbent improves, rather than at every iteration of the heuristic.

For any given threshold $\beta$, the GRASP iterations can be interrupted when $F_X^k(UB^k)$ becomes less than or equal to $\beta$, i.e., as soon as the probability of finding a solution at least as good as the incumbent in the next iteration becomes smaller than or equal to the threshold $\beta$. The probability value $F_X^k(UB^k)$ can be used to estimate the number of iterations that must be performed by the algorithm to find a new solution that is at least as good as the incumbent. Since the user is able to account for the average time taken by each GRASP iteration, the threshold defining the stopping criterion can either be fixed or determined online so as to limit the computation time when the probability of finding an improving solution becomes very small.

### 5.5.2 Gaussian approximation for GRASP iterations

Computational experiments and statistical tests have shown that the solution values obtained by GRASP heuristics for a number of combinatorial optimization problems fit a normal distribution. If $f_1, \ldots, f_N$ denote a sample formed by all solution values obtained along $N$ GRASP iterations, the null hypothesis stating that the sample $f_1, \ldots, f_N$ follows a normal distribution usually cannot be rejected with 90% confidence level by the chi-square test after relatively few iterations are performed.

We illustrate below that the solution values obtained along the GRASP iterations fit a normal distribution with numerical results obtained for four instances of the 2-path network design problem. The chi-square test shows that, already after as few as 50 iterations, the solution values obtained by the heuristic fit a normal distribution very closely. Table 5.1 lists the mean, standard deviation, skewness, and kurtosis for these four instances for $N = 50, 100, 500, 1,000, 5,000$, and $10,000$ GRASP iterations. Skewness measures the symmetry of the original data, while kurtosis measures the shape of the fitted distribution. Ideally, they should be equal to 0 and 3, respectively, in the case of a perfect normal fit. This table shows that the mean consistently converges very quickly to a steady-state value when the number of iterations increases. Furthermore, the mean after 50 iterations is already very close to that of the normal fit after 10,000 iterations. The skewness values are consistently very close to 0, while the measured kurtosis of the sample is always close to 3.

**Table 5.1** Statistics for normal fittings for a heuristic to the 2-path network design problem.

| Instance | Iterations | Mean | Std. dev. | Skewness | Kurtosis |
|---|---|---|---|---|---|
| | 50 | 372.920000 | 7.583772 | 0.060352 | 3.065799 |
| | 100 | 373.550000 | 7.235157 | -0.082404 | 2.897830 |
| 2pndp50 | 500 | 373.802000 | 7.318661 | -0.002923 | 2.942312 |
| | 1,000 | 373.854000 | 7.192127 | 0.044952 | 3.007478 |
| | 5,000 | 374.031400 | 7.442044 | 0.019068 | 3.065486 |
| | 10,000 | 374.063500 | 7.487167 | -0.010021 | 3.068129 |
| | 50 | 540.080000 | 9.180065 | 0.411839 | 2.775086 |
| | 100 | 538.990000 | 8.584282 | 0.314778 | 2.821599 |
| 2pndp70 | 500 | 538.334000 | 8.789451 | 0.184305 | 3.146800 |
| | 1,000 | 537.967000 | 8.637703 | 0.099512 | 3.007691 |
| | 5,000 | 538.576600 | 8.638989 | 0.076935 | 3.016206 |
| | 10,000 | 538.675600 | 8.713436 | 0.062057 | 2.969389 |
| | 50 | 698.100000 | 9.353609 | -0.020075 | 2.932646 |
| | 100 | 700.790000 | 9.891709 | -0.197567 | 2.612179 |
| 2pndp90 | 500 | 701.766000 | 9.248310 | -0.035663 | 2.883188 |
| | 1,000 | 702.023000 | 9.293141 | -0.120806 | 2.753207 |
| | 5,000 | 702.281000 | 9.149319 | 0.059303 | 2.896096 |
| | 10,000 | 702.332600 | 9.196813 | 0.022076 | 2.938744 |
| | 50 | 1,599.240000 | 13.019309 | 0.690802 | 3.311439 |
| | 100 | 1,600.060000 | 14.179436 | 0.393329 | 2.685849 |
| 2pndp200 | 500 | 1,597.626000 | 13.052744 | 0.157841 | 3.008731 |
| | 1,000 | 1,597.727000 | 12.828035 | 0.083604 | 3.009355 |
| | 5,000 | 1,598.313200 | 13.017984 | 0.057133 | 3.002759 |
| | 10,000 | 1,598.366100 | 13.066900 | 0.008450 | 3.019011 |

Figure 5.12 displays the normal distribution fit for each instance and for each value of $N$. Together with the statistics reported in Table 5.1, these plots illustrate the robustness of the normal fits to the solution values obtained along the iterations of the GRASP heuristic.

Since the null hypothesis cannot be rejected with a 90% confidence level, we can approximate the solution values obtained by a GRASP heuristic with a normal distribution whose fit is progressively better as more iterations are accounted for.

### 5.5.3 Stopping rule implementation

Recall that $X$ is a random variable representing the value of the objective function for the local minimum obtained by each GRASP iteration. We illustrated in the previous section that the distribution of $X$ can be approximated by a normal

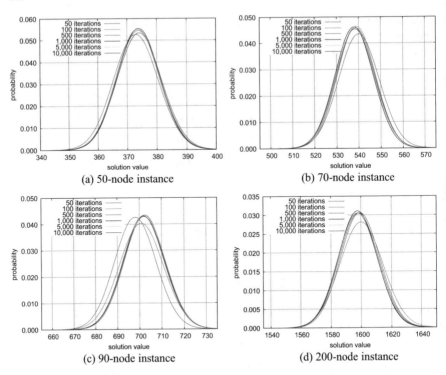

**Fig. 5.12** Normal distribution fits for four instances of the 2-path network design problem.

distribution $N(m^k, S^k)$ with mean $m^k = (1/k) \cdot \sum_{j=1}^{k} f_j$ and standard deviation $S^k = [(1/(k-1)) \cdot \sum_{j=1}^{k} (f_j - m^k)^2]^{1/2}$, whose probability density function and cumulative probability distribution are, respectively, $f_X^k(\cdot)$ and $F_X^k(\cdot)$.

The pseudo-code in Figure 5.13 extends the previous template of a GRASP heuristic for minimization presented in Figure 5.5, implementing the termination rule based on stopping the GRASP iterations whenever the probability $F_X^k(UB^k)$ of improving the best known solution value becomes smaller than or equal to the threshold $\beta$. Lines 14 and 15 update the sample $f_1, \ldots, f_k$ and the best known solution value $UB^k = f^*$ at each iteration $k$. The mean and the standard deviation of the fitted normal distribution in iteration $k$ are computed in line 16. The probability of finding a solution whose value is better than the currently best known solution value is computed in line 17.

This approach also makes it possible to apply stopping rules based on estimating the number of iterations needed to improve the value of the best solution found by each percentage point. For example, consider instance 2pndp90 of the 2-path network design problem, with the threshold $\beta = 10^{-3}$. Figure 5.14 plots the expected number of additional iterations needed to find a solution that improves the best known solution value by each percentage point that might be sought. For instance, the expected number of iterations needed to improve the best solution value

```
begin GRASP+STOP(β);
1   f* ← ∞;
2   k ← 0;
3   prob ← 1;
4   while prob > β do
5       S ← SEMI-GREEDY;
6       if S is not feasible then
7           S ← Repair(S);
8       end-if;
9       S ← LOCAL-SEARCH(S);
10      if f(S) < f* then
11          S* ← S;
12          f* ← f(S);
13      end-if;
14      k ← k + 1;
15      fₖ ← f(S);
16      Update the means and the standard deviations of f₁,...,fₖ;
17      prob ← Fₓᵏ(f*) = ∫₋∞^{f*} fₓᵏ(τ)dτ;
18  end-while;
19  return S*, f(S*);
end GRASP+STOP.
```

**Fig. 5.13** Pseudo-code of a basic GRASP with probabilistic stopping rule.

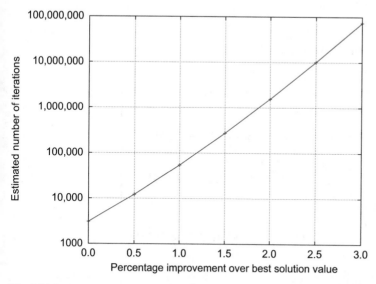

**Fig. 5.14** Estimated number of additional iterations needed to improve the best solution value.

found at termination by 0.5% is 12,131. If one seeks a percentage improvement of 1%, then the expected number of additional iterations to be performed increases to 54,153.

## 5.6 GRASP for multiobjective optimization

In this section, we consider a multiobjective optimization problem in which one seeks a solution $S^*$ belonging to the set of feasible solutions $F$ that optimizes a set of $k$ objective functions $f_j(S) = \sum_{i \in S} c_i^j$, for $j = 1, \ldots, k$. Specifically, we want to determine the set $\mathbb{P}$ of efficient points (usually called the *efficient Pareto frontier*). In its minimization form, a point or solution $S^* \in F$ is said to be *efficient* if there is no other solution $S \in F$ such that $f_i(S) \leq f_i(S^*)$ for all $i = 1, \ldots, k$ and $f_j(S) < f_j(S^*)$ for at least one $j \in \{1, \ldots, k\}$. In summary, efficiency requires that a solution to a multiobjective function be such that no single objective can be improved without deteriorating some other objective. In this context, we say that a solution $S^*$ *dominates* another solution $S \in F$ if $S^*$ is not worse than $S$ for all the objectives, and is better for at least one objective. Similarly, we say that $S^*$ *weakly dominates* $S$ if it is not worse than $S$ with respect to all objectives.

Since GRASP is a heuristic, instead of computing an exact Pareto efficient set, we will determine an approximation of this set. The pseudo-code in Figure 5.15 illustrates a multiobjective GRASP. In line 1, the Pareto efficient set $\mathbb{P}$ is initialized empty. The iterations take place in the loop from line 2 to 14. The steps are similar to those of the standard single-objective GRASP, except that multiobjective versions of the construction and local search algorithms are used and the approximate Pareto optimal set has to be managed.

---

**begin** MULTIOBJECTIVE-GRASP;
1   $\mathbb{P} \leftarrow \varnothing$;
2   **while** stopping criterion not satisfied **do**
3       $S \leftarrow$ MULTIOBJECTIVE-CONSTRUCTION;
4       **if** $\mathbb{P} = \varnothing$ **then** $\mathbb{P} \leftarrow \{S\}$;
5       **else**
6           **if** $S$ is not dominated by any solution in $\mathbb{P}$ **then**
7               $\mathbb{P} \leftarrow \mathbb{P} \cup \{S\}$;
8               **forall** $S' \in \mathbb{P} \setminus \{S\}$ **do**
9                   **if** $S'$ is dominated by $S$ **then** $\mathbb{P} \leftarrow \mathbb{P} \setminus \{S'\}$;
10              **end-forall**;
11          **end-if**;
12      **end-if**;
13      $(S, \mathbb{P}) \leftarrow$ MULTIOBJECTIVE-IMPROVEMENT$(S, \mathbb{P})$;
14  **end-while**;
15  **return** $\mathbb{P}$;
**end** MULTIOBJECTIVE-GRASP.

---

**Fig. 5.15** Pseudo-code of a multiobjective GRASP heuristic.

Multiobjective construction takes place in line 3, where a solution $S$ is built. If the Pareto efficient set $\mathbb{P}$ is empty, then the constructed solution $S$ is added to it in line 4. Otherwise, if $S$ is not dominated by any solution in $\mathbb{P}$, then it is added to $\mathbb{P}$ in line 7 and any other solution in $\mathbb{P}$ that is dominated by $S$ is removed from $\mathbb{P}$ in line 9. In line 13, multiobjective improvement takes as input the constructed solution $S$ and the current Pareto efficient set $\mathbb{P}$ and returns an improved solution. Instead of verifying only if the solution returned by the improvement procedure should be included in the Pareto efficient set, multiobjective improvement also verifies every solution visited along the search. The algorithm returns an approximate Pareto efficient set in line 15.

## 5.7 Bibliographical notes

Early proposals for what we called multistart methods in Section 5.1 can be found in the domains of heuristic scheduling (Muth and Thompson, 1963; Crowston et al., 1963), the traveling salesman problem (Held and Karp, 1970; Lawler et al., 1985), and the knapsack problems with single and multiple constraints (Senju and Toyoda, 1968; Wyman, 1973; Kochenberger et al., 1974). The survey by Martí et al. (2013a) gives an account of multistart methods and briefly sketches historical developments that have motivated the field. Focusing on contributions that define the current state of the art, two categories of multistart methods are considered: memory-based and memoryless procedures.

The semi-greedy multistart algorithm (without local search) discussed in Section 5.2 was proposed by Hart and Shogan (1987) and was independently developed by Feo and Resende (1989). In that paper, Feo and Resende described for the first time a GRASP heuristic, not referring to either the name GRASP or to greedy randomized adaptive search procedures, but simply calling the algorithm a probabilistic heuristic.

GRASP as a metaheuristic was presented and discussed in Section 5.3. This acronym was first introduced in the technical report by Feo et al. (1989) that appeared later as a journal paper in Feo et al. (1994). GRASP, as a general-purpose metaheuristic, was introduced by Feo and Resende (1995). Other tutorials on the method were authored by Pitsoulis and Resende (2002), Resende and Ribeiro (2003b; 2010; 2014), Ribeiro (2002), and Resende and Silva (2011). Annotated bibliographies of GRASP appeared in Festa and Resende (2002; 2009a;b).

The GRASP for the maximum covering problem, for which Figures 5.8, 5.9, and 5.11 were originally produced, appeared in Resende (1998). The GRASP for the maximum weighted satisfiability problem, for which Figures 5.4, 5.6, 5.7, and 5.10 were produced, was presented in Resende et al. (1997). See also Resende and Feo (1996) for the first proposal of a GRASP for the weighted satisfiability problem and Festa et al. (2006) for an improved GRASP for the maximum weighted satisfiability problem.

Filtering strategies discussed in Section 5.4 to accelerate the local search phase were originally proposed and applied by Feo et al. (1994) and Martins et al. (2000).

Proposed Bayesian stopping rules were not followed by sufficient computational studies to validate their effectiveness or to give evidence of their efficiency (Bartkutė et al., 2006; Bartkutė and Sakalauskas, 2009; Boender and Kan, 1987; Dorea, 1990; Hart, 1998). Orsenigo and Vercellis (2006) developed a Bayesian framework for stopping rules aimed at controlling the number of iterations in a GRASP heuristic. Stopping rules have also been discussed by Duin and Voss (1999) and Voss et al. (2005) in another context. The statistical estimation of optimal values for combinatorial optimization problems as a way to evaluate the performance of heuristics was also addressed by Rardin et al. (2001) and Serifoglu and Ulusoy (2004). Ribeiro et al. (2011; 2013) proposed and developed the probabilistic stopping rules described in Section 5.5. The 2-path network design problem was introduced and proved to be *NP*-hard by Dahl and Johannessen (2004). The GRASP heuristic used in the computational experiments with the 2-path network design problem was proposed by Ribeiro and Rosseti (2002; 2007).

Martí et al. (2015) surveyed applications of GRASP to multiobjective optimization, such as the multiobjective knapsack problem (Vianna and Arroyo, 2004), the multicriteria minimum spanning tree problem (Arroyo et al., 2008), the multiobjective quadratic assignment problem (Li and Landa-Silva, 2009), a learning classification problem (Ishida et al., 2009), the biobjective path dissimilarity and biorienteering problems (Martí et al., 2015), environmental investment decision making (Higgins et al., 2008), partial classification of databases (Reynolds and de la Iglesia, 2009; Reynolds et al., 2009), flow shop scheduling (Davoudpour and Ashrafi, 2009), biobjective set packing (Delorme et al., 2010), biobjective commercial territory design (Salazar-Aguilar et al., 2013), path dissimilarity (Martí et al., 2009), line balancing (Chica et al., 2010), and locating and sizing capacitors for reactive power compensation (Antunes et al., 2014), among others.

# Chapter 6
# Runtime distributions

Runtime distributions or time-to-target plots display on the ordinate axis the probability that an algorithm will find a solution at least as good as a given target value within a given running time, shown on the abscissa axis. They provide a very useful tool to characterize the running times of stochastic algorithms for combinatorial optimization problems and to compare different algorithms or strategies for solving a given problem. They have been widely used as a tool for algorithm design and comparison.

## 6.1 Time-to-target plots

Let $\mathscr{P}$ be an optimization problem and $\mathscr{H}$ a randomized heuristic for this problem. Furthermore, let $\mathscr{I}$ be a specific instance of $\mathscr{P}$ and let *look4* be a solution cost *target value* for this instance.

Heuristic $\mathscr{H}$ is run $N$ times on the fixed instance $\mathscr{I}$ and the algorithm is made to stop as soon as a solution whose objective function is at least as good as the given target value *look4* is found. For each of the $N$ runs, the random number generator used in the implementation of the heuristic is initialized with a distinct seed and, therefore, the runs are assumed to be independent. The solution time of each run is recorded and saved. To compare their empirical and theoretical distributions, we follow a standard graphical methodology for data analysis. This methodology is used to produce the time-to-target plots (TTT-plots) and is described next.

After concluding the $N$ independent runs, solution times are sorted in increasing order. We associate with the $i$-th sorted solution time $t_i$ a probability $p_i = (i - 1/2)/N$, and plot the points $z_i = (t_i, p_i)$, for $i = 1, \ldots, N$. We comment on this choice of $p_i$ later in Section 6.2. Figure 6.1 illustrates this estimated cumulative probability distribution plot for problem $\mathscr{P}$, a GRASP heuristic $\mathscr{H}$, instance $\mathscr{I}$, and target *look4*. We can see that the probability that the heuristic finds a solution at least as good as the target value in at most 416 seconds is about 50%, in at most 1064 seconds is about 80%, and in at most 1569 seconds is about 90%.

© Springer Science+Business Media New York 2016
M.G.C. Resende, C.C. Ribeiro, *Optimization by GRASP*,
DOI 10.1007/978-1-4939-6530-4_6

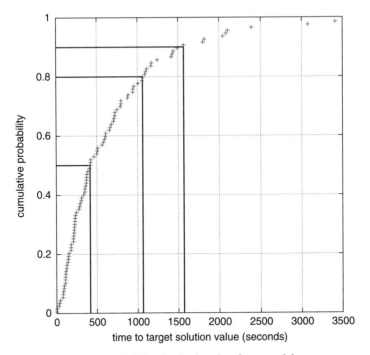

**Fig. 6.1** Cumulative probability distribution plot of measured data.

## 6.2 Runtime distribution of GRASP

The plot in Figure 6.1 appears to fit an exponential distribution, or more generally, a shifted exponential distribution. To estimate the parameters of this two-parameter exponential distribution, we first draw the theoretical quantile-quantile plot (or Q-Q plot) for the data. To describe Q-Q plots, we recall that the cumulative distribution function for the two-parameter exponential distribution is given by $F(t) = 1 - e^{-(t-\mu)/\lambda}$, where $\lambda$ is the mean of the distribution data (and also is the standard deviation of the data) and $\mu$ is the shift of the distribution with respect to the ordinate axis.

The *quantiles* of the data of an empirical distribution are derived from the (sorted) raw data, which in our case are $N$ measured (sorted) running times. Quantiles are cutpoints that group a set of sorted observations into classes of equal (or approximately equal) size. For each value $p_i$, $i = 1, \ldots, N$, we associate a $p_i$-quantile $q(p_i)$ of the theoretical distribution. For each $p_i$-quantile we have, by definition, that $F((q(p_i)) = p_i$. Hence, $q(p_i) = F^{-1}(p_i)$ and therefore, for the two-parameter exponential distribution, we have $q(p_i) = -\lambda \cdot \ln(1 - p_i) + \mu$. Note that if we were to use $p_i = 1/N$, for $i = 1, \ldots, N$, then $q(p_N)$ would be undefined.

A theoretical quantile-quantile plot (or theoretical Q-Q plot) is obtained by plotting the quantiles of the data of an empirical distribution against the quantiles of a

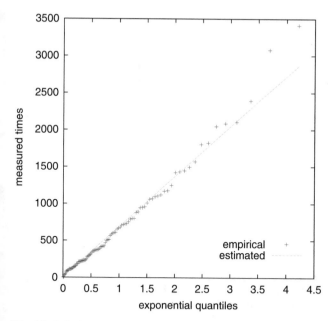

**Fig. 6.2** Q-Q plot showing fitted line.

theoretical distribution. This involves three steps. First, the data (in this case, the measured solution times) are sorted in ascending order. Second, the quantiles of the theoretical exponential distribution are obtained. Finally, a plot of the data against the theoretical quantiles is made.

In a situation where the theoretical distribution is a close approximation of the empirical distribution, the points in the Q-Q plot will have a nearly straight configuration. In a plot of the data against a two-parameter exponential distribution with $\lambda = 1$ and $\mu = 0$, the points would tend to follow the line $y = \hat{\lambda} \cdot x + \hat{\mu}$. Consequently, parameters $\lambda$ and $\mu$ of the two-parameter exponential distribution can be estimated, respectively, by the slope $\hat{\lambda}$ and the intercept $\hat{\mu}$ of the line depicted in the Q-Q plot.

The Q-Q plot shown in Figure 6.2 is obtained by plotting the measured times in the ordinate against the quantiles of a two-parameter exponential distribution with $\lambda = 1$ and $\mu = 0$ in the abscissa, given by $q(p_i) = -\ln(1 - p_i)$, for $i = 1, \dots, n$. To avoid possible distortions caused by outliers, we do not estimate the distribution mean with the data mean or by linear regression on the points of the Q-Q plot. Instead, we estimate the slope $\hat{\lambda}$ of the line $y = \lambda \cdot x + \mu$ using the upper quartile $q_u$ and lower quartile $q_l$ of the data. The *upper quartile* $q_u$ and *lower quartile* $q_l$ are, respectively, the $q(1/4)$ and $q(3/4)$ quantiles. We take $\hat{\lambda} = (z_u - z_l)/(q_u - q_l)$ as an estimate of the slope, where $z_u$ and $z_l$ are the $u$-th and $l$-th points of the ordered measured times, respectively. This informal estimation of the distribution of the measured data mean is robust since it will not be distorted by a few outliers. Consequently, the estimate for the shift is $\hat{\mu} = z_l - \hat{\lambda} q_l$.

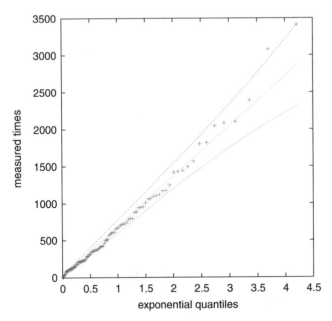

**Fig. 6.3** Q-Q plot with variability information.

To analyze the straightness of the Q-Q plots, we superimpose them with variability information. For each plotted point, we show plus and minus one standard deviation in the vertical direction from the line fitted to the plot. An estimate of the standard deviation for point $z_i$, $i = 1, \ldots, n$, of the Q-Q plot is $\hat{\sigma} = \hat{\lambda}[p_i/(1 - p_i)n]^{\frac{1}{2}}$. Figure 6.3 shows an example of a Q-Q plot with superimposed variability information.

When observing a theoretical quantile-quantile plot with superimposed standard deviation information, one should avoid turning such information into a formal test. One important fact that must be kept in mind is that the natural variability of the data generates departures from the straightness, even if the model of the distribution is valid. The most important reason for portraying standard deviation is that it gives us a sense of the relative variability of the points in the different regions of the plot. However, since one is trying to make simultaneous inferences from many individual inferences, it is difficult to use standard deviations to judge departures from the reference distribution. For example, the probability that a particular point deviates from the reference line by more than two standard deviations is small. However, the probability that any of the points deviates from the line by two standard deviations is probably much greater. In order statistics, this is made more difficult by the high correlation that exists between neighboring points. If one plotted point deviates by more than one standard deviation, there is a good chance that a whole bunch of them will too. Another point to keep in mind is that standard deviations vary substantially in the Q-Q plot. As one can observe in the Q-Q plot in Figure 6.3, the standard deviation of the points near the high end is substantially larger than the standard deviation of the points near the other end.

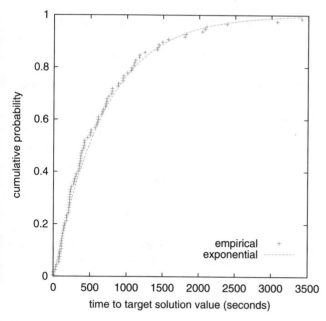

**Fig. 6.4** Superimposed empirical runtime distribution and best exponential fit.

Once the two parameters of the distribution have been estimated, a superimposed plot of the empirical and theoretical distributions can be made. Figure 6.4 depicts the superimposed empirical and theoretical distributions corresponding to the Q-Q plot in Figure 6.3.

The runtime distribution of a pure GRASP heuristic has been shown experimentally to behave as a random variable that fits an exponential distribution. Later in this book, we will discuss the implication of this observation with respect to parallel implementations of GRASP and restart strategies.

However, in the case of more elaborate heuristics where setup times are not negligible, the runtimes fit a two-parameter or shifted exponential distribution.

Therefore, the probability density function of the time-to-target random variable is given by $f(t) = (1/\lambda) \cdot e^{-t/\lambda}$ in the first case (exponential distribution) and by $f(t) = (1/\lambda) \cdot e^{-(t-\mu)/\lambda}$ in the second (shifted exponential distribution), with the parameters $\lambda \in \mathbb{R}^+$ and $\mu \in \mathbb{R}^+$ being associated with the shape and the shift of the exponential function, respectively. Figure 6.4 illustrates this result, depicting the superimposed empirical and theoretical distributions observed for an instance of the maximum covering problem where one wants to choose 500 out of 1000 facility locations such that, of the 10,000 customers, the sum of the weights of those that are covered is maximized. The best known solution for this instance is 33,343,542 and the target solution value used was 33,339,175 (about 0.01% off of the best known solution).

However, if path-relinking is applied as an intensification step at the end of each GRASP iteration (see Chapter 9 in this book), then the iterations are no longer independent and the memoryless characteristic of GRASP is destroyed. This also happens in the case of cooperative parallel implementations of GRASP (see also Chapter 10 in this book). Consequently, the time-to-target random variable may not fit an exponential distribution in such situations. This result is illustrated by two implementations of GRASP with bidirectional path-relinking. The first is an application to the 2-path network design problem. The runtime distribution and the corresponding quantile-quantile plot for an instance with 80 nodes and 800 origin-destination pairs are depicted in Figure 6.5. The second is an application to the three-index assignment problem. Runtime distributions and the corresponding quantile-quantile plots for Balas and Saltzman problems 22.1 (target value set to 8) and 24.1 (target value set to 7) are shown in Figures 6.6 and 6.7, respectively. For both heuristics and these three example instances, we observe that points steadily deviate by more than one standard deviation from the estimate for the upper quantiles in the quantile-quantile plots (i.e., many points associated with large computation times fall outside the plus or minus one standard deviation bounds). Therefore, we cannot say that these runtime distributions are exponentially distributed.

## 6.3 Comparing algorithms with exponential runtime distributions

We assume the existence of two randomized algorithms $A_1$ and $A_2$ for the approximate solution of some optimization problem. Furthermore, we assume that their solution times fit exponential (or shifted exponential) distributions. We denote by $X_1$ (resp. $X_2$) the continuous random variable representing the time needed by algorithm $A_1$ (resp. $A_2$) to find a solution as good as a given target value:

$$X_1 \mapsto \begin{cases} 0, & \tau < T_1 \\ \lambda_1 \cdot e^{-\lambda_1(\tau-T_1)}, & \tau \geq T_1 \end{cases}$$

and

$$X_2 \mapsto \begin{cases} 0, & \tau < T_2 \\ \lambda_2 \cdot e^{-\lambda_2(\tau-T_2)}, & \tau \geq T_2 \end{cases}$$

where $T_1$, $\lambda_1$, $T_2$, and $\lambda_2$ are parameters ($\lambda_1$ and $\lambda_2$ define the shape of each shifted exponential distribution, whereas $T_1$ and $T_2$ denote by how much each of them is shifted). The cumulative probability distribution and the probability density function of $X_1$ are depicted in Figure 6.8.

Since both algorithms stop when they find a solution at least as good as the target, we can say that algorithm $A_1$ performs better than $A_2$ if the former stops before the latter. Therefore, we must evaluate the probability $Pr(X_1 \leq X_2)$ that the random variable $X_1$ takes a value smaller than or equal to $X_2$. Conditioning on the value of $X_2$ and applying the total probability theorem, we obtain

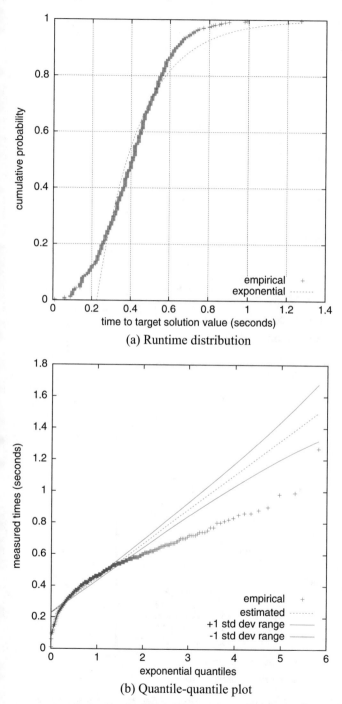

(a) Runtime distribution

(b) Quantile-quantile plot

**Fig. 6.5** Runtime distribution and quantile-quantile plot for GRASP with bidirectional path-relinking of an instance of the 2-path network design problem with 80 nodes and 800 origin-destination pairs, with target set to 588.

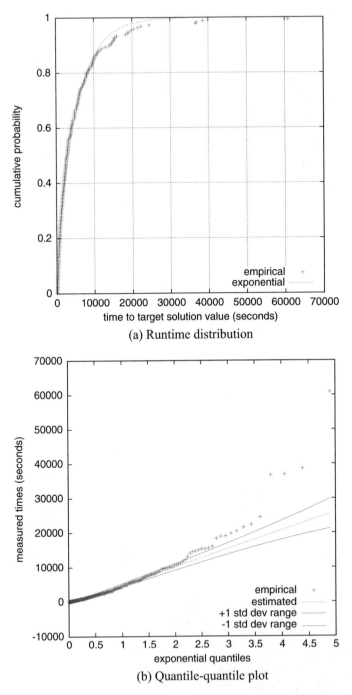

(a) Runtime distribution

(b) Quantile-quantile plot

**Fig. 6.6** Runtime distribution and quantile-quantile plot for GRASP with bidirectional path-relinking on Balas and Saltzman problem 22.1, with the target value set to 8.

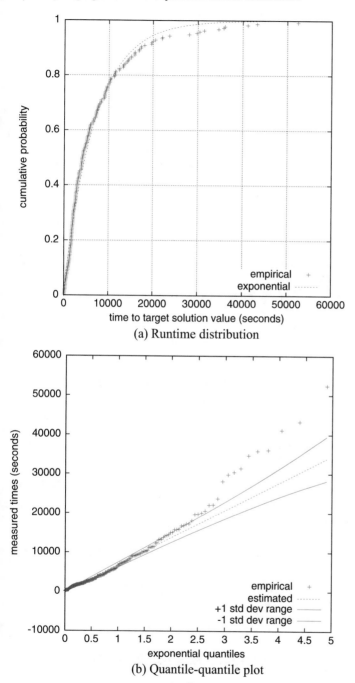

(a) Runtime distribution

(b) Quantile-quantile plot

**Fig. 6.7** Runtime distribution and quantile-quantile plot for GRASP with bidirectional path-relinking on Balas and Saltzman problem 24.1, with the target value set to 7.

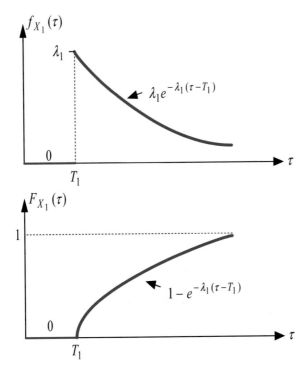

**Fig. 6.8** Probability density function and cumulative probability distribution of the random variable $X_1$.

$$Pr(X_1 \leq X_2) = \int_{-\infty}^{\infty} Pr(X_1 \leq X_2 | X_2 = \tau) \cdot f_{X_2}(\tau) \cdot d\tau =$$

$$= \int_{T_2}^{\infty} Pr(X_1 \leq X_2 | X_2 = \tau) \cdot \lambda_2 \cdot e^{-\lambda_2(\tau - T_2)} \cdot d\tau = \int_{T_2}^{\infty} Pr(X_1 \leq \tau) \cdot \lambda_2 \cdot e^{-\lambda_2(\tau - T_2)} \cdot d\tau.$$

Let $v = \tau - T_2$. Then, $dv = d\tau$ and

$$Pr(X_1 \leq X_2) = \int_0^{\infty} Pr(X_1 \leq (v + T_2)) \cdot \lambda_2 \cdot e^{-\lambda_2 v} \cdot dv. \qquad (6.1)$$

Using the formula of cumulative probability function of the random variable $X_1$ (see Figure 6.8), we obtain

$$Pr(X_1 \leq (v + T_2)) = 1 - e^{-\lambda_1(v + T_2 - T_1)}. \qquad (6.2)$$

Replacing (6.2) in (6.1) and solving the integral, we conclude that

$$Pr(X_1 \leq X_2) = 1 - e^{-\lambda_1(T_2 - T_1)} \cdot \frac{\lambda_2}{\lambda_1 + \lambda_2}. \qquad (6.3)$$

This result can be better interpreted by rewriting expression (6.3) as

$$Pr(X_1 \leq X_2) = (1 - e^{-\lambda_1(T_2 - T_1)}) + e^{-\lambda_1(T_2 - T_1)} \cdot \frac{\lambda_1}{\lambda_1 + \lambda_2}. \qquad (6.4)$$

The first term of the right-hand side of equation (6.4) is the probability that $0 \leq X_1 \leq T_2$, in which case $X_1$ is clearly less than or equal to $X_2$. The second term is given by the product of the factors $e^{-\lambda_1(T_2 - T_1)}$ and $\lambda_1/(\lambda_1 + \lambda_2)$, in which the former corresponds to the probability that $X_1 \geq T_2$ and the latter to the probability that $X_1$ be less than or equal to $X_2$, given that $X_1 \geq T_2$.

To illustrate the above result, we consider two algorithms for solving the server replication for reliable multicast problem. Algorithm $A_1$ is an implementation of pure GRASP with $\alpha = 0.2$, while algorithm $A_2$ is a pure GRASP heuristic with $\alpha = 0.9$. The runs were performed on an Intel Core2 Quad with 2.40 GHz of clock speed and 4 GB of RAM memory. Figure 6.9 depicts the runtime distributions of each algorithm, obtained after 500 runs with different seeds of an instance with the target value set at 2830. The parameters of the two distributions are $\lambda_1 = 0.524422349$, $T_1 = 0.36$, $\lambda_2 = 0.190533895$, and $T_2 = 0.51$. Applying expression (6.3), we get $Pr(X_1 \leq X_2) = 0.684125$. This probability is consistent with Figure 6.10, in which we superimposed the runtime distributions of the two pure GRASP heuristics for the same instance. The plots in this figure show that the pure GRASP with $\alpha = 0.2$ outperforms one with $\alpha = 0.9$, since the runtime distribution of the former is to the left of the runtime distribution of the latter.

If the solution times do not fit exponential (or two-parameter shifted exponential) distributions, as for the case of GRASP with path-relinking heuristics, then the the closed form result established in expression (6.3) does not hold. Algorithms in this situation cannot be compared by this approach. The next section extends this approach to general runtime distributions.

## 6.4 Comparing algorithms with general runtime distributions

Let $X_1$ and $X_2$ be two continuous random variables, with cumulative probability distributions $F_{X_1}(\tau)$ and $F_{X_2}(\tau)$ and probability density functions $f_{X_1}(\tau)$ and $f_{X_2}(\tau)$, respectively. Then,

$$Pr(X_1 \leq X_2) = \int_{-\infty}^{\infty} Pr(X_1 \leq \tau) \cdot f_{X_2}(\tau) \cdot d\tau = \int_{0}^{\infty} Pr(X_1 \leq \tau) \cdot f_{X_2}(\tau) \cdot d\tau,$$

since $f_{X_1}(\tau) = f_{X_2}(\tau) = 0$ for any $\tau < 0$. For an arbitrary small real number $\varepsilon$, the above expression can be rewritten as

$$Pr(X_1 \leq X_2) = \sum_{i=0}^{\infty} \int_{i \cdot \varepsilon}^{(i+1) \cdot \varepsilon} Pr(X_1 \leq \tau) \cdot f_{X_2}(\tau) \cdot d\tau. \qquad (6.5)$$

Since $Pr(X_1 \leq i \cdot \varepsilon) \leq Pr(X_1 \leq \tau) \leq Pr(X_1 \leq (i+1) \cdot \varepsilon)$ for $i \cdot \varepsilon \leq \tau \leq (i+1) \cdot \varepsilon$, then replacing $Pr(X_1 \leq \tau)$ by $Pr(X_1 \leq i \cdot \varepsilon)$ and by $Pr(X_1 \leq (i+1) \cdot \varepsilon)$ in (6.5) leads to

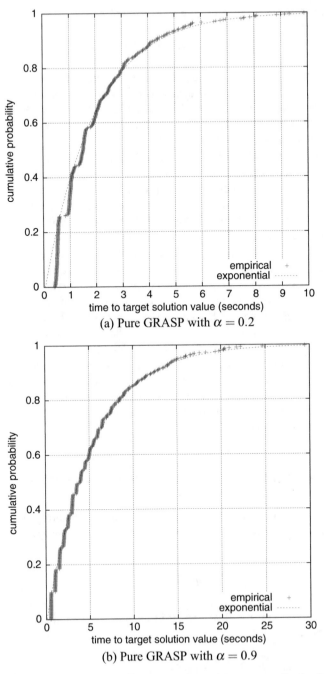

**Fig. 6.9** Runtime distributions of an instance of the server replication for reliable multicast problem with $m = 25$ and the target value set at 2830.

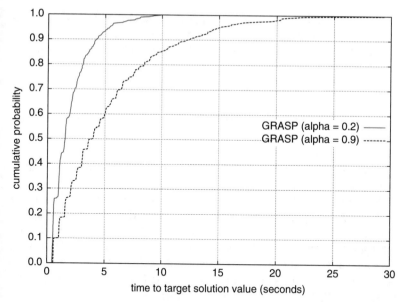

**Fig. 6.10** Superimposed runtime distributions of pure GRASP with $\alpha = 0.2$ and pure GRASP with $\alpha = 0.9$.

$$\sum_{i=0}^{\infty} F_{X_1}(i \cdot \varepsilon) \int_{i \cdot \varepsilon}^{(i+1) \cdot \varepsilon} f_{X_2}(\tau) \cdot d\tau \leq Pr(X_1 \leq X_2) \leq \sum_{i=0}^{\infty} F_{X_1}((i+1) \cdot \varepsilon) \int_{i \cdot \varepsilon}^{(i+1) \cdot \varepsilon} f_{X_2}(\tau) \cdot d\tau.$$

Let $L(\varepsilon)$ and $R(\varepsilon)$ be the value of the left- and right-hand sides of the above expression, respectively, with $\Delta(\varepsilon) = R(\varepsilon) - L(\varepsilon)$ being the difference between the upper and lower bounds to $Pr(X_1 \leq X_2)$. Then, we have that

$$\Delta(\varepsilon) = \sum_{i=0}^{\infty} [F_{X_1}((i+1) \cdot \varepsilon) - F_{X_1}(i \cdot \varepsilon)] \int_{i \cdot \varepsilon}^{(i+1) \cdot \varepsilon} f_{X_2}(\tau) \cdot d\tau. \tag{6.6}$$

Let $\delta = \max_{\tau \geq 0} \{f_{X_1}(\tau)\}$. Since $|F_{X_1}((i+1) \cdot \varepsilon) - F_{X_1}(i \cdot \varepsilon)| \leq \delta \cdot \varepsilon$ for $i \geq 0$, expression (6.6) turns out to be

$$\Delta(\varepsilon) \leq \sum_{i=0}^{\infty} \delta \cdot \varepsilon \int_{i \cdot \varepsilon}^{(i+1) \cdot \varepsilon} f_{X_2}(\tau) \cdot d\tau = \delta \cdot \varepsilon \int_0^{\infty} f_{X_2}(\tau) \cdot d\tau = \delta \cdot \varepsilon.$$

Consequently,

$$\Delta(\varepsilon) \leq \delta \cdot \varepsilon, \tag{6.7}$$

i.e., the difference $\Delta(\varepsilon)$ between the upper and lower bounds to $Pr(X_1 \leq X_2)$ (or the absolute error in the integration) is smaller than or equal to $\delta\varepsilon$. Therefore, this difference can be made as small as desired by choosing a sufficiently small value for $\varepsilon$.

In order to numerically evaluate a good approximation to $Pr(X_1 \leq X_2)$, we select the appropriate value of $\varepsilon$ such that the resulting approximation error $\Delta(\varepsilon)$ is sufficiently small. Next, we compute $L(\varepsilon)$ and $R(\varepsilon)$ to obtain the approximation

$$Pr(X_1 \leq X_2) \approx \frac{L(\varepsilon) + R(\varepsilon)}{2}. \tag{6.8}$$

In practice, the above probability distributions are unknown. Instead of the distributions, the information available is limited to a sufficiently large number $N_1$ (resp. $N_2$) of observations of the random variable $X_1$ (resp. $X_2$). Since the value of $\delta = \max_{\tau \geq 0}\{f_{X_1}(\tau)\}$ is also unknown beforehand, the appropriate value of $\varepsilon$ cannot be estimated. Then, we proceed iteratively as follows.

Let $t_1(j)$ (resp. $t_2(j)$) be the value of the $j$-th smallest observation of the random variable $X_1$ (resp. $X_2$), for $j = 1, \ldots, N_1$ (resp. $N_2$). We set the bounds $a = \min\{t_1(1), t_2(1)\}$ and $b = \max\{t_1(N_1), t_2(N_2)\}$ and choose an arbitrary number $h$ of integration intervals to compute an initial value $\varepsilon = (b-a)/h$ for each integration interval. For sufficiently small values of the integration interval $\varepsilon$, the probability density function $f_{X_1}(\tau)$ in the interval $[i \cdot \varepsilon, (i+1) \cdot \varepsilon]$ can be approximated by $\hat{f}_{X_1}(\tau) = (\hat{F}_{X_1}((i+1) \cdot \varepsilon) - \hat{F}_{X_1}(i \cdot \varepsilon))/\varepsilon$, where

$$\hat{F}_{X_1}(i \cdot \varepsilon) = |\{t_1(j), j = 1, \ldots, N_1 : t_1(j) \leq i \cdot \varepsilon\}|. \tag{6.9}$$

The same approximations hold for random variable $X_2$.

Finally, the value of $Pr(X_1 \leq X_2)$ can be computed as in expression (6.8), using the estimates $\hat{f}_{X_1}(\tau)$ and $\hat{f}_{X_2}(\tau)$ in the computation of $L(\varepsilon)$ and $R(\varepsilon)$. If the approximation error $\Delta(\varepsilon) = R(\varepsilon) - L(\varepsilon)$ becomes sufficiently small, then the procedure stops. Otherwise, the value of $\varepsilon$ is halved and the above steps are repeated until convergence.

## 6.5 Numerical applications to sequential algorithms

We illustrate next an application of the procedure described in the previous section for the comparison of randomized algorithms (running on the same instance) on three problems: server replication for reliable multicast, routing and wavelength assignment, and 2-path network design.

### 6.5.1 DM-D5 and GRASP algorithms for server replication

Multicast communication consists of simultaneously delivering the same information to many receivers, from single or multiple sources. Network services specially designed for multicast are needed. The scheme used in current multicast services creates a delivery tree, whose root represents the sender, whose leaves represent the receivers, and whose internal nodes represent network routers or relaying servers. Transmission is performed by creating copies of the data at split points of the tree. An important issue regarding multicast communication is how to provide reliable service, ensuring the delivery of all packets from the sender to receivers. A successful technique to provide reliable multicast service is the server replication approach, in which data is replicated at some of the multicast-capable relaying hosts (also called replicated or repair servers) and each of them is responsible for the retransmission of packets to receivers in its group. The problem consists in selecting the best subset of the multicast-capable relaying hosts to act as replicated servers in a multicast scenario. It is a special case of the $p$-median problem.

DM-GRASP is a hybrid version of GRASP described in Section 7.8 of this book, which incorporates a data mining process. We compare two heuristics for the server replication problem: algorithm $A_1$ is an implementation of the DM-D5 version of DM-GRASP, in which the mining algorithm is periodically applied, while $A_2$ is a pure GRASP heuristic. We present results for two instances using the same network scenario, with $m = 25$ and $m = 50$ replication servers.

Each algorithm was run 200 times with different seeds. The target was set at 2,818.925 (the best known solution value is 2,805.89) for the instance with $m = 25$ and at 2,299.07 (the best known solution value is 2,279.84) for the instance with $m = 50$. Figures 6.11 and 6.12 depict runtime distributions and quantile-quantile plots for DM-D5, for the instances with $m = 25$ and $m = 50$, respectively. Running times of DM-D5 did not fit exponential distributions for any of the instances. GRASP solution times were exponential for both.

The empirical runtime distributions of DM-D5 and GRASP are superimposed in Figure 6.13. Algorithm DM-D5 outperformed GRASP, since the runtime distribution of the DM-D5 is to the left of the distribution for GRASP on the both instances, with $m = 25$ and $m = 50$. Consistently, the computations show that $Pr(X_1 \leq X_2) = 0.619763$ (with $L(\varepsilon) = 0.619450$, $R(\varepsilon) = 0.620075$, $\Delta(\varepsilon) = 0.000620$, and $\varepsilon = 0.009552$) and $Pr(X_1 \leq X_2) = 0.854113$ (with $L(\varepsilon) = 0.853800$, $R(\varepsilon) = 0.854425$, $\Delta(\varepsilon) = 0.000625$, and $\varepsilon = 0.427722$) for the instances with $m = 25$ and $m = 50$, respectively.

We also investigate the convergence of the proposed measure with the sample size (i.e., with the number of independent runs of each algorithm). Convergence with the sample size is illustrated next for the same $m = 25$ instance of the server replication problem, with the same target 2,818.925 already used in the previous experiment. Once again, algorithm $A_1$ is the DM-D5 version of DM-GRASP and algorithm $A_2$ is the pure GRASP heuristic. The estimation of $Pr(X_1 \leq X_2)$ is computed for $N = 100, 200, 300, 400, 500, 600, 700, 800, 900, 1000, 2000, 3000, 4000,$ and 5000 independent runs of each algorithm. Table 6.1 shows the results obtained, which are also displayed in Figure 6.14. We notice that the estimation of $Pr(X_1 \leq X_2)$ stabilizes as the sample size $N$ increases.

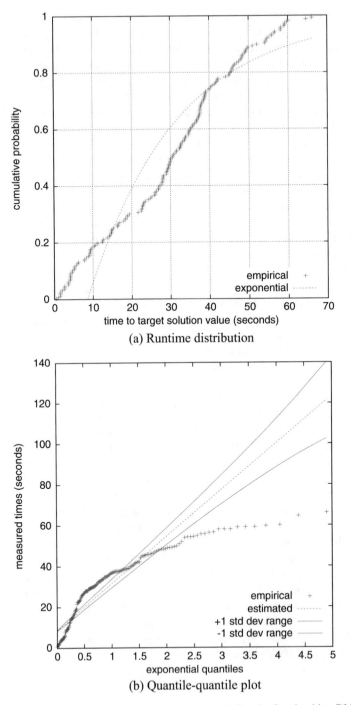

(a) Runtime distribution

(b) Quantile-quantile plot

**Fig. 6.11** Runtime distribution and quantile-quantile plot for algorithm DM-D5 on the instance with $m = 25$ and the target value set at 2,818.925.

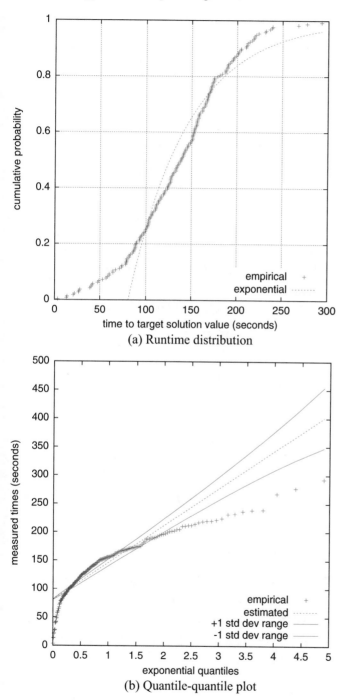

(a) Runtime distribution

(b) Quantile-quantile plot

**Fig. 6.12** Runtime distribution and quantile-quantile plot for algorithm DM-D5 on the instance with $m = 50$ and the target value set at 2,299.07.

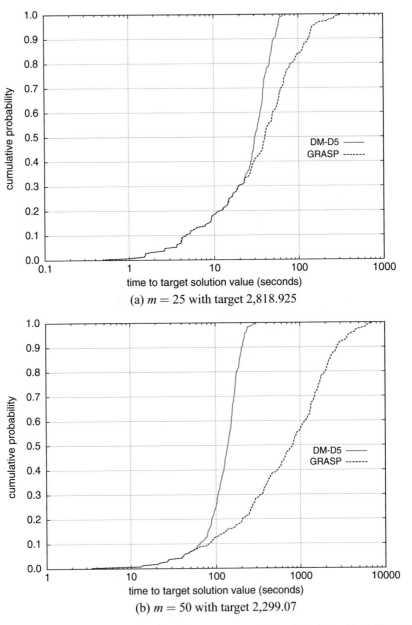

**Fig. 6.13** Superimposed runtime distributions of DM-D5 and GRASP: (a) $Pr(X_1 \leq X_2) = 0.619763$, and (b) $Pr(X_1 \leq X_2) = 0.854113$.

**Table 6.1** Convergence of the estimation of $Pr(X_1 \leq X_2)$ with the sample size for the $m = 25$ instance of the server replication problem.

| $N$ | $L(\varepsilon)$ | $Pr(X_1 \leq X_2)$ | $R(\varepsilon)$ | $\Delta(\varepsilon)$ | $\varepsilon$ |
|---|---|---|---|---|---|
| 100 | 0.655900 | 0.656200 | 0.656500 | 0.000600 | 0.032379 |
| 200 | 0.622950 | 0.623350 | 0.623750 | 0.000800 | 0.038558 |
| 300 | 0.613344 | 0.613783 | 0.614222 | 0.000878 | 0.038558 |
| 400 | 0.606919 | 0.607347 | 0.607775 | 0.000856 | 0.038558 |
| 500 | 0.602144 | 0.602548 | 0.602952 | 0.000808 | 0.038558 |
| 600 | 0.596964 | 0.597368 | 0.597772 | 0.000808 | 0.038558 |
| 700 | 0.591041 | 0.591440 | 0.591839 | 0.000798 | 0.038558 |
| 800 | 0.593197 | 0.593603 | 0.594009 | 0.000812 | 0.042070 |
| 900 | 0.593326 | 0.593719 | 0.594113 | 0.000788 | 0.042070 |
| 1000 | 0.594849 | 0.595242 | 0.595634 | 0.000785 | 0.042070 |
| 2000 | 0.588913 | 0.589317 | 0.589720 | 0.000807 | 0.047694 |
| 3000 | 0.583720 | 0.584158 | 0.584596 | 0.000875 | 0.047694 |
| 4000 | 0.582479 | 0.582912 | 0.583345 | 0.000866 | 0.047694 |
| 5000 | 0.584070 | 0.584511 | 0.584953 | 0.000882 | 0.050604 |

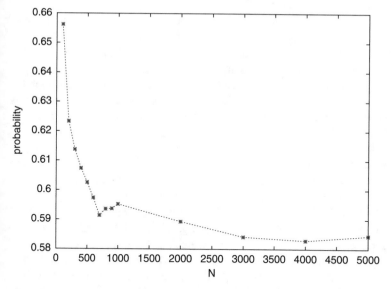

**Fig. 6.14** Convergence of the estimation of $Pr(X_1 \leq X_2)$ with the sample size for the $m = 25$ instance of the server replication problem.

### 6.5.2 Multistart and tabu search algorithms for routing and wavelength assignment

A point-to-point connection between two endnodes of an optical network is called a lightpath. Two lightpaths may use the same wavelength, provided they do not share any common link. The routing and wavelength assignment problem is that of routing a set of lightpaths and assigning a wavelength to each of them, minimizing the number of wavelengths needed. A decomposition strategy is compared with a multistart greedy heuristic. Two networks are used for benchmarking. The first has 27 nodes representing the capitals of the 27 states of Brazil, with 70 links connecting them. There are 702 lightpaths to be routed. Instance Finland is formed by 31 nodes and 51 links, with 930 lightpaths to be routed. Each algorithm was run 200 times with different seeds. The target was set at 24 (the best known solution value) for instance Brazil and at 50 for instance Finland (the best known solution value is 47). Algorithm $A_1$ is the multistart heuristic, while $A_2$ is the tabu search decomposition scheme. The multistart solution times fit exponential distributions for both instances. Figures 6.15 and 6.16 display runtime distributions and quantile-quantile plots for instances Brazil and Finland, respectively.

The empirical runtime distributions of the decomposition and multistart strategies are superimposed in Figure 6.17. The direct comparison of the two approaches shows that decomposition clearly outperformed the multistart strategy for instance Brazil, since $Pr(X_1 \leq X_2) = 0.13$ in this case (with $L(\varepsilon) = 0.129650$, $R(\varepsilon) = 0.130350$, $\Delta(\varepsilon) = 0.000700$, and $\varepsilon = 0.008163$). However, the situation changes, for instance Finland. Although both algorithms have similar performances, multistart is slightly better with respect to the measure proposed in this work, since $Pr(X_1 \leq X_2) = 0.536787$ (with $L(\varepsilon) = 0.536525$, $R(\varepsilon) = 0.537050$, $\Delta(\varepsilon) = 0.000525$, and $\varepsilon = 0.008804$).

As done for the server replication problem in Section 6.5.1, we also investigate the convergence of the proposed measure with the sample size (i.e., with the number of independent runs of each algorithm). Convergence with the sample size is illustrated next for the Finland instance of the routing and wavelength assignment problem, with the target set at 49. Once again, algorithm $A_1$ is the multistart heuristic and algorithm $A_2$ is the tabu search decomposition scheme. The estimation of $Pr(X_1 \leq X_2)$ is computed for $N = 100, 200, 300, 400, 500, 600, 700, 800, 900, 1000, 2000, 3000, 4000$, and $5000$ independent runs of each algorithm. Table 6.2 shows the results obtained, which are also displayed in Figure 6.18. Once again, we notice that the estimation of $Pr(X_1 \leq X_2)$ stabilizes as the sample size $N$ increases.

### 6.5.3 GRASP algorithms for 2-path network design

Given a connected undirected graph with non-negative weights associated with its edges, together with a set of origin-destination nodes, the 2-path network design

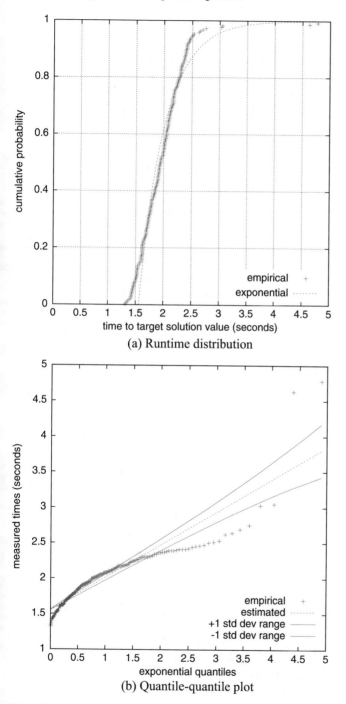

**Fig. 6.15** Runtime distribution and quantile-quantile plot for tabu search on Brazil instance with the target value set at 24.

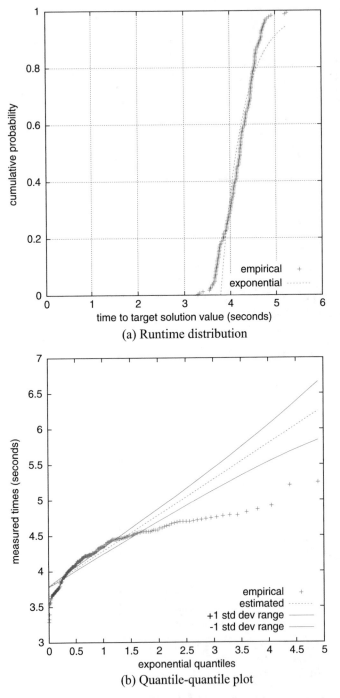

(a) Runtime distribution

(b) Quantile-quantile plot

**Fig. 6.16** Runtime distribution and quantile-quantile plot for tabu search on Finland instance with the target value set at 50.

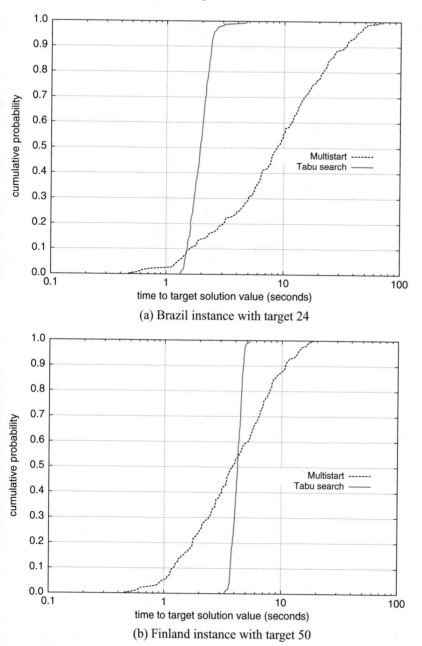

**Fig. 6.17** Superimposed runtime distributions of multistart and tabu search: (a) $Pr(X_1 \leq X_2) = 0.13$, and (b) $Pr(X_1 \leq X_2) = 0.536787$.

**Table 6.2** Convergence of the estimation of $Pr(X_1 \leq X_2)$ with the sample size for the Finland instance of the routing and wavelength assignment problem.

| $N$ | $L(\varepsilon)$ | $Pr(X_1 \leq X_2)$ | $R(\varepsilon)$ | $\Delta(\varepsilon)$ | $\varepsilon$ |
|---|---|---|---|---|---|
| 100 | 0.000001 | 0.000200 | 0.000400 | 0.000400 | 1.964844 |
| 200 | 0.000100 | 0.004875 | 0.009650 | 0.009550 | 0.000480 |
| 300 | 0.006556 | 0.012961 | 0.019367 | 0.012811 | 0.000959 |
| 400 | 0.007363 | 0.013390 | 0.019425 | 0.012063 | 0.000959 |
| 500 | 0.007928 | 0.014694 | 0.021460 | 0.013532 | 0.000610 |
| 600 | 0.006622 | 0.013069 | 0.019517 | 0.012894 | 0.000610 |
| 700 | 0.005722 | 0.011261 | 0.016800 | 0.011078 | 0.000610 |
| 800 | 0.005033 | 0.011667 | 0.018302 | 0.013269 | 0.000610 |
| 900 | 0.004556 | 0.010461 | 0.016367 | 0.011811 | 0.000610 |
| 1000 | 0.004100 | 0.009425 | 0.014750 | 0.010650 | 0.000610 |
| 2000 | 0.006049 | 0.011580 | 0.017112 | 0.011063 | 0.000610 |
| 3000 | 0.007802 | 0.014395 | 0.020987 | 0.013185 | 0.000610 |
| 4000 | 0.007408 | 0.013698 | 0.019988 | 0.012580 | 0.000610 |
| 5000 | 0.006791 | 0.013090 | 0.019389 | 0.012598 | 0.000623 |

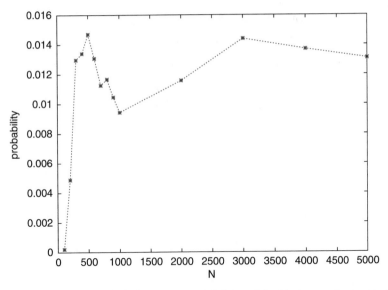

**Fig. 6.18** Convergence of the estimation of $Pr(X_1 \leq X_2)$ with the sample size for the Finland instance of the routing and wavelength assignment problem.

problem consists in finding a minimum weighted subset of edges containing a path formed by at most two edges between every origin-destination pair. Applications can be found in the design of communication networks, in which paths with few edges are sought to enforce high reliability and small delays.

### 6.5.3.1 Instance with 90 nodes

We first compare four GRASP heuristics for solving an instance of the 2-path network design problem with 90 nodes. The first heuristic is a pure GRASP (algorithm $A_1$). The others integrate different path-relinking strategies (see Chapters 8 and 9) for search intensification at the end of each GRASP iteration: forward path-relinking (algorithm $A_2$), bidirectional path-relinking (algorithm $A_3$), and backward path-relinking (algorithm $A_4$).

Each algorithm was run 500 independent times on the benchmark instance with 90 nodes and 900 origin-destination pairs, with the solution target value set at 673 (the best known solution value is 639). The runtime distributions and quantile-quantile plots for the different versions of GRASP with path-relinking are shown in Figures 6.19 to 6.21.

The empirical runtime distributions of the four algorithms are superimposed in Figure 6.22. Algorithm $A_2$ (as well as $A_3$ and $A_4$) performs much better than $A_1$, as indicated by $Pr(X_2 \leq X_1) = 0.986604$ (with $L(\varepsilon) = 0.986212$, $R(\varepsilon) = 0.986996$, $\Delta(\varepsilon) = 0.000784$, and $\varepsilon = 0.029528$). Algorithm $A_3$ outperforms $A_2$, as illustrated by the fact that $Pr(X_3 \leq X_2) = 0.636000$ (with $L(\varepsilon) = 0.630024$, $R(\varepsilon) = 0.641976$, $\Delta(\varepsilon) = 0.011952$, and $\varepsilon = 1.354218 \times 10^{-6}$). Finally, we observe that algorithms $A_3$ and $A_4$ behave very similarly, although $A_4$ performs slightly better for this instance, since $Pr(X_4 \leq X_3) = 0.536014$ (with $L(\varepsilon) = 0.528560$, $R(\varepsilon) = 0.543468$, $\Delta(\varepsilon) = 0.014908$, and $\varepsilon = 1.001358 \times 10^{-6}$).

As for the problems considered in Sections 6.5.1 and 6.5.2, we also investigate the convergence of the proposed measure as a function of sample size (i.e., with the number of independent runs of each algorithm). Convergence with the sample size is illustrated next for the 90-node instance of the 2-path network design problem, with the same target 673 previously used. We recall that algorithm $A_1$ is the GRASP with backward path-relinking heuristic, while algorithm $A_2$ is the GRASP with bidirectional path-relinking heuristic. The estimation of $Pr(X_1 \leq X_2)$ is computed for $N = 100, 200, 300, 400, 500, 600, 700, 800, 900, 1000, 2000, 3000, 4000$, and 5000 independent runs of each algorithm. Table 6.3 shows the results, which are also displayed in Figure 6.23. Once again, the estimation of $Pr(X_1 \leq X_2)$ stabilizes as the sample size $N$ increases.

### 6.5.3.2 Instance with 80 nodes

We next compare five GRASP heuristics for the 2-path network design problem, with and without path-relinking, for solving an instance with 80 nodes and 800 origin-destination pairs, with target value set at 588 (the best known solution value is 577). In this example, the first algorithm is a pure GRASP (algorithm $A_1$). The other heuristics integrate different path-relinking strategies at the end of each GRASP iteration (see Chapters 8 and 9): forward path-relinking (algorithm $A_2$), bidirectional path-relinking (algorithm $A_3$), backward path-relinking (algorithm $A_4$), and mixed path-relinking (algorithm $A_5$). As before, each heuristic was run independently 500 times.

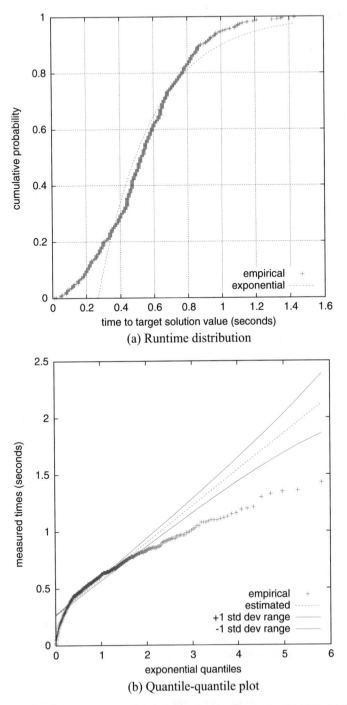

(a) Runtime distribution

(b) Quantile-quantile plot

**Fig. 6.19** Runtime distribution and quantile-quantile plot for GRASP with forward path-relinking on 90-node instance with the target value set at 673.

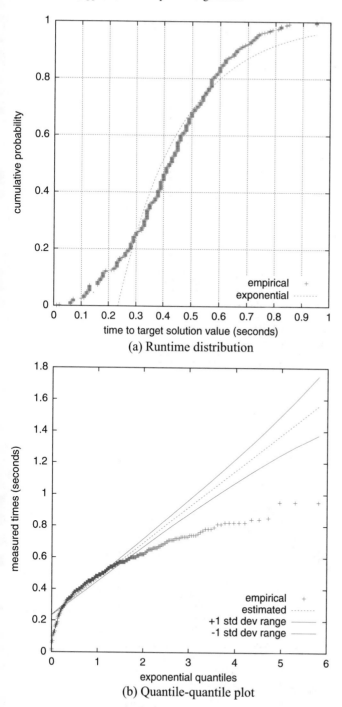

(a) Runtime distribution

(b) Quantile-quantile plot

**Fig. 6.20** Runtime distribution and quantile-quantile plot for GRASP with bidirectional path-relinking on 90-node instance with the target value set at 673.

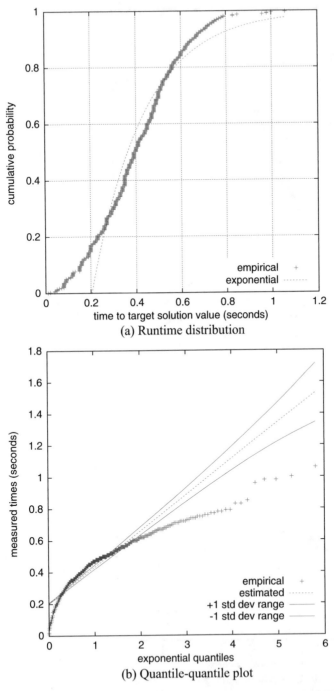

**Fig. 6.21** Runtime distribution and quantile-quantile plot for GRASP with backward path-relinking on 90-node instance with the target value set at 673.

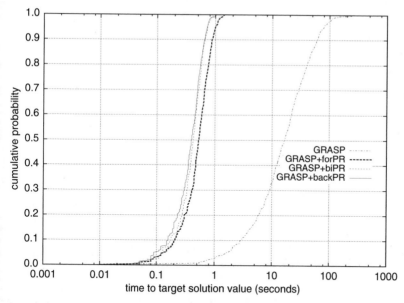

**Fig. 6.22** Superimposed runtime distributions of pure GRASP and three versions of GRASP with path-relinking.

**Table 6.3** Convergence of the estimation of $Pr(X_1 \leq X_2)$ with the sample size for the 90-node instance of the 2-path network design problem.

| $N$ | $L(\varepsilon)$ | $Pr(X_1 \leq X_2)$ | $R(\varepsilon)$ | $\Delta(\varepsilon)$ | $\varepsilon$ |
|---|---|---|---|---|---|
| 100 | 0.553300 | 0.559150 | 0.565000 | 0.011700 | $4.387188 \times 10^{-7}$ |
| 200 | 0.553250 | 0.553850 | 0.554450 | 0.001199 | $4.501629 \times 10^{-7}$ |
| 300 | 0.551578 | 0.557483 | 0.563389 | 0.011811 | $4.501629 \times 10^{-7}$ |
| 400 | 0.545244 | 0.551241 | 0.557238 | 0.011994 | $4.730511 \times 10^{-7}$ |
| 500 | 0.546604 | 0.552420 | 0.558236 | 0.011632 | $5.035686 \times 10^{-7}$ |
| 600 | 0.538867 | 0.544749 | 0.550631 | 0.011764 | $5.073833 \times 10^{-7}$ |
| 700 | 0.536320 | 0.542181 | 0.548041 | 0.011720 | $5.073833 \times 10^{-7}$ |
| 800 | 0.537533 | 0.543298 | 0.549064 | 0.011531 | $5.073833 \times 10^{-7}$ |
| 900 | 0.533912 | 0.539671 | 0.545430 | 0.011517 | $5.073833 \times 10^{-7}$ |
| 1000 | 0.531595 | 0.537388 | 0.543180 | 0.011585 | $5.073833 \times 10^{-7}$ |
| 2000 | 0.528224 | 0.533959 | 0.539698 | 0.011469 | $5.722427 \times 10^{-7}$ |
| 3000 | 0.530421 | 0.536128 | 0.541835 | 0.011414 | $6.027603 \times 10^{-7}$ |
| 4000 | 0.532695 | 0.538364 | 0.544033 | 0.011338 | $6.027603 \times 10^{-7}$ |
| 5000 | 0.530954 | 0.536566 | 0.542178 | 0.011225 | $6.027603 \times 10^{-7}$ |

The empirical runtime distributions of the five algorithms are superimposed in Figure 6.24. Algorithm $A_2$ (as well as $A_3$, $A_4$, and $A_5$) performs much better than $A_1$, as indicated by $Pr(X_2 \leq X_1) = 0.970652$ (with $L(\varepsilon) = 0.970288$, $R(\varepsilon) = 0.971016$, $\Delta(\varepsilon) = 0.000728$, and $\varepsilon = 0.014257$). Algorithm $A_3$ outperforms $A_2$, as shown by the fact that $Pr(X_3 \leq X_2) = 0.617278$ (with $L(\varepsilon) = 0.610808$, $R(\varepsilon) = 0.623748$, $\Delta(\varepsilon) = 0.012940$, and $\varepsilon = 1.220703 \times 10^{-6}$). Algorithm $A_4$ performs slightly better than $A_3$ for this instance, since $Pr(X_4 \leq X_3) = 0.537578$ (with $L(\varepsilon) = 0.529404$,

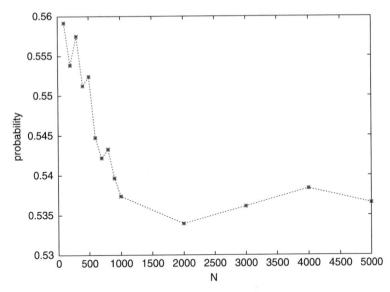

**Fig. 6.23** Convergence of the estimation of $Pr(X_1 \leq X_2)$ with the sample size for the 90-node instance of the 2-path network design problem.

$R(\varepsilon) = 0.545752$, $\Delta(\varepsilon) = 0.016348$, and $\Delta(\varepsilon) = 1.201630 \times 10^{-6}$). Algorithms $A_5$ and $A_4$ also behave very similarly, but $A_5$ is slightly better for this instance since $Pr(X_5 \leq X_4) = 0.556352$ (with $L(\varepsilon) = 0.547912$, $R(\varepsilon) = 0.564792$, $\Delta(\varepsilon) = 0.016880$, and $\varepsilon = 1.001358 \times 10^{-6}$).

## 6.6 Comparing and evaluating parallel algorithms

We conclude this chapter by describing the use of the runtime distribution methodology to evaluate and compare parallel implementations of stochastic local search algorithms. Once again, the 2-path network design problem is used to illustrate this application.

Figures 6.25 and 6.26 superimpose the runtime distributions of, respectively, cooperative and independent parallel implementations of GRASP with bidirectional path-relinking for the same problem on 2, 4, 8, 16, and 32 processors, on an instance with 100 nodes and 1000 origin-destination pairs, using 683 as target value. Each algorithm was run independently 200 times. We denote by $A_1^k$ (resp. $A_2^k$) the cooperative (resp. independent) parallel implementation running on $k$ processors, for $k = 2, 4, 8, 16, 32$.

Table 6.4 shows the probability that the cooperative parallel implementation performs better than the independent implementation on 2, 4, 8, 16, and 32 processors. We observe that the independent implementation performs better than the cooper-

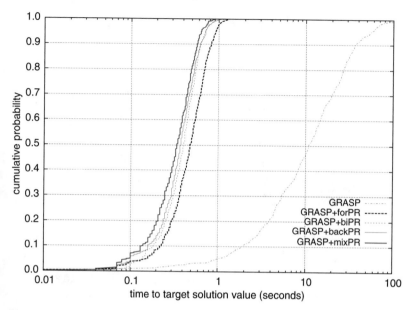

**Fig. 6.24** Superimposed empirical runtime distributions of pure GRASP and four GRASP with path-relinking heuristics.

ative implementation on two processors. In that case, the cooperative implementation does not benefit from the availability of two processors, since only one of them performs iterations, while the other acts as the master. However, as the number of processors increases from two to 32, the cooperative implementation performs progressively better than the independent implementation, since more processors are devoted to perform GRASP iterations. The proposed methodology is clearly consistent with the relative behavior of the two parallel versions for any number of processors. Furthermore, it illustrates that the cooperative implementation becomes progressively better than the independent implementation when the number of processors increases.

Table 6.5 displays the probability that each of the two parallel implementations performs better on $2^{j+1}$ than on $2^j$ processors, for $j = 1, 2, 3, 4$. Both implementations scale appropriately as the number of processors grows. Once again, we can see that the performance measure appropriately describes the relative behavior of the two parallel strategies and provides insight on how parallel algorithms scale with the number of processors. The table shows numerical evidence to evaluate the trade-offs between computation times and the number of processors in parallel implementations.

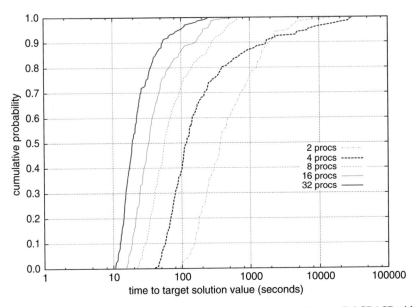

**Fig. 6.25** Superimposed empirical runtime distributions of cooperative parallel GRASP with bidirectional path-relinking running on 2, 4, 8, 16, and 32 processors.

**Table 6.4** Comparing cooperative (algorithm $A_1$) and independent (algorithm $A_2$) parallel implementations.

| Processors $(k)$ | $Pr(X_1^k \leq X_2^k)$ |
|:---:|:---:|
| 2 | 0.309784 |
| 4 | 0.597253 |
| 8 | 0.766806 |
| 16 | 0.860864 |
| 32 | 0.944938 |

**Table 6.5** Comparing the parallel implementations on $2^{j+1}$ (algorithm $A_1$) and $2^j$ (algorithm $A_2$) processors, for $j = 1, 2, 3, 4$.

| Processors $(a)$ | Processors $(b)$ | $Pr(X_1^a \leq X_1^b)$ | $Pr(X_2^a \leq X_2^b)$ |
|:---:|:---:|:---:|:---:|
| 4 | 2 | 0.766235 | 0.651790 |
| 8 | 4 | 0.753904 | 0.685108 |
| 16 | 8 | 0.724398 | 0.715556 |
| 32 | 16 | 0.747531 | 0.669660 |

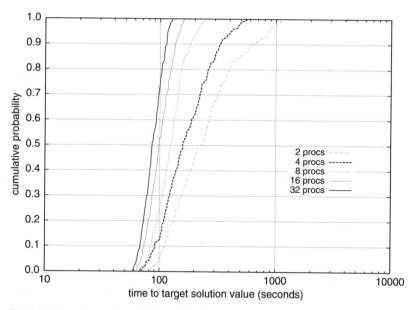

**Fig. 6.26** Superimposed empirical runtime distributions of independent parallel GRASP with bidirectional path-relinking running on 2, 4, 8, 16, and 32 processors.

## 6.7 Bibliographical notes

The time-to-target, or runtime distribution, plots introduced in Section 6.1 were first used by Feo et al. (1994). They have been also advocated by Hoos and Stützle (1998b;a) as a way to characterize the execution times of stochastic algorithms for combinatorial optimization. Aiex et al. (2007) developed a perl program to create time-to-target plots for measured times that are assumed to fit a shifted exponential distribution, following closely the work of Aiex et al. (2002).

Section 6.2 reports that the runtime distributions of GRASP heuristics follow exponential distributions and shows how the best fittings can be obtained. In fact, Aiex et al. (2002), Battiti and Tecchiolli (1992), Dodd (1990), Eikelder et al. (1996), Hoos (1999), Hoos and Stützle (1999), Osborne and Gillett (1991), Selman et al. (1994), Taillard (1991), Verhoeven and Aarts (1995), and others observed that in many implementations of randomized heuristics, such as simulated annealing, genetic algorithms, iterated local search, tabu search, and GRASP, the random variable time-to-target value (i.e., the runtime) is exponentially distributed or fits a shifted exponential distribution. Hoos and Stützle (1998c; 1999) conjectured that this is true for all methods for combinatorial optimization based on stochastic local search.

Aiex et al. (2002) used TTT-plots to show experimentally that the running times of GRASP heuristics fit shifted exponential distributions, reporting computational results for 2400 runs of GRASP heuristics for each of five different problems: maximum stable set (Feo et al., 1994; Resende et al., 1998), quadratic assignment

(Li et al., 1994; Resende et al., 1996), graph planarization (Resende and Ribeiro, 1997; Ribeiro and Resende, 1999; Resende and Ribeiro, 2001), maximum weighted satisfiability (Resende et al., 2000), and maximum covering (Resende, 1998). To compare the empirical and the theoretical runtime distributions, a standard graphical methodology for data analysis was used (Chambers et al., 1983). Experiments with instances of the 2-path network design problem and of the three-index assignment problem were reported to show that implementations of GRASP with path-relinking may not follow exponential distributions. The 2-path network design problem was introduced and proved to be *NP*-hard by Dahl and Johannessen (2004). The GRASP heuristics used in the computational experiments with this problem were proposed by Ribeiro and Rosseti (2002; 2007). The three-index assignment problem considered in the experiments reported by Aiex et al. (2005) was studied by Balas and Saltzman (1991), from where problem instances 22.1 and 24.1 were taken.

The closed form result developed in Section 6.3 to compare two exponential algorithms and the iterative procedure proposed in Section 6.4 to compare two algorithms following generic distributions were first presented by Ribeiro et al. (2009). This work was extended by Ribeiro et al. (2012) and was also applied in the comparison of parallel heuristics. Ribeiro and Rosseti (2015) developed the code to compare runtime distributions of randomized algorithms.

Different problems and algorithms were used in Sections 6.5 and 6.6 to illustrate the application of the iterative procedure to compare generic runtime distributions of two algorithms. Algorithms for solving the server replication for reliable multicast problem were described by Fonseca et al. (2008) and Santos et al. (2008). The DM-GRASP hybrid version of GRASP that incorporates a data mining process appeared in Santos et al. (2008). Its basic principle consisted in mining for patterns found in high-quality solutions to guide the construction of new solutions. Variant DM-D5 appeared in Fonseca et al. (2008).

Noronha and Ribeiro (2006) proposed a decomposition heuristic for solving the routing and wavelength assignment problem. First, a set of possible routes is precomputed for each lightpath. Next, one of the precomputed routes and a wavelength are assigned to each lightpath by a tabu search heuristic solving an instance of the partition coloring problem. Manohar et al. (2002) developed the multistart greedy heuristic for the same problem. The Finland instance of the routing and wavelength assignment problem came from Hyytiä and Virtamo (1998).

# Chapter 7
# Extended construction heuristics

In Chapter 3, we considered cardinality-based and quality-based adaptive greedy algorithms as a generalization of greedy algorithms. Next, we presented semi-greedy algorithms that are obtained by randomizing adaptive greedy algorithms and constitute the main foundation for developing the construction phase of GRASP heuristics. In this chapter, we consider enhancements, extensions, and variants of greedy randomized adaptive construction procedures such as Reactive GRASP, the probabilistic choice of the construction parameter $\alpha$, random plus greedy and sampled greedy constructions, cost perturbations, bias functions, principles of intelligent construction based on memory and learning, the proximate optimality principle and local search applied to partially constructed solutions, and pattern-based construction strategies using vocabulary building or data mining.

## 7.1 Reactive GRASP

The choice of the parameter $\alpha$ of a quality-based semi-greedy algorithm used in the GRASP construction phase determines the blend of greediness and randomness that is used in the construction. One basic strategy is to use a fixed value for $\alpha$. Another strategy consists in using a different value chosen at random in each iteration. Reactive GRASP is a strategy in which the algorithm progressively learns and updates the best value of $\alpha$. It was the first proposal to incorporate a learning mechanism in the otherwise memoryless construction phase of GRASP.

In the context of *Reactive GRASP*, the value of the restricted candidate list parameter $\alpha$ is not fixed, but instead is randomly selected at each iteration from a discrete set of possible values. This selection is guided by the solution values found during previous iterations. One way to accomplish this is to use a rule that considers a set $\Psi = \{\alpha_1, \ldots, \alpha_m\}$ of possible values for $\alpha$. The probabilities associated with the choice of each value are all initially made equal to $p_i = 1/m$, for $i = 1, \ldots, m$. Furthermore, let $f^*$ be the incumbent solution value and let $A_i$ be the average value of all solutions found using $\alpha = \alpha_i$, for $i = 1, \ldots, m$. The selection probabilities are

© Springer Science+Business Media New York 2016
M.G.C. Resende, C.C. Ribeiro, *Optimization by GRASP*,
DOI 10.1007/978-1-4939-6530-4_7

periodically reevaluated by taking $p_i = q_i / \sum_{j=1}^{m} q_j$, with $q_i = f^*/A_i$ for $i = 1,\ldots,m$. For the case of minimization, the value of $q_i$ will be larger for values of $\alpha = \alpha_i$ leading to the best solutions on average. Larger values of $q_i$ correspond to more suitable values for the parameter $\alpha$. Therefore, the probabilities associated with the more appropriate values increase when they are reevaluated.

The reactive approach leads to improvements over the basic GRASP in terms of robustness and solution quality, due to greater diversification and less reliance on parameter tuning.

## 7.2 Probabilistic choice of the RCL parameter

The computational results obtained by Reactive GRASP show that using a single fixed value for the restricted candidate parameter $\alpha$ very often hinders finding high-quality solutions, which could be found if other values were used. These results motivated the study of the behavior of GRASP using alternative strategies for the variation of the value of the restricted candidate list parameter $\alpha$:

(R) $\alpha$ self tuned according with the Reactive GRASP procedure;
(E) $\alpha$ randomly chosen from a uniform discrete probability distribution;
(H) $\alpha$ randomly chosen from a decreasing nonuniform discrete probability distribution; and
(F) fixed value of $\alpha$, close to the purely greedy choice.

We summarize the results obtained in experiments incorporating these four strategies into the GRASP procedures developed for four different optimization problems: (P-1) matrix decomposition for traffic assignment in communication satellites, (P-2) set covering, (P-3) weighted MAX-SAT, and (P-4) graph planarization. Let $\Psi = \{\alpha_1,\ldots,\alpha_m\}$ be the set of possible values for the parameter $\alpha$ for the first three strategies. The strategy for choosing and self-tuning the value of $\alpha$ in the case of the Reactive GRASP procedure (R) is the one described in Section 7.1. In the case of the strategy (E) based on using a discrete uniform distribution, all choice probabilities are equal to $1/m$. The third case corresponds to a hybrid strategy (H), in which the following probabilities are considered: $p(\alpha = 0.1) = 0.5$, $p(\alpha = 0.2) = 0.25$, $p(\alpha = 0.3) = 0.125$, $p(\alpha = 0.4) = 0.03$, $p(\alpha = 0.5) = 0.03$, $p(\alpha = 0.6) = 0.03$, $p(\alpha = 0.7) = 0.01$, $p(\alpha = 0.8) = 0.01$, $p(\alpha = 0.9) = 0.01$, and $p(\alpha = 1.0) = 0.005$. In the last strategy (F), the value of $\alpha$ is fixed as tuned and recommended in the original references reporting results for these problems. A subset of instances from the literature was considered for each class of test problems. Numerical results are reported in Table 7.1. For each problem, we first list the number of instances considered. Next, for each strategy, we give the number of instances for which it found the best solution (hits), as well as the average computation time (in seconds) on an IBM 9672 model R34. The number of GRASP iterations was fixed at 10,000.

**Table 7.1** Computational results for different strategies for the variation of parameter $\alpha$.

| Problem Instances | R | | E | | H | | F | |
|---|---|---|---|---|---|---|---|---|
| | Hits | Time (s) | Hits | Time (s) | Hits | Time (s) | Hits | Time (s) |
| P-1 | 36 | 34 | 579.0 | 35 | 358.2 | 32 | 612.6 | 24 | 642.8 |
| P-2 | 7 | 7 | 1346.8 | 6 | 1352.0 | 6 | 668.2 | 5 | 500.7 |
| P-3 | 44 | 22 | 2463.7 | 23 | 2492.6 | 16 | 1740.9 | 11 | 1625.2 |
| P-4 | 37 | 28 | 6363.1 | 21 | 7292.9 | 24 | 6326.5 | 19 | 5972.0 |
| Total | 124 | 91 | | 85 | | 78 | | 59 | |

Strategy (F) presented the shortest average computation times for three of the four problem types. It was also the one with the least variability in the constructed solutions and, as a consequence, found the best solution the fewest times. The reactive strategy (R) is the one which most often found the best solutions, however, at the cost of computation times that are longer than those of some of the other strategies. The high number of hits observed by strategy (E) also illustrates the effectiveness of strategies based on the variation of $\alpha$, the parameter that defines the size of the restricted candidate list.

## 7.3 Random plus greedy and sampled greedy

In Section 3.4 of Chapter 3, we described the semi-greedy scheme used in the construction phase of GRASP to build solutions that serve as starting points for local search. Two alternative randomized greedy approaches that run faster than the semi-greedy algorithm are introduced next. Both have been originally applied to the $p$-median problem.

Instead of combining greediness and randomness at each step of the construction procedure, the *random plus greedy* scheme applies randomness during the first $p$ construction steps to produce a random partial solution. Next, the algorithm completes the solution with one or more pure adaptive greedy construction steps. The resulting solution is randomized greedy. One can control the balance between greediness and randomness in the construction by changing the value of the parameter $p$. Larger values of $p$ are associated with solutions that are more random, while smaller values result in greedier solutions.

Similar to the random plus greedy procedure, the *sampled greedy* construction also combines randomness and greediness, but in a different way. This procedure is also controlled by a parameter $p$. At each step of the construction process, the procedure builds a restricted candidate list by sampling $\min\{p, |C|\}$ elements of the candidate set $C$ of elements that can be added to the current partial solution. Each element of the restricted candidate list is evaluated by the greedy function. The element with the smallest greedy function value is added to the partial solution. This two-step process is repeated until there are no more candidate elements. The resulting solution is also randomized greedy. Once again, the balance between greediness and randomness can be controlled by changing the value of the parameter $p$, i.e., the number of candidate elements that are sampled. Small sample sizes lead to more random solutions, while large sample sizes lead to greedier solutions.

## 7.4 Cost perturbations

The idea of introducing noise into the original costs to change the objective function adds more flexibility to algorithm design. Furthermore, in circumstances where the construction algorithm is not very sensitive to randomization, it can also be more effective than the greedy randomized adaptive construction of the basic GRASP procedure. This is indeed the case for the shortest path heuristic used as one of the main building blocks of the construction phase of GRASP heuristics for the Steiner problem in graphs.

The cost perturbation methods used in a hybrid algorithm for the Steiner problem in graphs incorporate learning mechanisms associated with intensification and diversification strategies. Three distinct weight randomization methods were applied to generate cost perturbations for the shortest path heuristic. At any given GRASP iteration, the modified weight of each edge is randomly selected from a uniform distribution in an interval which depends on the selected weight randomization method applied at that iteration. The different weight randomization methods use frequency information and are used to enforce intensification and diversification strategies. Experimental results show that the strategy combining these three perturbation methods is more robust than any of them used in isolation, leading to the best overall results on a quite broad mix of test instances with different characteristics. The GRASP heuristic using this cost perturbation strategy was among the most effective heuristics available for the Steiner problem in graphs at the time of its development.

Another situation where cost perturbations can be very effective appears when no greedy algorithm is available for straightforward randomization. A typical situation is the case of a hybrid GRASP developed for the prize-collecting Steiner tree problem, which makes use of a primal-dual approximation algorithm to build initial solutions using perturbed costs. A new solution is built at each iteration using node prizes updated by a perturbation function, according to the structure of the current solution. Two different prize perturbation schemes were used. In *perturbation by eliminations*, the primal-dual algorithm used in the construction phase is driven to build a new solution without some of the nodes that appeared in the solution constructed in the previous iteration. In *perturbation by prize changes*, noise is introduced into the node prizes to change the objective function.

## 7.5 Bias functions

In the construction phase of the basic GRASP heuristic, the next element to be introduced in the solution is chosen at random from the elements in the restricted candidate list. The elements of the restricted candidate list are assigned equal probabilities of being chosen. However, as was the case for Reactive GRASP, any probability distribution can be used to bias the selection towards some particular candidates. Another selection mechanism used in the construction phase makes use of a family of such probability distributions. They are based on the rank $r(\sigma)$ assigned to each candidate element $\sigma$, according to its greedy function value. Several bias functions can be used:

- random bias: $\texttt{bias}(r(\sigma)) = 1$;
- linear bias: $\texttt{bias}(r(\sigma)) = 1/r(\sigma)$;
- log bias: $\texttt{bias}(r(\sigma)) = \log^{-1}(r(\sigma) + 1)$;
- exponential bias: $\texttt{bias}(r(\sigma)) = e^{-r(\sigma)}$; and
- polynomial bias of order $n$: $\texttt{bias}(r(\sigma)) = r(\sigma)^{-n}$.

Consider that any one of the above bias functions is being used. Once the rank $r(\sigma)$ and its corresponding bias, $\texttt{bias}(r(\sigma))$, are evaluated for all elements in the candidate set $C$, the probability $\pi(\sigma)$ of selecting element $\sigma$ is given by

$$\pi(\sigma) = \frac{\texttt{bias}(r(\sigma))}{\sum_{\sigma' \in C} \texttt{bias}(r(\sigma'))}.$$

The basic GRASP heuristic uses a random bias function. The evaluation of these bias functions can be applied to all candidate elements or can be limited to the elements of the restricted candidate list.

## 7.6 Memory and learning

Flexible and adaptive memory techniques have been the source of a number of developments to improve multistart methods, which otherwise would simply resort to random restarts.

The basic memoryless GRASP heuristic does not make use of information gathered in previously performed iterations. As in tabu search and other multistart heuristics, a long-term memory strategy can be used to add memory to GRASP.

An *elite solution* is a high-quality solution found during the iterations of a search algorithm. *Long-term memory* can be implemented by maintaining a *pool* or *set* of elite solutions. To become an elite solution, a solution must be either better than the best member of the pool, or better than its worst member and sufficiently different from the other solutions in the pool. For example, one can count identical solution attributes and set a threshold for rejection. In Chapter 9, we revisit elite sets.

A *strongly determined variable* is one that cannot be changed without eroding the objective or changing significantly other variables. A *consistent variable* is one that receives a particular value in a large portion of the elite solution set. Let $I(e)$ be a measure of the strong determination and consistency features of a solution element $e$ of the ground set $E$. Then, $I(e)$ becomes larger as $e$ appears more often in the pool of elite solutions. The *intensity function* $I(e)$ is used in the construction phase as follows. Recall that $g(e)$ is the greedy function, i.e., the incremental cost associated with the incorporation of element $e \in E$ into the solution under construction. Let $K(e) = F(g(e), I(e))$ be a composite function of the greedy and the intensification functions. For example, $K(e) = \lambda \cdot g(e) + I(e)$. The intensification scheme biases selection from the restricted candidate list RCL to those elements $e \in E$ with a high value of $K(e)$ by setting its selection probability to be $p(e) = K(e)/\sum_{s \in \text{RCL}} K(s)$. Function $K(e)$ can vary with time by changing the value of $\lambda$. For example, $\lambda$ can be set to a large value when diversification is required and can be gradually decreased as intensification is called for.

## 7.7 Proximate optimality principle in construction

The proximate optimality principle is based on the idea that good solutions at one level of the search are likely to be found close to good solutions at an adjacent level. A GRASP interpretation of this principle suggests that imperfections introduced during steps of the GRASP construction can be ironed-out by applying local search during (and not only at the end of) the construction phase.

Because of efficiency considerations, a practical implementation of the proximate optimality principle to GRASP consists in applying local search a few times during the construction phase, but not necessarily at every construction iteration.

## 7.8 Pattern-based construction

Different strategies have been devised to intelligently explore adaptive memory information. Their main underlying principle consists in exploring information collected and updated along the search to improve the performance of different construction and local search methods.

Vocabulary building is an intensification strategy for creating new solutions from good fragments of high-quality solutions previously found and stored in an elite set. Data mining refers to the automatic extraction of new and potentially useful knowledge from data sets. The extraction of frequent items is often at the core of data mining methods. Frequent items extracted from the elite set represent patterns appearing in high-quality solutions that may be used as building blocks in an adapted construction phase.

Both *data mining* and *vocabulary building* can be combined into more efficient implementations of GRASP or other multistart procedures. Since data mining has been further explored in this context, we focus our presentation of pattern-based construction strategies in the hybridization GRASP with data mining.

The main kinds of rules and patterns mined from data sets are frequent items, association rules, sequential patterns, classification rules, and data clusters. In the context of a data set formed by solutions obtained by GRASP, the extracted frequent items represent patterns that are common to high-quality solutions, i.e., subsets of variables that often appear in the elite set.

Let $I = \{i_1, \ldots, i_n\}$ be a set of items. A *transaction t* is a subset of $I$ and a *data set* $\mathscr{T}$ is a set of transactions. An *item set F*, with support $s \in [0, 1]$, is a subset of $I$ which occurs in at least $s \cdot |\mathscr{T}|$ transactions of $\mathscr{T}$. An item set $F$ is said to be *frequent* if its support $s$ is greater than or equal to a minimum threshold $\underline{s}$ specified by the user. The frequent item mining problem consists in extracting all frequent item sets from a data set $\mathscr{T}$ with a minimum support $\underline{s}$ specified as a parameter. A frequent item set is maximal if it has no superset that is also frequent. Maximal frequent item sets are useful to avoid mining and using patterns that are subsets of larger patterns.

The principle behind the incorporation of a data mining process in GRASP is that patterns or frequent item sets found in high-quality solutions obtained in earlier iterations can be used in later iterations to improve the search procedure.

The DM-GRASP (Data Mining GRASP) heuristic starts with the elite set generation phase, which consists of executing $n_{\text{iter}}$ pure GRASP iterations to obtain an elite set $\mathscr{E}$ formed by the $n_{\mathscr{E}}$ best distinct solutions found along these iterations. Next, a data mining process is applied to extract a set $P$ of common patterns from the solutions in the elite set $\mathscr{E}$. These patterns are subsets of attributes that frequently appear in solutions of the elite set $\mathscr{E}$ (or, equivalently, variables that are set at persistent values in these solutions). A frequent pattern mined from the elite set with support $s \in [0, 1]$ represents a subset of attributes that occur in $s \cdot n_{\mathscr{E}}$ elite solutions. The hybrid phase is the last to be performed. An additional $n_{\text{iter}}$ slightly modified GRASP iterations are executed. The construction phase of each of these modified iterations starts from a pattern selected from the set of mined patterns $P$, and not from scratch. Therefore, DM-GRASP spends the first half of its iterations in the elite set generation phase and the second half in the hybrid phase, which makes use of the mined frequent patterns. We observe that the number of iterations $n_{\text{iter}}$ may be replaced by any other stopping criterion.

A pseudo-code of the DM-GRASP hybridization for a minimization problem is illustrated in Figure 7.1. The best solution value $f^*$ and the elite set $\mathscr{E}$ are initialized in lines 1 and 2, respectively. The $n_{\text{iter}}$ pure GRASP iterations of the first phase are carried out in the while loop in lines 3 to 14. A solution $S$ is constructed with a semi-greedy algorithm in line 4. Since a semi-greedy algorithm cannot always generate a feasible solution, a repair procedure may have to be invoked in line 6 to make changes in $S$ so that it becomes feasible (alternatively, the solution $S$ may be simply discarded and followed by a new run of the semi-greedy algorithm, until a feasible solution is built). Local search is then applied starting from the feasible solution provided by the semi-greedy algorithm or by the repair procedure. If the objective function value $f(S)$ of the local minimum produced in line 8 is better than the value $f^*$ of the incumbent, then the local minimum is made the incumbent and its objective function value is placed in $f^*$ in lines 10 and 11. The elite set $\mathscr{E}$ is updated in line 13: if the new solution $S$ is added to $\mathscr{E}$, then a previously obtained elite solution is discarded. Algorithm UPDATE-ELITE-SET described in Section 9.2 receives as inputs the local optimum $S$ and the current elite set $\mathscr{E}$ and returns the updated elite set. The data mining algorithm extracts the set of frequent patterns $P$ from the elite set $\mathscr{E}$ in line 15. The loop from line 16 to 27 corresponds to the hybridization phase and runs until an additional $n_{iter}$ iterations are performed. Each iteration starts in line 17 by the selection of a pattern $p \in P$. An adapted construction procedure based on the SemiGreedy algorithm is performed in line 18, starting from a partial solution defined by pattern $p$ as a starting point, and not from scratch. The feasibility of solution $S$ is tested in line 19. A repair procedure may have to be invoked in line 20 to make changes in $S$ so that it becomes feasible (as before, the solution $S$ may also be simply discarded and followed by a new run of the adapted semi-greedy algorithm, until a feasible solution is built). Local search is applied in line 22 to solution $S$. If the objective function value $f(S)$ of the local minimum $S$ is better than the value $f^*$ of the incumbent, then this new local minimum is made the incumbent in line 24 and its objective function value is placed in $f^*$ in line 25. The best solution $S^*$ and its cost $f(S^*)$ are returned in line 28.

To illustrate the improvements brought by DM-GRASP to the basic GRASP procedure, we summarize below some computational results obtained for the problem of server replication for reliable multicast, which was introduced in Section 6.5.1 of Chapter 6.

In computational experiments performed to compare the performance of GRASP and DM-GRASP both heuristics were implemented in C++ and were run ten times for each problem instance, with different random seeds. The GRASP parameter $\alpha$ was set to 0.7. Each run consisted of 500 iterations. In the hybrid DM-GRASP, the elite set generation phase made use of $n_{iter} = 250$ iterations and the hybrid phase performed the remaining $n_{iter} = 250$ iterations. The size of the elite set $\mathscr{E}$ was set at ten, as well as that of the set $P$ of mined patterns. The pattern extraction algorithm used a support value such that a set of nodes may be considered as a frequent pattern if it appears at least in two elite solutions.

The computational results are shown in Table 7.2. The first two columns summarize the characteristics of the problem instances, showing the multicast scenario and the number $m$ of nodes to be set as replicated servers. The next three columns contain the results obtained by a previously developed GRASP heuristic (best solution value, average solution value, and computation time in seconds), while the last three columns depict the same results for DM-GRASP. The best solution obtained by DM-GRASP improved on the solution obtained by GRASP in 12 out of the 20 instances in this table, while GRASP never obtained a better solution. The best results are indicated in boldface. DM-GRASP obtained better average solution values for 13 out of the 20 instances, while GRASP obtained better average values for only four instances. Furthermore, DM-GRASP was considerably faster for all instances.

```
begin DM-GRASP;
1   f* ← ∞;
2   𝓔 ← ∅;
3   for i = 1,...,n_iter do
4       S ← SEMI-GREEDY;
5       if S is not feasible then
6           S ← Repair(S);
7       end-if;
8       S ← LOCAL-SEARCH(S);
9       if f(S) < f* then
10          S* ← S;
11          f* ← f(S);
12      end-if;
13      UPDATE-ELITE-SET(S, 𝓔);
14  end-for;
15  P ← Mine(𝓔);
16  for i = 1,...,n_iter do
17      Select a pattern p ∈ P;
18      Build a solution S using an adapted version of the semi-greedy algorithm
            starting from pattern p;
19      if S is not feasible then
20          S ← Repair(S);
21      end-if;
22      S ← LOCAL-SEARCH(S);
23      if f(S) < f* then
24          S* ← S;
25          f* ← f(S);
26      end-if;
27  end-for;
28  return S*, f(S*);
end DM-GRASP.
```

**Fig. 7.1** Pseudo-code of a DM-GRASP heuristic for minimization.

The last column shows the average reduction in time obtained by DM-GRASP with respect to GRASP. On average, DM-GRASP ran in 36.8% time less than GRASP.

## 7.9 Lagrangean GRASP heuristics

### 7.9.1 Lagrangean relaxation and subgradient optimization

Lagrangean relaxation can be used to provide lower bounds for combinatorial optimization problems. However, the primal solutions produced by the algorithms used to solve the Lagrangean dual problem are not necessarily feasible. Lagrangean heuristics exploit dual multipliers to generate primal feasible solutions.

**Table 7.2** Comparison between GRASP and DM-GRASP for the reliable multicast problem.

| Instances | | GRASP | | | DM-GRASP | | | Time |
|---|---|---|---|---|---|---|---|---|
| Scenario | $m$ | Best | Average | Time (s) | Best | Average | Time (s) | reduction |
| CONF_1 | 5 | **63762.2** | **63762.2** | 28231.5 | **63762.2** | **63762.2** | **20292.0** | 28.1% |
| | 10 | **44480.7** | **44480.7** | 43826.0 | **44480.7** | **44480.7** | **30881.8** | 29.5% |
| | 15 | **31328.6** | 31347.2 | 43374.8 | **31328.6** | **31328.6** | **30058.0** | 30.7% |
| | 20 | **23625.2** | 23775.9 | 43314.2 | **23625.2** | **23763.7** | **26831.3** | 38.0% |
| CONF_2 | 10 | **11894.1** | **11894.1** | 3083.9 | **11894.1** | **11894.1** | **2631.8** | 14.7% |
| | 20 | 10076.3 | 10076.3 | 5239.0 | **10047.1** | **10047.1** | **3280.1** | 37.4% |
| | 30 | **9207.8** | **9208.7** | 7211.5 | **9207.8** | 9211.7 | **4196.7** | 41.9% |
| | 40 | 8668.5 | 8676.6 | 6787.9 | **8642.3** | **8646.3** | **4418.2** | 34.9% |
| CONF_3 | 20 | 11130.1 | 11177.4 | 7518.6 | **11114.5** | **11114.5** | **5661.6** | 24.7% |
| | 40 | 9631.7 | 9652.9 | 15077.0 | **9584.3** | **9596.5** | **8724.9** | 42.1% |
| | 60 | 8855.9 | **8869.3** | 19683.5 | **8848.1** | 8869.4 | **11312.0** | 42.5% |
| | 80 | **8550.4** | **8557.8** | 17747.8 | **8550.4** | 8559.1 | **10628.1** | 40.1% |
| BROAD_1 | 25 | 2818.9 | 2818.9 | 1555.1 | **2807.2** | **2807.2** | **1004.8** | 35.4% |
| | 50 | 2296.6 | 2299.0 | 3709.2 | **2281.8** | **2287.4** | **2301.7** | 38.0% |
| | 75 | 2039.3 | 2045.9 | 5530.9 | **2020.9** | **2030.8** | **3366.7** | 39.1% |
| | 100 | **1873.6** | **1877.5** | 7183.0 | **1873.6** | 1877.9 | **4160.2** | 42.0% |
| BROAD_2 | 50 | 2444.0 | 2444.2 | 5096.8 | **2425.6** | **2431.4** | **2994.0** | 41.3% |
| | 100 | 2019.0 | 2020.2 | 9246.2 | **2018.9** | **2020.1** | **5188.1** | 43.9% |
| | 150 | 1836.3 | 1837.0 | 11482.1 | **1836.2** | **1836.9** | **6098.7** | 46.9% |
| | 200 | 1727.9 | 1729.6 | 14047.3 | **1726.5** | **1729.3** | **7705.9** | 45.1% |
| | | | | | | Average reduction in time: | | 36.8% |

Given a mathematical programming problem $\mathscr{P}$ formulated as

$$f^* = \min f(x) \tag{7.1}$$

$$g_i(x) \leq 0, \quad i = 1, \ldots, m, \tag{7.2}$$

$$x \in X, \tag{7.3}$$

its Lagrangean relaxation is obtained by associating dual multipliers $\lambda_i \in \mathbb{R}_+$ to each inequality (7.2), for $i = 1, \ldots, m$. This results in the following *Lagrangean relaxation problem LRP($\lambda$)*

$$\min f'(x) = f(x) + \sum_{i=1}^{m} \lambda_i \cdot g_i(x) \tag{7.4}$$

$$x \in X, \tag{7.3}$$

whose optimal solution $x(\lambda)$ gives a lower bound $f'(x(\lambda))$ to the optimal value of the original problem $\mathscr{P}$ defined by (7.1) to (7.3). The best (dual) lower bound is given by the solution of the *Lagrangean dual problem* $\mathscr{D}$

$$f_\mathscr{D} = f'(x(\lambda^*)) = \max_{\lambda \in \mathbb{R}_+^m} f'(x(\lambda)). \tag{7.5}$$

Subgradient optimization is used to solve the dual problem $\mathscr{D}$ defined by (7.5). Subgradient algorithms start from any feasible set of dual multipliers, such as $\lambda_i = 0$, for $i = 1, \ldots, m$, and iteratively generate updated multipliers.

At any iteration $q$, let $\lambda^q$ be the current vector of multipliers and let $x(\lambda^q)$ be an optimal solution to problem $LRP(\lambda^q)$, whose optimal value is $f'(x(\lambda^q))$. Furthermore, let $\bar{f}$ be a known upper bound to the optimal value of problem $\mathscr{P}$ defined by (7.1) to (7.3). Additionally, let $g^q \in \mathbb{R}^m$ be a subgradient of $f'(x)$ at $x = x(\lambda^q)$, with $g_i^q = g_i(x(\lambda^q))$ for $i = 1, \ldots, m$. To update the Lagrangean multipliers, the algorithm makes use of a step size

$$d^q = \frac{\eta \cdot (\bar{f} - f'(x(\lambda^q)))}{\sum_{i=1}^m (g_i^q)^2}, \tag{7.6}$$

where $\eta \in (0, 2]$. Multipliers are then updated as

$$\lambda_i^{q+1} = \max\{0; \lambda_i^q + d^q \cdot g_i^q\}, \quad i = 1, \ldots, m, \tag{7.7}$$

and the subgradient algorithm proceeds to iteration $q + 1$.

### 7.9.2 A template for Lagrangean heuristics

We describe next a template for Lagrangean heuristics that make use of the dual multipliers $\lambda^q$ and of the optimal solution $x(\lambda^q)$ to each problem $LRP(\lambda^q)$ to build feasible solutions to the original problem $\mathscr{P}$ defined by (7.1) to (7.3). In the following, we assume that the objective function and all constraints are linear functions, i.e., $f(x) = \sum_{i=1}^n c_j x_j$ and $g_i(x) = \sum_{j=1}^n d_{ij} x_j - e_i$, for $i = 1, \ldots, m$.

Let $\mathscr{H}$ be a primal heuristic that builds a feasible solution $x$ to $\mathscr{P}$, starting from the initial solution $x^0 = x(\lambda^q)$ at every iteration $q$ of the subgradient algorithm. Heuristic $\mathscr{H}$ is first applied using the original costs $c_j$, i.e., using the cost function $f(x)$. In any subsequent iteration $q$ of the subgradient algorithm, $\mathscr{H}$ uses either Lagrangean reduced costs $c'_j = c_j - \sum_{i=1}^m \lambda_i^q d_{ij}$ or complementary costs $\bar{c}_j = (1 - x_j(\lambda^q)) \cdot c_j$.

Let $x^{\mathscr{H}, \gamma}$ be the solution obtained by heuristic $\mathscr{H}$, using a generic cost vector $\gamma$ corresponding to either one of the above modified cost schemes or to the original cost vector. Its cost can be used to update the upper bound $\bar{f}$ to the optimal value of the original problem (7.1) to (7.3). This upper bound can be further improved by local search and is used to adjust the step size defined by equation (7.6).

The algorithm in Figure 7.2 shows the pseudo-code of a Lagrangean heuristic. Lines 1 to 4 initialize the upper and lower bounds, the iteration counter, and the dual multipliers. The iterations of the subgradient algorithm are performed along the loop in lines 5 to 24. The reduced costs are computed in line 6 and the Lagrangean relaxation problem is solved in line 7. In the first iteration of the Lagrangean heuristic, the original cost vector is assigned to $\gamma$ in line 9, while in subsequent iterations a modified cost vector is assigned to $\gamma$ in line 11. Heuristic $\mathcal{H}$ is applied in line 13 at the first iteration and after every $H$ iterations thereafter (i.e., whenever the iteration counter $q$ is a multiple of the input parameter $H$) to produce a feasible solution $x^{\mathcal{H},\gamma}$ to problem (7.1) to (7.3). If the cost of this solution is smaller than the current upper bound, then the best solution and its cost are updated in lines 14 to 18. If the lower bound $f'(x(\lambda^q))$ is greater than the current lower bound $f_{\mathscr{D}}$, then $f_{\mathscr{D}}$ is updated in line 19. Line 20 computes a subgradient at $x(\lambda^q)$ and line 21 computes the step size. The dual multipliers are updated in line 22 and the iteration counter is incremented in line 23. The best solution found and its cost are returned in line 24.

---

**begin** LAGRANGEAN-HEURISTIC($H$);
1   $\bar{f} \leftarrow +\infty$;
2   $f_{\mathscr{D}} \leftarrow -\infty$;
3   $q \leftarrow 0$;
4   $\lambda_i^q \leftarrow 0$, $i = 1, \ldots, m$;
5   **repeat**
6       Compute reduced costs: $c_j' \leftarrow c_j - \sum_{i=1}^{m} \lambda_i^q d_{ij}$, $j = 1, \ldots, n$;
7       Solve $LRP(\lambda^q)$ to obtain a solution $x(\lambda^q)$;
8       **if** $q = 0$ **then**
9           $\gamma \leftarrow c$;
10      **else**
11          Set $\gamma$ to the modified cost vector $c'$ or $\bar{c}$;
12      **end-if**;
13      **if** $q$ is a multiple of $H$ **then** apply heuristic $\mathcal{H}$ with cost vector $\gamma$ to obtain $x^{\mathcal{H},\gamma}$;
14      **if** $f(x^{\mathcal{H},\gamma}) < \bar{f}$
15      **then do**;
16          $x^* \leftarrow x^{\mathcal{H},\gamma}$;
17          $\bar{f} \leftarrow f(x^{\mathcal{H},\gamma})$;
18      **end-if**;
19      **if** $f'(x(\lambda^q)) > f_{\mathscr{D}}$ **then** $f_{\mathscr{D}} \leftarrow f'(x(\lambda^q))$;
20      Compute a subgradient: $g_i^q \leftarrow g_i(x(\lambda^q))$, $i = 1, \ldots, m$;
21      Compute the step size: $d^q \leftarrow \eta \cdot (\bar{f} - f'(x(\lambda^q))) / \sum_{i=1}^{m} (g_i^q)^2$;
22      Update the dual multipliers: $\lambda_i^{q+1} \leftarrow \max\{0, \lambda_i^q - d^q g_i^q\}$, $i = 1, \ldots, m$;
23      $q \leftarrow q + 1$;
24  **until** stopping criterion satisfied;
25  **return** $x^*, f(x^*)$;
**end** LAGRANGEAN-HEURISTIC.

---

**Fig. 7.2** Pseudo-code of a template for a Lagrangean heuristic.

Different choices for the initial solution $x^0$, for the modified costs $\gamma$, and for the primal heuristic $\mathcal{H}$ itself lead to different variants of the above algorithm. The integer parameter $H$ defines the frequency in which $\mathcal{H}$ is applied. The smaller the value of $H$, the greater the number of times $\mathcal{H}$ is applied. Therefore, the computation time increases as the value of $H$ decreases. In particular, one should set $H = 1$ if the primal heuristic $\mathcal{H}$ is to be applied at every iteration.

### 7.9.3 Lagrangean GRASP

Different choices for the primal heuristic $\mathscr{H}$ in the template of Algorithm 7.2 lead to distinct Lagrangean heuristics. We consider two variants: the first makes use of a greedy algorithm with local search, while in the second a GRASP with path-relinking heuristic is used.

*Greedy heuristic*: This heuristic repairs the solution $x(\lambda^q)$ produced in line 7 of the Lagrangean heuristic described in Algorithm 7.2 to make it feasible for problem $\mathscr{P}$. It makes use of the modified costs ($c'$ or $\bar{c}$). Local search can be applied to the resulting solution, using the original cost vector $c$. We refer to this approach as a *greedy Lagrangean heuristic* (GLH).

*GRASP heuristic*: Instead of simply performing one construction step followed by local search as for GLH, this variant applies a GRASP heuristic to repair the solution $x(\lambda^q)$ produced in line 7 of the Lagrangean heuristic to make it feasible for problem $\mathscr{P}$.

Although the GRASP heuristic produces better solutions than the greedy heuristic, the greedy heuristic is much faster. To appropriately address this trade-off, we adapt line 10 of Algorithm 7.2 to use the GRASP heuristic with probability $\beta$ and the greedy heuristic with probability $1 - \beta$, where $\beta$ is a parameter of the algorithm.

We note that this strategy involves three main parameters: the number $H$ of iterations after which the basic heuristic is always applied, the number $Q$ of iterations performed by the GRASP heuristic when it is chosen as the primal heuristic, and the probability $\beta$ of choosing the GRASP heuristic as $\mathscr{H}$. We shall refer to the Lagrangean heuristic that uses this hybrid strategy as LAGRASP($\beta,H,Q$).

We next summarize computational results obtained for 135 instances of the set $k$-covering problem. These instances have up to 400 constraints and 4000 binary variables.

The first experiment with the GRASP Lagrangean heuristic established the relationship between running times and solution quality for different parameter settings. Parameter $\beta$, the probability of GRASP being applied as the heuristic $\mathscr{H}$, was set to $0, 0.25, 0.50, 0.75$, and 1. Parameter $H$, the number of iterations between successive calls to the heuristic $\mathscr{H}$, was set to $1, 5, 10$, and 50. Parameter $Q$, the number of iterations carried out by the GRASP heuristic, was set to $1, 5, 10$, and 50. By combining some of these parameter values, 68 variants of the hybrid LAGRASP($\beta,H,Q$) heuristic were created. Each variant was applied eight times to a subset formed by 21 instances, with different initial seeds being given to the random number generator.

The plot in Figure 7.3 summarizes the results for all variants evaluated, displaying points whose coordinates are the values of the average deviation from the best known solution value and the total time in seconds for processing the eight runs on all instances, for each combination of parameter values. Eight variants of special interest are identified and labeled with the corresponding parameters $\beta$, $H$, and $Q$, in this order. These variants correspond to selected Pareto points in the plot in Figure 7.3. Setting $\beta = 0$ and $H = 1$ corresponds to the greedy Lagrangean heuristic (GLH) or, equivalently, to LAGRASP($0,1,-$), whose average deviation from the best

value amounted to 0.12% in 4,859.16 seconds of total running time. Table 7.3 shows
the average deviation from the best known solution value and the total time for each
of the eight selected variants.

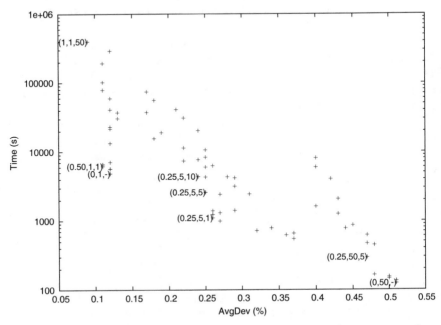

**Fig. 7.3** Average deviation from the best value and total running time for 68 different variants of
LAGRASP on a reduced set of 21 instances of the set $k$-covering problem: each point represents a
unique combination of parameters $\beta$, $H$, and $Q$.

**Table 7.3** Summary of the numerical results obtained with the selected variants of the GRASP
Lagrangean heuristic on a reduced set of 21 instances of the set $k$-covering problem. These values
correspond to the coordinates of the selected variants in Figure 7.3. The total time is given in
seconds.

| Heuristic | Average deviation | Total time (s) |
|---|---|---|
| LAGRASP(1,1,50) | 0.09 % | 399,101.14 |
| LAGRASP(0.50,1,1) | 0.11 % | 6,198.46 |
| LAGRASP(0,1,-) | 0.12 % | 4,859.16 |
| LAGRASP(0.25,5,10) | 0.24 % | 4,373.56 |
| LAGRASP(0.25,5,5) | 0.25 % | 2,589.79 |
| LAGRASP(0.25,5,1) | 0.26 % | 1,101.64 |
| LAGRASP(0.25,50,5) | 0.47 % | 292.95 |
| LAGRASP(0,50,-) | 0.51 % | 124.26 |

In another experiment, all 135 test instances were considered for the comparison of the above selected eight variants of LAGRASP. Table 7.4 summarizes the
results obtained by the eight selected variants. It shows that LAGRASP(1,1,50)

found the best solutions, with their average deviation from the best values amounting to 0.079%. It also found the best known solutions in 365 executions, again with the best performance when the eight variants are evaluated side by side, although at the cost of the longest running times. On the other hand, the smallest running times were observed for LAGRASP(0,50,-), which was over 3000 times faster than LAGRASP(1,1,50) but found the worst-quality solutions among the eight variants considered.

**Table 7.4** Summary of the numerical results obtained with the selected variants of the GRASP Lagrangean heuristic on the full set of 135 instances of the set $k$-covering problem. The total time is given in seconds.

| Heuristic | Average deviation | Hits | Total time (s) |
|---|---|---|---|
| LAGRASP(1,1,50) | 0.079 % | 365 | 1,803,283.64 |
| LAGRASP(0.50,1,1) | 0.134 % | 242 | 30,489.17 |
| LAGRASP(0,1,-) | 0.135 % | 238 | 24,274.72 |
| LAGRASP(0.25,5,10) | 0.235 % | 168 | 22,475.54 |
| LAGRASP(0.25,5,5) | 0.247 % | 163 | 11,263.80 |
| LAGRASP(0.25,5,1) | 0.249 % | 164 | 5,347.78 |
| LAGRASP(0.25,50,5) | 0.442 % | 100 | 1,553.35 |
| LAGRASP(0,50,-) | 0.439 % | 97 | 569.30 |

Figure 7.4 illustrates the merit of the proposed approach for one of the test instances. We first observe that all variants reach the same lower bounds, which is expected since they depend exclusively on the common subgradient algorithm. However, as the lower bound appears to stabilize, the upper bound obtained by LAGRASP(0,1,-) (or GLH) also seems to freeze. On the other hand, the other variants continue to make improvements in discovering better upper bounds, since the randomized GRASP construction help them to escape from locally optimal solutions and find new, improved upper bounds.

Finally, we also report on the comparison of the performance of GRASP with backward path-relinking and LAGRASP when the same time limits are used as the stopping criterion for all heuristics and variants running on all 135 test instances. Eight runs were performed for each heuristic and each instance, using different initial seeds for the random number generator. The results in Table 7.5 show that all variants of LAGRASP outperformed GRASP with backward path-relinking and were able to find solutions whose costs are very close to or as good as the best known solution values, while GRASP with backward path-relinking found solutions whose costs are on average 4.05% larger than the best known solution values.

Figure 7.5 displays for one test instance the typical behavior of these heuristics. As opposed to the GRASP with path-relinking heuristic, the Lagrangean heuristics are able to escape from local optima for longer and keep on improving the solutions to obtain the best results.

To conclude, we note that an important feature of Lagrangean heuristics is that they provide not only a feasible solution (which gives an upper bound, in the case of a minimization problem), but also a lower bound that may be used to give an estimate of the optimality gap that may be considered as a stopping criterion.

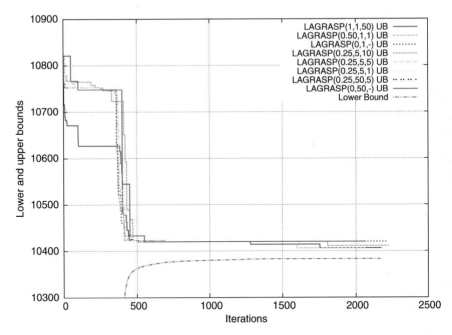

**Fig. 7.4** Evolution of lower and upper bounds over the iterations for different variants of LA-GRASP. The number of iterations taken by each LAGRASP variant depends on the step size, which in turn depends on the upper bounds produced by each heuristic.

**Table 7.5** Summary of results for the best variants of LAGRASP and GRASP.

| Heuristic | Average deviation | Hits |
|---|---|---|
| LAGRASP(1,1,50) | 3.30 % | 0 |
| LAGRASP(0.50,1,1) | 0.35 % | 171 |
| LAGRASP(0,1,-) | 0.35 % | 173 |
| LAGRASP(0.25,5,10) | 0.45 % | 138 |
| LAGRASP(0.25,5,5) | 0.45 % | 143 |
| LAGRASP(0.25,5,1) | 0.46 % | 137 |
| LAGRASP(0.25,50,5) | 0.65 % | 97 |
| LAGRASP(0,50,-) | 0.65 % | 93 |
| GRASP with backward path-relinking | 4.05 % | 0 |

## 7.10 Bibliographical notes

The Reactive GRASP approach considered in Section 7.1 was developed by Prais and Ribeiro (2000a) in the context of a traffic assignment problem in communication satellites. It has been widely explored and used in a number of successful applications. Computational experiments on this traffic assignment problem, reported in Prais and Ribeiro (2000a), showed that Reactive GRASP found better solutions than the basic algorithm for many test instances. In addition to the applications in Prais and Ribeiro (1999; 2000a;b), this approach was used in power

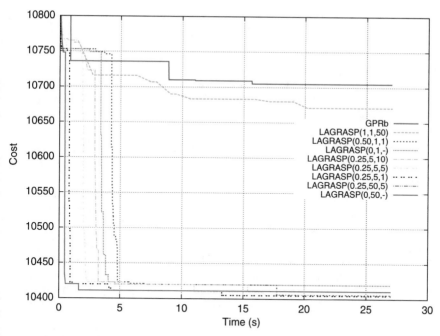

**Fig. 7.5** Evolution of solution costs with time for the best variants of LAGRASP and GRASP with backward path-relinking.

transmission network expansion planning (Bahiense et al., 2001; Binato and Oliveira, 2002), job shop scheduling (Binato et al., 2002), parallel machine scheduling with setup times (Kampke et al., 2009), balancing reconfigurable transfer lines (Essafi et al., 2012), container loading (Parreño et al., 2008), channel assignment in mobile phone networks (Gomes et al., 2001), broadcast scheduling (Butenko et al., 2004; Commander et al., 2004), just-in-time scheduling (Alvarez-Perez et al., 2008), single machine scheduling (Armentano and Araujo, 2006), examination scheduling (Casey and Thompson, 2003), semiconductor manufacturing (Deng et al., 2010), rural road network development (Scaparra and Church, 2005), maximum diversity (Duarte and Martí, 2007; Santos et al., 2005; Silva et al., 2004), max-min diversity (Resende et al., 2010a), capacitated location (Delmaire et al., 1999), locating emergency services (Silva and Serra, 2007), point-feature cartographic label placement (Cravo et al., 2008), set packing (Delorme et al., 2004), strip-packing (Alvarez-Valdes et al., 2008b), biclustering of gene expression data (Dharan and Nair, 2009; Das and Idicula, 2010), constrained two-dimensional nonguillotine cutting (Alvarez-Valdes et al., 2004), capacitated clustering (Deng and Bard, 2011), capacitated multi-source Weber problem (Luis et al., 2011), capacitated location routing (Prins et al., 2005), vehicle routing (Repoussis et al., 2007), family traveling salesperson (Morán-Mirabal et al., 2014), driver scheduling (Leone et al., 2011), portfolio optimization (Anagnostopoulos et al., 2010), automated test case prioritization (Maia et al., 2010), Golomb ruler search (Cotta and Fernández,

2004), commercial territory design motivated by a real-world application in a beverage distribution firm (Ríos-Mercado and Fernández, 2009), combined production-distribution (Boudia et al., 2007), and therapist routing and scheduling (Bard et al., 2014), among others.

The use of probabilistically determined values of the construction parameter $\alpha$ reported in Section 7.2 originally appeared in Prais and Ribeiro (1999; 2000b). The four tested strategies were incorporated into basic GRASP heuristics implemented for matrix decomposition for traffic assignment in communication satellites (Prais and Ribeiro, 2000a), set covering (Feo and Resende, 1989), weighted MAX-SAT (Resende et al., 1997; 2000), and graph planarization (Resende and Ribeiro, 1997; Ribeiro and Resende, 1999).

The two alternative randomized greedy approaches described in Section 7.3 were originally proposed in Resende and Werneck (2004) and compared with the semi-greedy algorithm for the $p$-median problem.

The idea of introducing perturbations into the original costs discussed in Section 7.4 is similar to that used in the so-called "noising" method of Charon and Hudry (1993; 2002). It was first applied in the context of GRASP to the shortest path heuristic of Takahashi and Matsuyama (1980), which is used as the main building block of the construction phase of the hybrid procedure proposed by Ribeiro et al. (2002) for the Steiner tree problem in graphs.

Another situation where cost perturbations can be very effective appears when no greedy algorithm is available for straightforward randomization. Canuto et al. (2001) made effective use of cost perturbations in their GRASP heuristic for the prize-collecting Steiner tree problem in graphs, for which no greedy algorithm was available to build starting solutions. In that case, the primal-dual algorithm of Goemans and Williamson (1996) was applied to build initial solutions, using different perturbed costs at each iteration of the hybrid GRASP procedure.

In the construction procedure of the basic GRASP, the next element to be introduced in the solution is chosen at random from the restricted candidate list. The elements of the restricted candidate list are assigned equal probabilities of being chosen. However, any probability distribution can be used to bias the selection towards some particular candidates. Bresina (1996) proposed a family of probability distributions to bias the selection mechanism in the construction phase of GRASP towards some particular candidates, as described in Section 7.5, instead of randomly choosing any element in the restricted candidate list (the basic GRASP heuristic uses a random bias function). Bresina's selection procedure applied to elements of the restricted candidate list was used in Binato et al. (2002).

Adaptive memory fundamentals and uses are reported by Rochat and Taillard (1995), Fleurent and Glover (1999), Patterson et al. (1999), Melián et al. (2004), and Martí et al. (2013a). Fleurent and Glover (1999) proposed the use of the long-term memory scheme described in Section 7.6 in multistart heuristics. The function $K(e)$ can vary with time by changing the value of $\lambda$. Procedures for changing the value of $\lambda$ were reported by Binato et al. (2002).

Glover and Laguna (1997) stated the proximate optimality principle as introduced in Section 7.7. Fleurent and Glover (1999) provided its interpretation in the context of GRASP and suggested the application of local search also during the construction phase. Local search was applied by Binato et al. (2002) after 40% and 80% of the construction moves have been taken, as well as at the end of the construction phase.

Section 7.8 presented alternative construction strategies based on the use of frequent patterns that appear in high-quality solutions previously detected. Vocabulary building and data mining can be combined into efficient pattern-based implementations of GRASP or other multistart procedures. Vocabulary building is an intensification strategy originally proposed in Glover and Laguna (1997) and Glover et al. (2000) for creating new solutions from good fragments of high-quality solutions previously found and stored in an elite set. See also Scholl et al. (1998) and Berger et al. (2000) for some successful applications. Aloise and Ribeiro (2011) developed a multistart procedure based on vocabulary building for multicommodity network design.

Data mining refers to the automatic extraction of new and potentially useful knowledge from data sets (Han et al., 2011; Witten et al., 2011). The extracted knowledge, expressed in terms of patterns or rules, represents important features of the data set at hand. The extraction of frequent items is one of the issues involved in the data mining process. Some algorithms exist to efficiently mine frequent items (Agrawal and Srikant, 1994; Han et al., 2000; Orlando et al., 2002; Goethals and Zaki, 2003). The patterns mined in the context of GRASP correspond to subsets of attributes that frequently appear in elite solutions.

The hybridization of GRASP with a data mining process was first introduced and applied to the set packing problem by Ribeiro et al. (2004; 2006). Afterwards, the method was evaluated in the context of three other applications, namely the maximum diversity problem (Santos et al., 2005), the server replication for reliable multicast problem (Santos et al., 2006), and the $p$-median problem (Plastino et al., 2009), with equally successful outcomes. The DM-GRASP hybrid heuristic, developed by Ribeiro et al. (2006), used a frequent item strategy that enhanced GRASP in terms of both solution quality and computation times. Frequent items extracted from the elite set represent patterns appearing in high-quality solutions, which are then used to perform an adapted construction phase which makes use of them. Frequent patterns are mined by the FPMax* algorithm (Grahne and Zhu, 2003). This hybridization strategy was also successfully applied to other combinatorial optimization problems (Santos et al., 2008; Plastino et al., 2009; 2011).

The MDM-GRASP variant, which performs data mining not only in the first phase, but also along the entire execution of the algorithm whenever the elite set changes, was developed by Barbalho et al. (2013) and Plastino et al. (2014). These references give numerical evidence that DM-GRASP and MDM-GRASP are able to improve not only the basic GRASP heuristic, but also implementations of Reactive GRASP and of GRASP with path-relinking.

Lagrangean relaxation (Beasley, 1993; Fisher, 2004) is a mathematical programming technique that can be used to provide lower bounds for minimization problems. Held and Karp (1970; 1971) were among the first to explore the use of the dual multipliers produced by Lagrangean relaxation to derive lower bounds, applying this idea in the context of the traveling salesman problem. Lagrangean heuristics further explore the use of different dual multipliers to generate feasible solutions. Beasley (1987; 1990b) described a Lagrangean heuristic for set covering, which can be extended to the set $k$-covering problem. The set multicovering or set $k$-covering problem is an extension of the classical set covering problem, in which each object is required to be covered at least $k$ times. The problem finds applications in the design of communication networks and in computational biology. Pessoa et al. (2011; 2013) proposed the hybridization of GRASP and Lagrangean relaxation leading to the Lagrangean GRASP heuristic described in Section 7.9. They generated 135 set $k$-covering instances from 45 set covering instances of the OR-Library (Beasley, 1990a), using three different coverage factors $k$. The experiments they performed were on a 2.33 GHz Intel Xeon E5410 Quadcore computer running Linux Ubuntu 8.04. All algorithms were implemented in C and compiled with gcc 4.1.2. They used the same strategy proposed by Held et al. (1974) for updating the dual multipliers from one iteration to the next. Beasley (1990b) reported as computationally useful the adjustment of components of the subgradients to zero whenever they do not effectively contribute to the update of the multipliers, i.e., arbitrarily setting $g_i^q = 0$ whenever $g_i^q > 0$ and $\lambda_i^q = 0$, for $i = 1, \ldots, m$.

# Chapter 8
# Path-relinking

Path-relinking is a search intensification strategy. As a major enhancement to heuristic search methods for solving combinatorial optimization problems, its hybridization with other metaheuristics has led to significant improvements in both solution quality and running times of hybrid heuristics. In this chapter, we review the fundamentals of path-relinking, implementation issues and strategies, and the use of randomization in path-relinking.

## 8.1 Template and mechanics of path-relinking

*Path-relinking* is an intensification strategy to explore trajectories connecting elite solutions (i.e., high-quality solutions) of combinatorial optimization problems. In this section, we focus on the path-relinking operator, including its template and mechanics.

As introduced in Chapter 4, we consider the search space graph $\mathscr{G} = (F, M)$ associated with a combinatorial optimization problem. The nodes of this graph correspond to the set $F$ of feasible solutions. There is an edge $(S, S') \in M$ if and only if $S \in F$, $S' \in F$, $S' \in N(S)$, and $S \in N(S')$, where $N(S) \subseteq F$ is the neighborhood of $S$. Path-relinking is usually carried out between two solutions in $F$: one is the *initial solution* $S^i$, while the other is the *guiding solution* $S^g$. One or more paths connecting these solutions in the search space graph can be explored by path-relinking in the search for better solutions. Local search is often applied to the best solution in each of these paths since there is no guarantee that this solution is locally optimal.

### 8.1.1 Restricted neighborhoods

Let $S \in F$ be any solution (i.e., a node) on a path in $\mathscr{G}$ leading from the initial solution $S^i \in F$ to the guiding solution $S^g \in F$. Not all solutions in the neighborhood

M.G.C. Resende, C.C. Ribeiro, *Optimization by GRASP*,
DOI 10.1007/978-1-4939-6530-4_8

$N(S)$ are allowed to follow $S$ on this path from $S^i$ to $S^g$. Path-relinking restricts its possible choices to the feasible solutions in $N(S)$ that are more similar to $S^g$ than $S$ is (measures of solution similarity will be discussed later). We denote by $N(S : S^g) \subseteq N(S)$ this *restricted neighborhood*, which is therefore defined exclusively by moves that introduce in $S$ attributes of the guiding solution $S^g$ that do not appear in $S$.

The elements of the ground set $E$ that appear in $S$ but not in $S^g$ are those that must be removed from the current solution $S$ in a path leading to $S^g$. Similarly, the elements of the ground set $E$ that appear in $S^g$ but not in $S$ are those that must be incorporated into $S$ in a path leading to $S^g$. The restricted neighborhood $N(S : S^g)$ is formed by all feasible solutions in $N(S)$ that may appear in a path from $S$ to $S^g$.

Therefore, path-relinking may be viewed as a strategy that seeks to incorporate attributes of a guiding solution (which is often a high-quality solution) into the current solution, by favoring these attributes in the selected moves. After evaluating each potential move leading to a feasible solution in $N(S : S^g)$, the most common strategy is a greedy approach, where one selects the move resulting in a best-quality restricted neighbor of $S$ that is closer to $S^g$ than $S$ is.

We next illustrate the restricted neighborhoods used by path-relinking with three examples.

## Minimum spanning tree problem – Restricted neighborhood

Consider the weighted graph depicted in Figure 8.1(a) and two of its spanning trees in Figures 8.1(b) and 8.1(c). Suppose the spanning tree in Figure 8.1(b) is the current solution $S$ and the one in Figure 8.1(c) is the guiding solution $S^g$. The total weight of solution $S$ is 35, while that of $S^g$ is 32.

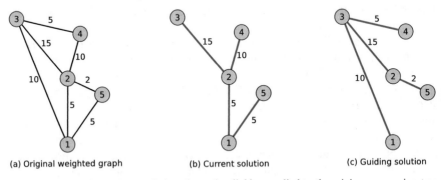

(a) Original weighted graph          (b) Current solution          (c) Guiding solution

**Fig. 8.1** Current and guiding solutions for path-relinking applied to the minimum spanning tree problem.

Edges $(3,4)$, $(1,3)$, and $(2,5)$ are present in $S^g$ but not in $S$, while edges $(2,4)$, $(1,2)$, and $(1,5)$ appear in $S$ but not in $S^g$. Therefore, to transform solution $S$ into $S^g$ it is necessary that all edges $(2,4)$, $(1,2)$, and $(1,5)$ be removed from $S$ and be replaced by the three edges $(3,4)$, $(1,3)$, and $(2,5)$ that originally appear only in $S^g$.

If $N$ is a swap neighborhood, then there are nine moves associated with ordered pairs of edges such that the first belongs to $S$ but not to $S^g$ (edge to be removed from $S$), while the second belongs to $S^g$ but not to $S$ (edge to be added to $S$). However, note that only four of the solutions resulting from these moves are feasible, corresponding to four ordered pairs of edges to be swapped: swap edge $(2,4)$ with $(3,4)$, edge $(1,2)$ with $(1,3)$, edge $(1,5)$ with $(2,5)$, and edge $(1,2)$ with $(2,5)$. Swapping, e.g., edge $(1,2)$ with $(3,4)$ would lead to an infeasible solution corresponding to a graph with two connected components (the first would be the cycle formed by nodes 2, 3, and 4, with the second being the edge connecting nodes 1 and 5).

This situation is illustrated in Figure 8.2, in which only four moves lead to the feasible solutions in the restricted neighborhood $N(S : S^g)$. If edge $(3,4)$ is added to $S$, then the cycle $(3,4) - (2,4) - (2,3)$ is created. This is followed by the removal of edge $(2,4)$, leading to solution A. If edge $(1,3)$ is added to $S$, then the cycle $(1,3) - (2,3) - (1,2)$ is created. In this case, edge $(1,2)$ is removed from $S$ and solution B is obtained. If edge $(2,5)$ is added to $S$, then the cycle $(2,5) - (1,5) - (1,2)$ is created. One possible edge to be removed is $(1,5)$, leading to solution C. Finally, we note that edge $(1,2)$ may also be removed following the addition of edge $(2,5)$, in which case the feasible solution D is obtained.

**Traveling salesman problem – Restricted neighborhood**

We now consider the traveling salesman problem associated with the weighted graph in Figure 8.3(a). Two of its Hamiltonian cycles are depicted in Figures 8.3(b) and 8.3(c). Suppose the tour in Figure 8.3(b) is the current solution represented by the linear permutation $S = (1,2,3,4,5)$, while that in Figure 8.3(c) is the guiding solution associated with the linear permutation $S^g = (1,3,5,2,4)$. We note that these two linear permutations correspond to two different circular permutations of the five cities, i.e., they are associated with two different tours. The total length of solution $S$ is 17, while that of $S^g$ is 18.

We observe that these two solutions only match in the first component of their associated linear permutations, i.e., both start from vertex 1: $S(1) = S^g(1) = 1$. Therefore, there are four misplaced cities between $S$ and $S^g$, each of them corresponding to a position in the tour for which the two linear permutations originally differ.

Let us suppose that neighborhood $N_2$ defined for the traveling salesman problem in Section 4.1 is being used. Each neighbor is obtained by a move consisting of the exchange of two cities in different positions of the current solution $S$. Therefore, solution $S = (1,2,3,4,5)$ has six neighbors if node 1 is fixed as the first: $(1,3,2,4,5)$, $(1,2,4,3,5)$, $(1,2,3,5,4)$, $(1,4,3,2,5)$, $(1,2,5,4,3)$, and $(1,5,3,4,2)$. Since $S^g = (1,3,5,2,4)$ is the guiding solution, four out of these six solutions in neighborhood $N_2(S)$ also belong to the restricted neighborhood $N(S : S^g)$, each of them making the position of a new city coincide in the new solution and in $S^g$: solution $A = (1,3,2,4,5)$ makes node 3 the second in the tour, solution $B = (1,2,5,4,3)$ makes node 5 the third in the tour, solution $C = (1,4,3,2,5)$ makes node 2 the fourth in the tour, and solution $D = (1,2,3,5,4)$ makes node 4 the last in the tour.

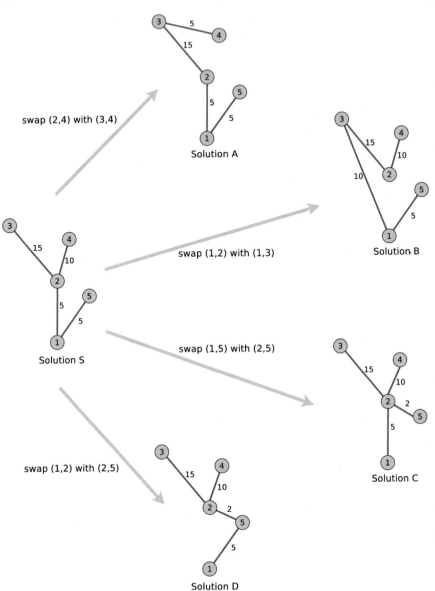

**Fig. 8.2** Minimum spanning tree problem: four swap moves from the current solution $S$ towards the guiding solution $S^g$.

The restricted neighborhood $N(S : S^g)$ is shown in Figure 8.4, where we denote by swap$(i, j)$ the move that swaps the cities in positions $i$ and $j$ of the linear permutation associated with the current solution $S$ and makes at least one of them coincide with its final position in the linear permutation corresponding to the guiding solution $S^g$. ∎

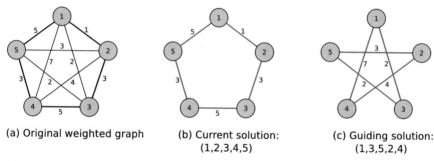

(a) Original weighted graph

(b) Current solution:
(1,2,3,4,5)

(c) Guiding solution:
(1,3,5,2,4)

**Fig. 8.3** Current solution $S = (1,2,3,4,5)$ and guiding solution $S^g = (1,3,5,2,4)$ for path-relinking applied to the traveling salesman problem.

## Knapsack problem – Restricted neighborhood

We consider the optimization version of the knapsack problem, as introduced in Section 1.2. In this problem, one has a set $I = \{1,\ldots,n\}$ of items to be placed in a knapsack. Integer numbers $a_i$ and $c_i$ represent, respectively, the weight and the utility of each item $i \in I$. We assume that each item fits in the knapsack by itself and denote by $b$ the maximum total weight that can be taken in the knapsack. We have seen in Section 4.1 that every solution $S$ of the knapsack problem can be represented by a binary vector $(x_1,\ldots,x_n)$, in which $x_i = 1$ if item $i$ is selected, $x_i = 0$ otherwise, for every $i = 1,\ldots,n$. A solution $S = (x_1,\ldots,x_n)$ is feasible if $\sum_{i \in I} a_i \cdot x_i \leq b$.

We recall the example in Figure 2.5, where four items are available to be placed in a knapsack of capacity 19. The weights of the yellow and green items are each equal to 10 and those of the blue and red items are both equal to 5. Therefore, only two of the four items fit together in the knapsack. The two heaviest items have utilities 20 and 10 to the hiker, while the two items with least weights have utilities 10 and 5. We consider the red, green, blue, and yellow items indexed by 1, 2, 3, and 4, respectively. The four items are illustrated in Figure 8.5(a) and two feasible solutions appear in Figures 8.5(b) and 8.5(c). Suppose the solution in Figure 8.5(b) is the current solution represented by vector $S = (1,1,0,0)$, while the solution in Figure 8.5(c) is the guiding solution associated with vector $S^g = (0,0,1,1)$.

These two solutions differ in all elements. Therefore, there are four moves in a path leading from $S$ to $S^g$, each of them corresponding to an item that appears in one solution, but not in the other. Following the solution representation proposed for the knapsack problem in Section 4.1, there are four possible neighbors in $N(S)$, each of them corresponding to flipping the value of one variable of the current solution $S$. We denote by flip$(j)$ the move that replaces the value of $x(j)$ by $1 - x(j)$, for $j = 1,\ldots,n$: flip(1) sets $x_1 = 0$, flip(2) sets $x_2 = 0$, flip(3) sets $x_3 = 1$, and flip(4) sets $x_4 = 1$. However, since the two last moves lead to infeasible solutions, the restricted neighborhood $N(S:S^g)$ illustrated in Figure 8.6 contains only two solutions: solution $A = (0,1,0,0)$ and solution $B = (1,0,0,0)$, corresponding, respectively, to the moves flip(1) that sets $x_1 = 0$ and flip(2) that sets $x_2 = 0$. ∎

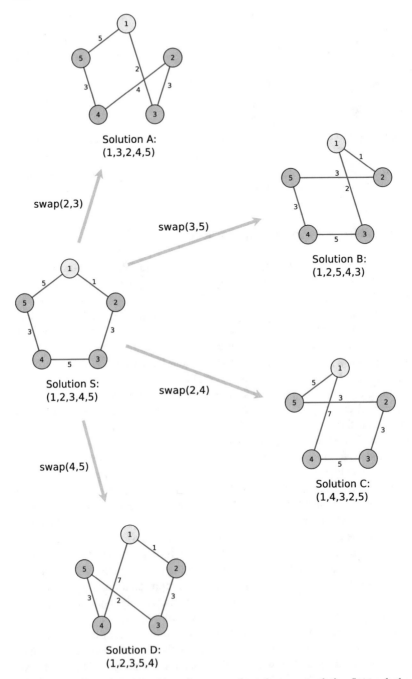

**Fig. 8.4** Traveling salesman problem: four moves from the current solution $S$ towards the guiding solution $S^g$.

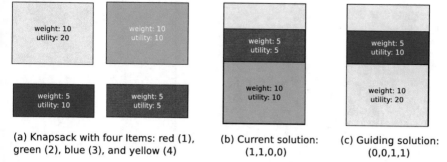

(a) Knapsack with four Items: red (1), green (2), blue (3), and yellow (4)

(b) Current solution: (1,1,0,0)

(c) Guiding solution: (0,0,1,1)

**Fig. 8.5** Current and guiding solutions for a knapsack problem with four items.

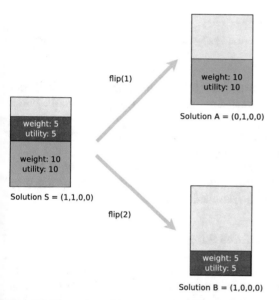

flip(1)

Solution A = (0,1,0,0)

Solution S = (1,1,0,0)

flip(2)

Solution B = (1,0,0,0)

**Fig. 8.6** Knapsack problem: two moves from the current solution $S$ towards the guiding solution $S^g$.

## 8.1.2 A template for forward path-relinking

The algorithm in Figure 8.7 is an implementation of *forward path-relinking* for a minimization problem, where $S^i \in F$ is the initial solution and $S^g \in F$ is the guiding solution. We assume that the guiding solution $S^g$ is at least as good as (and possibly better than) the initial solution $S^i$, hence the qualification of this strategy as forward. The current solution, the incumbent, and their cost are initialized in lines 1 to 3.

In line 4, the algorithm checks if the restricted neighborhood $N(S : S^g) \subseteq N(S)$ contains at least one feasible solution. In most cases, the restricted neighborhood $N(S : S^g)$ does not have to be explicitly computed and stored: instead, its elements may only be implicitly enumerated on-the-fly.

```
begin FORWARD-PR(S^i, S^g);
1   S ← S^i;
2   S* ← S;
3   f* ← f(S);
4   while |N(S : S^g)| ≥ 1 do
5       S ← argmin{f(S') : S' ∈ N(S : S^g)};
6       if f(S) < f* then
7           S* ← S;
8           f* ← f(S);
9       end-if;
10  end-while;
11  Apply local search to improve the best solution S*;
12  return S*, f(S*);
end FORWARD-PR(S^i, S^g).
```

**Fig. 8.7** Pseudo-code for a template of a forward path-relinking algorithm for minimization problems.

As the algorithm traverses the path from $S$ to $S^g$, the best restricted neighbor solution of the current solution is selected at each iteration. A path from the initial solution $S^i$ to the guiding solution $S^g$ is created in the loop going from line 4 to 10. The best restricted neighbor $S$ is selected in line 5. Lines 6 to 9 update the best solution $S^*$ and its cost if a new best-quality solution is found.

Since the best solution found along the path from $S^i$ to $S^g$ may not be locally optimal, local search is applied to it in line 11 and the final solution obtained by forward path-relinking and its cost are returned in line 12.

### Knapsack problem – Forward path-relinking

Figure 8.8 illustrates the full application of forward path-relinking to the same instance of the knapsack problem with four items that was used in the last example presented in the previous section. As before, we consider the red, green, blue, and yellow items indexed by 1, 2, 3, and 4, respectively, and we denote by flip($j$) the move that replaces the current value of $x_j$ by $1 - x_j$. The initial solution is $S^i = (1,1,0,0)$ and the guiding solution is $S^g = (0,0,1,1)$. We recall that the knapsack capacity is 19.

The first iteration of path-relinking corresponds to the example in Section 8.1. The initial solution is $S = S^i = (1,1,0,0)$ and there are two possible moves in the restricted neighborhood: flip(1) sets $x_1 = 0$ and leads to solution A = $(0,1,0,0)$, whose utility is 10, while flip(2) sets $x_2 = 0$ and leads to solution B = $(1,0,0,0)$, whose utility is 5. Since we are facing a maximization problem, move flip(1) is selected. Item 1 is removed from the knapsack and path-relinking proceeds from solution A. The second iteration begins with two possible moves to incorporate attributes of the guiding solution $S^g = (0,0,1,1,)$ into the current solution A: flip(2) sets $x_2 = 0$ and leads to solution C = $(0,0,0,0)$, whose utility is 0, while flip(3) sets $x_3 = 1$ and leads to solution D = $(0,1,1,0)$, whose utility is 20. Move flip(3) is selected, item 3 is included in the knapsack, and path-relinking moves to solution D. At this time, there is only one possible remaining move to be applied to solution

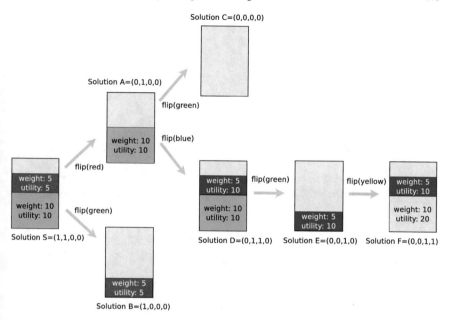

**Fig. 8.8** Example of forward path-relinking applied to an instance of the knapsack problem with four items: the path from the initial solution $S^i = (1,1,0,0)$ to the guiding solution $S^g = (0,0,1,1)$ has exactly four moves, corresponding to the number of items by which the initial and guiding solutions differ. In this example, the best solution along the generated path coincides with the guiding solution $S^g$.

D that makes the resulting solution closer, or more similar, to the guiding solution: flip(2) sets $x_2 = 0$ and leads to solution $E = (0,0,1,0)$, whose utility is 10. Finally, once again there is only one possible move to be performed at the fourth and last iteration: flip(4) sets $x_4 = 1$ and leads to solution $F = (0,0,1,1)$, which coincides with $S^g$ and whose utility is 30.

The initial and guiding solutions differ by all four elements, each of them corresponding to a move that would make the current solution closer to the guiding solution. As expected, path-relinking reaches the guiding solution after exactly four moves. In this particular example, the best solution along the generated path coincided with the guiding solution $S^g$. However, in many cases, the best solution found improves both the initial and the guiding solutions, as will be illustrated later in this chapter. ∎

## 8.2 Other implementation strategies for path-relinking

Path-relinking can be implemented using different strategies, as illustrated in Figure 8.9. These include not only forward path-relinking, as seen in Section 8.1.2, but also backward, back-and-forward, mixed, truncated, greedy randomized adaptive,

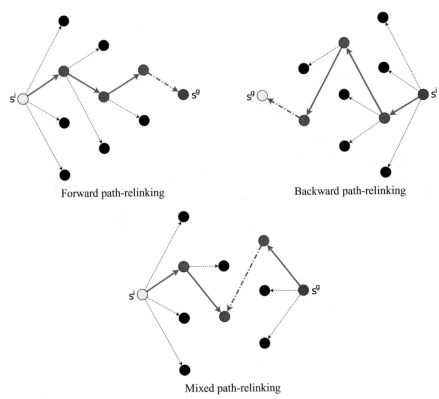

Forward path-relinking                                Backward path-relinking

Mixed path-relinking

**Fig. 8.9** Different implementations of path-relinking: (a) Forward path-relinking: a path is traversed from the initial solution $S^i$ to a guiding solution $S^g$ at least as good as $S^i$. (b) Backward
path-relinking: a path is traversed from the initial solution $S^i$ to a guiding solution $S^g$ that is not
better than $S^i$. (c) Mixed path-relinking: two subpaths are traversed, one starting at $S^i$ and the other
at $S^g$, which eventually meet in the middle of the trajectory connecting $S^i$ and $S^g$.

external, and evolutionary path-relinking, together with their hybrids. All these
strategies involve trade-offs between computation time and solution quality.

## 8.2.1 Backward and back-and-forward path-relinking

Suppose that path-relinking is applied to a minimization problem between two solutions $S^1$ and $S^2$ such that $f(S^1) \leq f(S^2)$, where $f(S)$ denotes the value of solution
$S$ for the objective function to be minimized. Path-relinking is always carried out
from an initial solution $S^i$ to a guiding solution $S^g$. We have seen in Section 8.1.2
that in the case of forward path-relinking, the initial and guiding solutions are set as
$S^i = S^2$ and $S^g = S^1$: in this case, the initial solution is not better than the guiding
solution.

Conversely, in *backward path-relinking*, we set $S^i = S^1$ and $S^g = S^2$: now, the guiding solution is not better than the initial solution. In *back-and-forward path-relinking*, backward path-relinking is applied first, followed by forward path-relinking. Path-relinking explores the restricted neighborhood of the initial solution more thoroughly than the restricted neighborhood of the guiding solution because, as it moves along the path, the size of the restricted neighborhood progressively decreases. If one of the solutions $S^1$ or $S^2$ is strictly better than the other, then backward path-relinking explores more thoroughly the restricted neighborhood of the solution which is the best among $S^1$ and $S^2$. Since it is more likely to find an improving solution in the restricted neighborhood of the better solution than in that of the worse, backward path-relinking usually tends to perform better than forward path-relinking. Back-and-forward path-relinking does at least as well as either backward or forward path-relinking, but takes about twice as long to compute, since two (usually distinct) paths of the same length are traversed. Computational experiments have confirmed that backward path-relinking usually outperforms forward path-relinking in terms of solution quality, while back-and-forward path-relinking finds solutions at least as good as forward or backward path-relinking, but at the expense of longer running times. Figure 8.10 illustrates this behavior on an instance of a routing problem in private virtual networks.

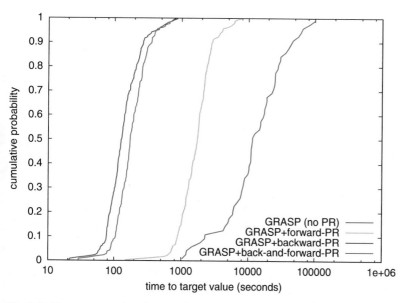

**Fig. 8.10** Time-to-target plots for pure GRASP and three variants of GRASP with path-relinking (forward, backward, and back-and-forward) on an instance of a routing problem in private virtual networks. The plots show that GRASP with backward path-relinking outperformed the other path-relinking variants as well as the pure GRASP heuristic, which was the slowest to find a solution whose value is at least as good as the target value.

## 8.2.2 Mixed path-relinking

In applying *mixed path-relinking* between two feasible solutions $S^i$ and $S^g$, the connecting path is explored from both extremities. At each iteration of path-relinking, the closest extremity to the new current solution alternates between the original initial solution $S^i$ and the original guiding solution $S^g$. The search behaves as if solutions in two different subpaths were visited alternately: the first of these subpaths leaves from the initial solution $S^i$ and leads to the guiding solution $S^g$, while the second emanates from $S^g$ and develops towards $S^i$. These two subpaths meet at some feasible solution in the middle of the trajectory, thus connecting $S^i$ and $S^g$ with a single path. We observe that, in this case, the qualification of a solution as being the initial or the guiding solution is meaningless, since the procedure behaves as if they keep permanently interchanging their role until the end. Figure 8.11 illustrates the steps of the application of mixed path-relinking to two solutions $S^i$ and $S^g$ for which the path connecting them is formed by five arcs. Moves alternate between the subpath leaving from the left and the subpath leaving from the right.

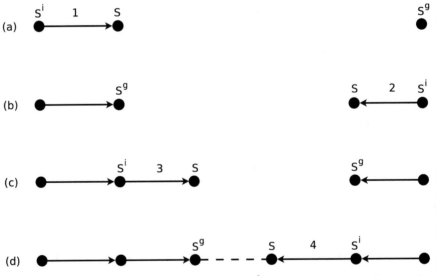

**Fig. 8.11** Mixed path-relinking between two solutions $S^i$ and $S^g$ for which the path connecting them is formed by five arcs: numbers above the arrows represent the order in which the moves are performed. Moves alternate between the subpath leaving from the left and the subpath leaving from the right.

Figure 8.12 shows the pseudo-code of a template for a mixed path-relinking algorithm between solutions $S^i$ and $S^g$ for a minimization problem. The pseudo-code of algorithm MIXED-PR is basically the same of algorithm FORWARD-PR, except for lines 10 to 12, in which the direction of the path is reversed by the exchange of the roles of the guiding and current solutions.

```
begin MIXED-PR(S^i, S^g);
1    S ← S^i;
2    S* ← S;
3    f* ← f(S*);
4    while |N(S : S^g)| ≥ 1 do
5        S ← argmin{f(S') : S' ∈ N(S : S^g)};
6        if f(S) < f* then
7            S* ← S;
8            f* ← f(S);
9        end-if;
10       S' ← S;
11       S ← S^g;
12       S^g ← S';
14   end-while;
14   Apply local search to improve the best solution S*;
15   return S*, f(S*);
end MIXED-PR.
```

**Fig. 8.12** Pseudo-code for a template of a mixed path-relinking algorithm for minimization problems.

While back-and-forward path-relinking thoroughly explores both restricted neighborhoods of $S^i$ and $S^g$, the mixed variant explores the entire restricted neighborhood of $S^i$ and all but one solution of the restricted neighborhood of $S^g$. This is in contrast with both forward and backward path-relinking, which each fully explore only one of the restricted neighborhoods.

Furthermore, mixed path-relinking explores half as many restricted neighbors as back-and-forward path-relinking and the same number of neighbors as either the backward or forward variants. Figure 8.13 illustrates the comparison of a pure GRASP heuristic with four of its variants combined with path-relinking and applied to an instance of the 2-path network design problem: forward, backward, back-and-forward, and mixed path-relinking. The time-to-target plots show that GRASP with mixed path-relinking has the best runtime profile among the variants compared.

## 8.2.3 Truncated path-relinking

One can expect to see most solutions produced by path-relinking to come from subpaths that are close to either the initial or the guiding solution.

Figure 8.14 illustrates this observation for 80 instances of the max-min diversity problem, where a GRASP with back-and-forward path-relinking was run on each instance for two minutes. In each application of path-relinking, the step which produced the best path-relinking solution was recorded. For each instance, the total numbers of best path-relinking solutions found in each tenth of the traversed paths were added up and the average numbers of solutions found in each tenth were

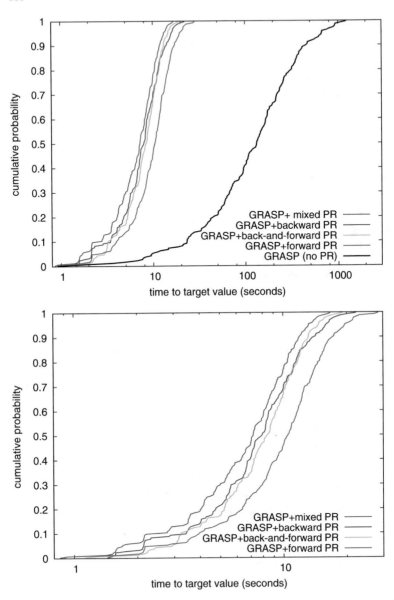

**Fig. 8.13** Time-to-target plots for pure GRASP and four variants of GRASP with path-relinking (forward, backward, back-and-forward, and mixed) on an instance of the 2-path network design problem. The plot on the bottom compares only the variants that include path-relinking.

computed. It was shown experimentally that exploring only the subpaths near the extremities often produces solutions as good as those found by exploring the entire path, since there is a higher concentration of better solutions close to the initial and guiding solutions explored by path-relinking. The figure shows that most of the best solutions obtained by path-relinking are found near the initial and guiding solutions, and only in 15% of the calls to path-relinking would it be necessary to exploit subpaths longer than 20% of the total number of moves.

It is straightforward to adapt path-relinking to explore only the restricted neighborhoods that are close to the extremities. *Truncated path-relinking* can be applied to either forward, backward, backward-and-forward, or mixed path-relinking: instead of exploring the entire path, it just explores a fraction of the path and, consequently, takes a fraction of the running time.

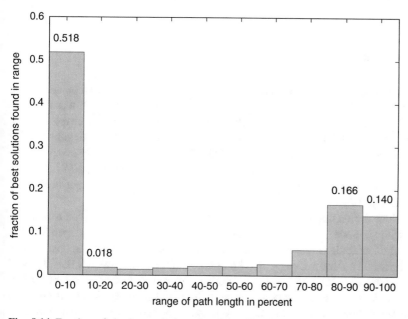

**Fig. 8.14** Fraction of the best solutions found by GRASP with backward-and-forward path-relinking that appear in each range of the path length from the initial to the guiding solutions on two-minute runs over 80 instances of the max-min diversity problem. Fifty four percent of the best solutions were found in subpaths that originate at the initial solutions and appear within the first 20% of the total number of moves performed, while 31% are close to the guiding solutions and appear in the last 20% of the moves performed in each path.

## 8.3 Minimum distance required for path-relinking

We assume that we want to connect two locally optimal solutions $S^1$ and $S^2$ with path-relinking. If $S^1$ and $S^2$ differ by only one of their components, then the path directly connects the two solutions and no solution, other than $S^1$ and $S^2$, is visited.

Since $S^1$ and $S^2$ are both local minima, then $f(S^1) \leq f(S)$ for all $S \in N(S^1)$ and $f(S^2) \leq f(S)$ for all $S \in N(S^2)$, where $N(S)$ denotes the neighborhood of solution $S$.

If $S^1$ and $S^2$ differ by exactly two moves, then any path between $S^1$ and $S^2$ visits exactly one intermediary solution $S \in N(S^1) \cap N(S^2)$. Consequently, solution $S$ cannot be better than either $S^1$ or $S^2$.

Likewise, if $S^1$ and $S^2$ differ by exactly three moves, then any path between them visits two intermediary solutions $S \in N(S^1)$ and $S' \in N(S^2)$ and, consequently, neither $S$ nor $S'$ can be better than both $S^1$ and $S^2$.

Therefore, things only get interesting when the two solutions $S^1$ and $S^2$ differ by at least four moves. Consequently, we can discard the application of path-relinking to pairs of solutions differing by less than four moves.

## 8.4 Dealing with infeasibilities in path-relinking

So far in our discussion about path-relinking, we assumed that at least one restricted neighbor of a solution $S$ with respect to a target guiding solution $S^g$ was feasible. Consider line 5 of both path-relinking templates shown earlier in this chapter, where we minimize $f(S)$, for $S \in F$. This step selects the best restricted neighbor of the current solution as $\arg\min\{f(S') : S' \in N(S : S^g)\}$. However, it may occur that all moves from the current solution $S$ lead to infeasible solutions, i.e., $N(S : S^g) = \varnothing$ and the result of the argmin operator is undefined. In this situation, path-relinking would have to stop.

Consider the example in Figures 8.15 to 8.17, where we are given a bipartite graph with six nodes, $A$, $B$, $C$, $D$, $E$, and $F$, and seek a maximum independent set, i.e., a set of mutually nonadjacent nodes of maximum cardinality. Suppose we are given the initial solution $S^i = \{A,B,C\}$ and the guiding solution $S^g = \{D,E,F\}$, and that we consider a neighborhood characterized by moves defined as swap($out,in$), where the node $out$ is replaced by the node $in$ in the solution. Since all nodes in the initial solution must be removed from it and all nodes in the guiding solution $S^g$ must be inserted, there are nine moves that might be applied to build a path from $S^i$ to $S^g$: swap($A,D$), swap($A,E$), swap($A,F$), swap($B,D$), swap($B,E$), swap($B,F$), swap($C,D$), swap($C,E$), and swap($C,F$). Applying any of these moves to $S^i$ results in one of nine infeasible solutions. Infeasibilities correspond to edges connecting pairs of nodes in a candidate independent set, i.e., they are associated with conflicting edges. In the figures, infeasibilities are indicated by edges in red. Of the nine moves, six lead to solutions that have a single infeasibility, while three lead to solutions that have two conflicting edges. In such situations, one possible strategy that might be applied is a greedy path-relinking operator that proceeds by moving to a least-infeasible solution.

In this example, suppose solution $\{B,C,D\}$ with a single infeasibility corresponding to edge $(B,D)$ is chosen, i.e., the algorithm moves to solution $S = \{B,C,D\}$. Now, there are only four moves that might be applied to $S$ to build a path to $S^g$: swap($B,E$), swap($B,F$), swap($C,E$), and swap($C,F$). Again, all moves lead to infeasible solutions, but two lead to a single infeasibility, while the others to two infeasibilities. Suppose the greedy choice is to apply move swap($B,E$) resulting

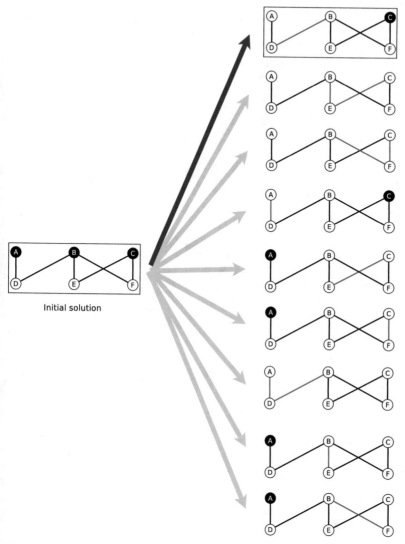

**Fig. 8.15** First iteration of path-relinking for a 6-node maximum independent set problem where all nine restricted neighbors of the initial solution are infeasible.

in solution $S' = \{C, D, E\}$, with a single infeasibility corresponding to edge $(C, E)$. There is now only the move swap$(C, F)$ leading from $S'$ to the guiding solution $S^g$. In this example, all restricted neighbors on all paths from the initial solution to the target solution are infeasible. In general, however, some may be feasible, some infeasible.

In a revised path-relinking operator that allows moves to infeasible solutions, each visited solution $S$ may be in either one of two possible situations: either at least one move from $S$ leads to a feasible solution, in which case $|N(S : S^g)| \geq 1$,

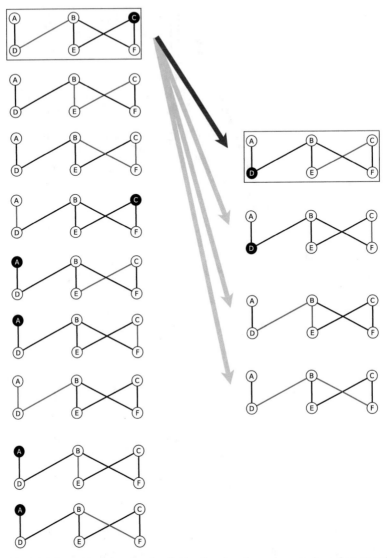

**Fig. 8.16** Second iteration of path-relinking for a 6-node maximum independent set problem where all four restricted neighbors of the current solution are infeasible.

or, alternatively, all restricted moves lead to infeasible solutions and the restricted neighborhood $N(S : S^g)$ becomes empty before the guiding solution is reached. In the first case, a greedy version of path-relinking selects a move that leads to a least cost feasible neighbor of $S$. Otherwise, the selected move is one that leads to an infeasible neighbor of $S$ with minimum infeasibility.

The pseudo-code in Figure 8.18 presents a revised mixed path-relinking procedure that allows feasible and infeasible moves. It is very similar to the template

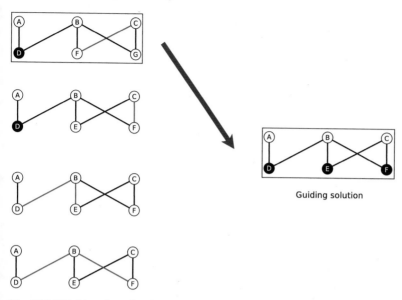

**Fig. 8.17** Third iteration of path-relinking for a 6-node maximum independent set problem, in which the path finally reaches the guiding solution.

presented in Figure 8.12, with the main difference corresponding to lines 4 to 10. Both feasible and infeasible moves are allowed in the neighborhood $N(S)$. As already observed for algorithm FORWARD-PR in Figure 8.7, the neighborhood $N(S)$ does not necessarily have to be explicitly enumerated. If the algorithm detects in line 5 that there is at least one move that once applied to $S$ leads to a feasible solution that is closer to $S^g$ than $S$ is, then the best restricted neighbor is selected in line 6. Otherwise, the restricted neighborhood is empty. Denoting by infeasibility$(S)$ a measure of the degree of infeasibility of a solution $S$, line 8 selects the best infeasible neighbor of the current solution, i.e., the one with the smallest measure of infeasibility. The updates in lines 11 and 12 are performed if the new incumbent solution $S$ is feasible and improves the best solution previously known.

## 8.5 Randomization in path-relinking

All previously described path-relinking strategies follow a greedy criterion to select the best move at each of their iterations. Therefore, path-relinking is limited to exploring a single path from a set of exponentially many paths between any pair of solutions. By adding randomization to path-relinking, *greedy randomized adaptive path-relinking* is not constrained to explore a single path. Instead of always selecting the move that results in the best solution, a restricted candidate list is constructed with the moves that result in promising solutions with costs in an interval

```
begin MIXED-PR-INFEASIBLE-MOVES(S^i, S^g);
1   S ← S^i;
2   S* ← S;
3   f* ← f(S*);
4   while |N(S)| > 1 do
5       if N(S : S^g) ≠ ∅ then
6           S ← argmin{f(S') : S' ∈ N(S : S^g)};
7       else
8           S ← argmin{infeasibility(S') : S' ∈ N(S)};
9       end-if;
10      if S is feasible and f(S) < f* then
11          S* ← S;
12          f* ← f(S);
13      end-if;
14      S' ← S;
15      S ← S^g;
16      S^g ← S';
17  end-while;
18  Apply local search to improve the best solution S*;
19  return S*, f(S*);
end MIXED-PR-INFEASIBLE-MOVES.
```

**Fig. 8.18** Pseudo-code for a revised template of a mixed path-relinking algorithm for minimization problems, with feasible and infeasible moves.

that depends on the values of the best and worst moves, as well as on a parameter in the interval $[0,1]$. A move is selected at random from this set to produce the next solution in the path.

By applying this strategy several times to the initial and guiding solutions, several paths can be explored. This strategy is useful when path-relinking is applied more than once to the same pair of solutions as it may occur in evolutionary path-relinking, which we will introduce in the next chapter.

## 8.6 External path-relinking and diversification

So far in this chapter, we have considered variants of path-relinking in which a path in the search space graph $\mathcal{G} = (F, M)$ connects two feasible solutions $S, T \in F$ by progressively introducing in one of them (the initial solution) attributes of the other (the guiding solution). Since attributes common to both solutions are not changed and all solutions visited belong to a path between the two solutions, we may also refer to this type of path-relinking as *internal path-relinking*.

*External path-relinking* extends any path connecting $S$ and $T$ in $\mathcal{G} = (F, M)$ beyond its extremities. To extend such a path beyond $S$, attributes not present in either $S$ or $T$ are introduced in $S$. Symmetrically, to extend it beyond $T$, attributes not present in either $S$ or $T$ are introduced in $T$. In its greedy variant, all moves are

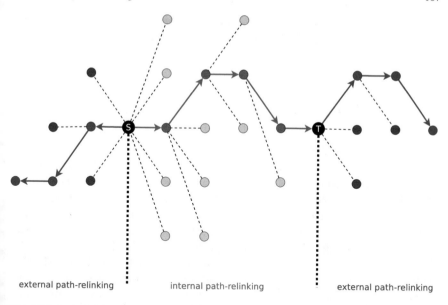

**Fig. 8.19** An internal path (red arcs, red nodes) from solution $S$ to solution $T$ and two external (blue arcs, blue nodes) paths, one emanating from solution $S$ and the other from solution $T$. These paths are produced by internal and external path-relinking.

evaluated and the solution chosen to be next in the path is one with best cost or, in case they are all infeasible, the one with least infeasibility. In either direction, the procedure stops when all attributes that do not appear in either $S$ or $T$ have been tested for extending the path. Once both paths are complete, local search may be applied to the best solution in each of them. The best of the two local minima is returned as the solution produced by the external path-relinking procedure.

Figure 8.19 illustrates internal and external path-relinking. The path with red nodes and edges is one resulting from internal path-relinking applied with $S$ as the initial solution and $T$ as the guiding solution. We observe that the orientation introduced by the arcs in this path is due only to the choice of the initial and guiding solutions. If the roles of solutions $S$ and $T$ were interchanged, it could have been computed and generated in the reverse direction. The same figure also illustrates two paths obtained by external path-relinking, one emanating from $S$ and the other from $T$, both represented with blue nodes and edges. The orientations of the arcs in each of these paths indicate that they necessarily emanate from either solution $S$ or $T$.

To conclude, we establish a parallel between internal and external path-relinking. Since internal path-relinking works by fixing all attributes common to the initial and guiding solutions and searches for paths between them satisfying this property, it is clearly an intensification strategy. Contrarily, external path-relinking progressively removes common attributes and replaces them by others that do not appear in either one of the initial or guiding solution. Therefore, it can be seen as a diversification strategy which produces solutions increasingly farther from both the initial and the guiding solutions. External path-relinking becomes therefore a tool for search diversification.

## 8.7 Bibliographical notes

Path-relinking, as introduced in Section 8.1, was originally proposed by Glover (1996b) as an intensification strategy to explore trajectories connecting elite solutions obtained by tabu search or scatter search (Glover and Laguna, 1997; Glover, 2000; Glover et al., 2000; 2003; 2004). Accounts and surveys of path-relinking, mostly in the context of GRASP applications, were authored by Resende and Ribeiro (2005a), Resende et al. (2010b), and Ribeiro and Resende (2012). Forward path-relinking corresponds to the original proposal for the implementation strategy.

Section 8.2 discussed other, more elaborate implementation strategies. The concepts of backward as well as of back-and-forward path-relinking appeared first in Ribeiro et al. (2002), with both names being later introduced in Aiex et al. (2005). Computational experiments reported in Ribeiro et al. (2002) and Resende and Ribeiro (2003a) were the first to show that backward path-relinking usually outperforms forward path-relinking, while back-and-forward path-relinking finds solutions at least as good as forward or backward path-relinking, but at the expense of longer running times.

Mixed path-relinking was suggested by Glover (1996b) and was first implemented and tested in the context of the 2-path network design problem by Rosseti (2003), followed by results in Resende and Ribeiro (2005a) and Ribeiro and Rosseti (2009), where it was shown that mixed path-relinking usually outperforms forward, backward, and back-and-forward path-relinking.

Resende et al. (2010a) showed empirically, for instances of the max-min diversity problem, that most of the best solutions obtained by path-relinking are found near the initial and guiding solutions, and that more are found near the best of these two solutions. Andrade and Resende (2007a) and Resende et al. (2010a) were the first to apply truncated path-relinking.

Sections 8.3 to 8.6 introduced several extensions of path-relinking. The requirement of a minimum distance between the initial and guiding solutions for the application of path-relinking appeared originally in Festa et al. (2005) and Festa et al. (2006). Infeasibility in path-relinking was first addressed by Mateus et al. (2011). Morán-Mirabal et al. (2013b) developed the approach to deal with infeasibility in path-relinking. Greedy randomized adaptive path-relinking was proposed by Faria Jr. et al. (2005). External path-relinking was introduced by Glover (2014) and first applied by Duarte et al. (2015) in a heuristic for differential dispersion minimization.

# Chapter 9
# GRASP with path-relinking

Path-relinking is a major enhancement to GRASP, adding a long-term memory mechanism to GRASP heuristics. GRASP with path-relinking implements long-term memory using an elite set of diverse high-quality solutions found during the search. In its most basic implementation, at each iteration the path-relinking operator is applied between the solution found at the end of the local search phase and a randomly selected solution from the elite set. The solution resulting from path-relinking is a candidate for inclusion in the elite set. In this chapter we examine elite sets, their integration with GRASP, the basic GRASP with path-relinking procedure, several variants of the basic scheme, including evolutionary path-relinking, and restart strategies for GRASP with path-relinking heuristics.

## 9.1 Memoryless GRASP

The basic GRASP heuristic, as presented in Chapter 5, searches the solution space by repeatedly applying independent searches in the solution space graph $\mathcal{G} = (F, M)$, each search starting from a different greedy randomized solution. Each independent search uses no information produced by any other search performed at previous iterations. The choices of starting solutions for local search are not influenced by information produced during the search. However, Reactive GRASP and adaptive memory techniques (introduced in Sections 7.1 and 7.6, respectively) do make use of information produced during the search. Reactive GRASP does so to select the blend of randomness and greediness used in the construction of the starting solutions for local search, while programming with adaptive memory determines the amount of intensification and diversification in the construction phase.

 The memoryless nature of basic, or pure, GRASP is in contrast with many successful metaheuristics, such as tabu search, genetic algorithms, and ant colony optimization, which make extensive use of information gathered during the search process to guide their choice of the region of the solution space to explore.

© Springer Science+Business Media New York 2016
M.G.C. Resende, C.C. Ribeiro, *Optimization by GRASP*,
DOI 10.1007/978-1-4939-6530-4_9

In this chapter, we show how path-relinking can be used with any GRASP heuristic to result in a hybrid procedure with a long-term memory mechanism. Given the same running time, this hybridization almost always produces better solutions than pure GRASP. Alternatively, given a target value, it almost always finds a solution at least as good as this target in less running time than pure GRASP.

## 9.2 Elite sets

An elite set $\mathscr{E}$ of solutions is a set formed by at most a fixed number $n_{\mathscr{E}}$ of diverse, high-quality solutions found during the run of a heuristic. The elite solutions should represent distinct promising regions of the solution space and therefore should not include solutions that are too similar, even if they are of high quality.

A basic scheme to maintain an elite set $\mathscr{E}$ for a minimization problem is outlined in the algorithm of Figure 9.1. The algorithm is given a candidate solution $S$ and determines if $S$ should be added to $\mathscr{E}$ and, if so, which solution, if any, should be removed from $\mathscr{E}$.

```
begin UPDATE-ELITE-SET(S, E);
1   if |E| < n_E then
2       if E = ∅ then
3           E ← E ∪ {S};
4       else
5           δ ← min{|Δ(S,S')| : S' ∈ E};
6           if δ > 0 then E ← E ∪ {S};
7       end-if;
8   else
9       f⁺ ← max{f(S') : S' ∈ E};
10      δ ← min{|Δ(S,S')| : S' ∈ E};
11      if f(S) < f⁺ and δ > 0 then
12          S⁻ ← argmin{|Δ(S,S')| : S' ∈ E such that f(S') ≥ f(S)};
13          E ← E ∪ {S} \ {S⁻};
14      end-if;
15  end-if;
16  return E;
end UPDATE-ELITE-SET.
```

**Fig. 9.1** Pseudo-code of a template for the maintenance of the elite set $\mathscr{E}$ of at most $n_{\mathscr{E}}$ elements in the context of a minimization problem.

If line 1 determines that the elite set $\mathscr{E}$ is not full, i.e., if $|\mathscr{E}| < n_{\mathscr{E}}$, then a candidate solution $S$ is always added to $\mathscr{E}$ if it is different from any solution currently in the set. This case is treated in lines 2 to 7 of the pseudo-code. In line 3, $S$ is added to $\mathscr{E}$ if the elite set is empty. Let the *symmetric difference* $\Delta(S,S')$ be formed by the ground set elements that belong to either $S$ or $S'$. In line 5, the minimum cardinality

$\delta$ among the symmetric differences between $S$ and the elements of $\mathscr{E}$ is computed. If $S$ is different from all elite solutions, then it is added to $\mathscr{E}$ in line 6.

Otherwise, if the elite set is full (i.e., if $|\mathscr{E}| = n_{\mathscr{E}}$), then any time a solution is added to the set, another solution must be removed from it, thus maintaining the size of $\mathscr{E}$ equal to $n_{\mathscr{E}}$. Our goal is to first improve the average quality of the elite set, and then maximize the diversity of its elements, which amounts to maximizing the cardinalities of the symmetric differences between all pairs of solutions in the set. This case is treated in lines 9 to 14. In line 9, the cost $f^+$ of the worst-valued elite set solution is computed, while in line 10 the minimum cardinality $\delta$ among the symmetric differences between $S$ and any element of $\mathscr{E}$ is determined. $S$ is added to $\mathscr{E}$ if it is better than the worst solution in the elite set and if it is different from all elite solutions, i.e., if $f(S) < f^+$ and $\delta > 0$ in line 11. This is accomplished in lines 12 and 13. Line 12 determines, among all elite set solutions valued no better than $S$, one which is most similar to $S$, i.e., one which minimizes the cardinality of its symmetric difference with respect to $S$. This solution, $S^-$, is removed from $\mathscr{E}$ in line 13. The new elite solution $S$ is inserted in the pool as a replacement for $S^-$ at the same line. The updated elite set is returned in line 16.

The algorithm in Figure 9.1 can be modified to increase the diversity of the elite set solutions by modifying lines 6 and 11, where condition $\delta > 0$ can be changed to $\delta \geq \underline{\delta}$, where $\underline{\delta} > 0$ is a parameter. In this case, instead of requiring that $S$ only be different from all other elite set solutions, we now require that it be sufficiently different by at least a given number of attributes.

## 9.3 Hybridization of GRASP with path-relinking

Path-relinking is a major enhancement to GRASP, equipping GRASP heuristics with a long-term memory mechanism and enabling search intensification beyond simple local search. In this section, we show how to hybridize path-relinking with GRASP.

To implement GRASP with path-relinking, we make use of an elite set $\mathscr{E}$, such as the one introduced in Section 9.2, to collect a diverse set of high-quality solutions found during the search. The elite set starts empty and is constrained to have at most $n_{\mathscr{E}}$ solutions. Each new locally optimal solution produced by the GRASP local search phase is relinked with one or more solutions from the elite set. Each solution resulting from path-relinking is considered as a candidate to be inserted in the elite set according to algorithm UPDATE-ELITE-SET of Figure 9.1.

The pseudo-code of Figure 9.2 outlines the main steps of a GRASP with path-relinking heuristic for minimization. This simple variant relinks the locally optimal solution produced in each GRASP iteration with a single, randomly chosen, solution from the elite set, following the forward path-relinking strategy described in Section 8.1.2. The output of the path-relinking operator is a candidate for inclusion in the elite set.

```
begin GRASP+PR;
1    𝓔 ← ∅;
2    while stopping criterion not satisfied do
3        S ← SEMI-GREEDY;
4        if S is not feasible then
5            S ← Repair(S);
6        end-if;
7        S ← LOCAL-SEARCH(S);
8        if |𝓔| > 0 then
9            Select an elite solution S′ at random from 𝓔;
10           S ← FORWARD-PR(S,S′);
11       end-if;
12       UPDATE-ELITE-SET(S,𝓔);
13 end-while;
14 return S* = argmin{f(S) : S ∈ 𝓔};
end GRASP+PR.
```

**Fig. 9.2** Pseudo-code of a template of a basic GRASP with path-relinking heuristic for minimization.

Line 1 of the pseudo-code initializes the elite set $\mathscr{E}$ as empty. The loop from line 2 to line 13 makes up the steps of GRASP with path-relinking. Lines 3 to 7 correspond to the semi-greedy construction, repair (in case of infeasibility), and local search phases of a basic GRASP heuristic. Forward path-relinking is performed in lines 9 and 10 in case the elite set is not empty: in line 9, an elite set solution $S'$ is selected at random from $\mathscr{E}$ while, in line 10, $S'$ is relinked with the locally optimal solution $S$ produced in line 7. The resulting solution, $S$, is tested for inclusion in the elite set in line 12, which updates $\mathscr{E}$ by applying algorithm UPDATE-ELITE-SET of Figure 9.1. The algorithm returns the best-valued elite solution in line 14, after a stopping criterion is met.

Enhancing GRASP with path-relinking almost always improves the performance of the heuristic. As an illustration, Figures 9.3 and 9.4 show time-to-target plots (or runtime distributions) for GRASP with and without path-relinking for four different applications. These plots show the empirical cumulative probability distributions of the time-to-target random variable, i.e., the time needed to find a solution at least as good as a given target value. For all problems, the plots show that GRASP with path-relinking is able to find target solutions faster than the memoryless basic algorithm.

## 9.4 Evolutionary path-relinking

As aforementioned, GRASP with path-relinking heuristics maintain an elite set of high-quality solutions. In the variant of GRASP with path-relinking introduced in Section 9.3, locally optimal solutions produced by local search are relinked with elite set solutions. Path-relinking can also be applied to pairs of elite set solutions to search for new high-quality solutions and to improve the quality of the elite set.

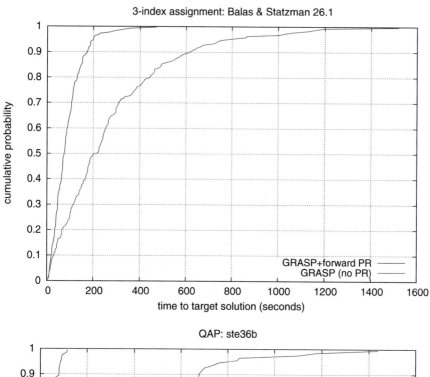

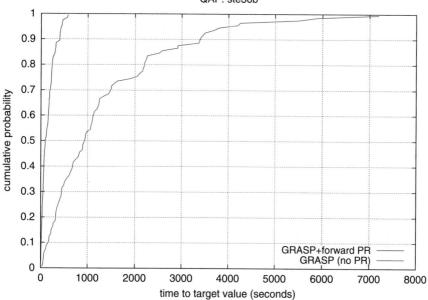

**Fig. 9.3** Time-to-target plots comparing running times of GRASP with and without path-relinking on distinct problems: three-index assignment and maximum satisfiability. Forward path-relinking was used in these two examples.

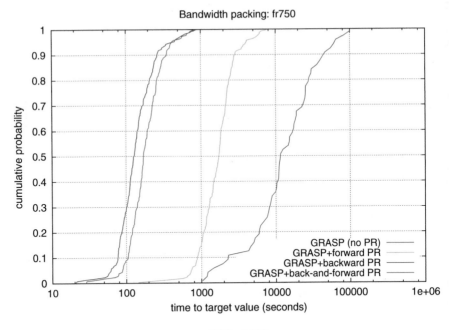

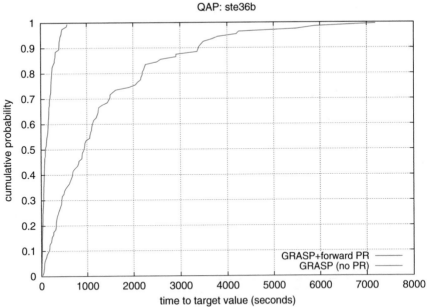

**Fig. 9.4** Time-to-target plots comparing running times of GRASP with and without path-relinking on distinct problems: bandwidth packing and quadratic assignment. Forward path-relinking was used in these two examples. In addition, on the bandwidth packing example, plots for GRASP with backward and back-and-forward path-relinking are also shown.

This procedure, called *evolutionary path-relinking* (EvPR), can be applied as a post-optimization phase of GRASP, after the main heuristic stops, or periodically, when the main heuristic is still running.

```
begin GRASP+EvPR;
1   𝓔 ← ∅;
2   f* ← ∞;
3   while stopping criterion not satisfied do
4       S ← SEMI-GREEDY;
5       if S is not feasible then
6           S ← Repair(S);
7       end-if;
8       S ← LOCAL-SEARCH(S);
9       if |𝓔| > 0 then
10          Select an elite solution S' at random from 𝓔;
11          S ← FORWARD-PR(S,S');
12      end-if;
13      UPDATE-ELITE-SET(S, 𝓔);
14  end-while;
15  𝓔 ← EvPR(𝓔);
16  return S* = argmin{f(S) : S ∈ 𝓔};
end GRASP+EvPR.
```

**Fig. 9.5** Pseudo-code of a template of a GRASP with evolutionary path-relinking heuristic where evolutionary path-relinking is applied at a post-processing step.

The pseudo-codes in Figures 9.5 and 9.6 correspond to the post-processing and periodic variants, respectively. The pseudo-code in Figure 9.5 is identical to that of the GRASP with path-relinking of Figure 9.2, with an additional step in line 15 where EvPR is applied.

The pseudo-code of Figure 9.6 adds lines 3 and 15 to 19 to manage the periodic application of EvPR. Line 3 initializes it2evPR, a counter of iterations to EvPR, with evPRfreq being the number of GRASP iterations between consecutive calls to EvPR. If evPRfreq iterations have passed without the application of EvPR, then in line 16 it is applied and the counter it2evPR is reinitialized in line 17. Finally, in line 19, it2evPR is decreased by one iteration.

Evolutionary path-relinking takes as input the elite set and returns either the same elite set or a renewed one with an improved average cost. This approach is outlined in the pseudo-code of Figure 9.7. While there exists a pair of solutions in the elite set for which path-relinking has not yet been applied, the two solutions are combined with path-relinking and the resulting solution is tested for membership in the elite set. If it is accepted, it then replaces the elite solution most similar to it among all solutions having worse cost. To explore more than one path connecting two solutions, evolutionary path-relinking can apply greedy randomized adaptive path-relinking a fixed number of times between each pair of elite solutions.

This strategy outperformed several other heuristics using GRASP with path-relinking, simulated annealing, tabu search, and a multistart strategy for the

```
begin GRASP+itEvPR(evPRfreq);
1   𝓔 ← ∅;
2   f* ← ∞;
3   it2evPR ← evPRfreq;
4   while stopping criterion not satisfied do
5       S ← SEMI-GREEDY;
6       if S is not feasible then
7           S ← Repair(S);
8       end-if;
9       S ← LOCAL-SEARCH(S);
10      if |𝓔| > 0 then
11          Select an elite solution S' at random from 𝓔;
12          S ← FORWARD-PR(S, S');
13      end-if;
14      UPDATE-ELITE-SET(S, 𝓔);
15      if it2evPR < 1 then
16          𝓔 ← EvPR(𝓔);
17          it2evPR ← evPRfreq + 1;
18      end-if;
19      it2evPR ← it2evPR − 1;
20  end-while;
21  return S* = argmin{f(S) : S ∈ 𝓔};
end GRASP+itEvPR.
```

**Fig. 9.6** Pseudo-code of a template of a GRASP with evolutionary path-relinking heuristic where evolutionary path-relinking is applied periodically during the search.

```
begin EvPR(𝓔);
1   while there exists solutions S¹, S² ∈ 𝓔 that have not yet been relinked, with S¹ ≠ S² do
2       S ← FORWARD-PR(S¹, S²);
3       UPDATE-ELITE-SET(S, 𝓔);
4   end-while;
5   return 𝓔;
end EvPR.
```

**Fig. 9.7** Pseudo-code of a template of the evolutionary path-relinking strategy.

max-min diversity problem. Figure 9.8 shows the evolution of the best solution found by the multistart strategy, pure GRASP, and GRASP with evolutionary path-relinking for a 500-element max-min diversity instance.

## 9.5 Restart strategies

Figure 9.9 shows a typical iteration count distribution for a GRASP with path-relinking heuristic. Observe in this example that for most of the independent runs whose iteration counts make up the plot, the algorithm finds a target solution in

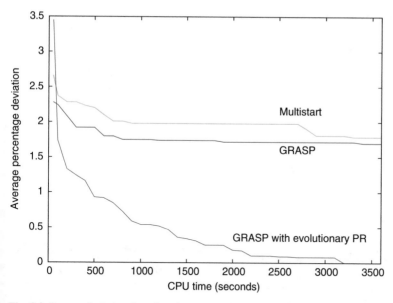

**Fig. 9.8** Percent deviation from best known solution value for GRASP with evolutionary path-relinking, pure GRASP, and a multistart algorithm for a 500-element instance of a max-min diversity problem with a time limit of 60 minutes.

relatively few iterations: about 25% of the runs take at most 101 iterations; about 50% take at most 192 iterations; and about 75% take at most 345. However, some runs take much longer: 10% take over 1000 iterations; 5% over 2000; and 2% over 9715 iterations. The longest run took 11607 iterations to find a solution at least as good as the target. These long tails contribute to a large average iteration count as well as to a high standard deviation. This section proposes strategies to reduce the tail of the distribution, consequently reducing the average iteration count and its standard deviation.

Consider again the distribution in Figure 9.9. The distribution shows that each run will take over 345 iterations with about 25% probability. Therefore, any time the algorithm is restarted, the probability that the new run will take over 345 iterations is also about 25%. By restarting the algorithm after 345 iterations, the new run will take more than 345 iterations with probability of also about 25%. Therefore, the probability that the algorithm will be still running after $345 + 345 = 690$ iterations is the probability that it takes more than 345 iterations multiplied by the probability that it takes more than 690 iterations given that it took more than 345 iterations, i.e., about $(1/4) \times (1/4) = (1/4)^2$. It follows by induction that the probability that the algorithm will still be running after $k$ periods of 345 iterations is $1/(4^k)$. In this example, the probability that the algorithm will be running after 1725 iterations will be about 0.1%, i.e., much less than the 5% probability that the algorithm will take over 2000 iterations without restart.

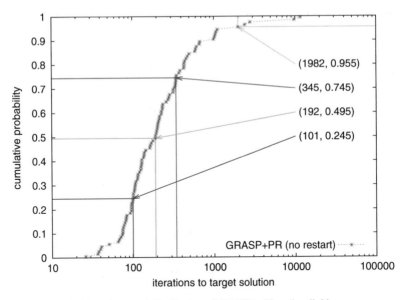

**Fig. 9.9** Typical iteration count distribution of GRASP with path-relinking.

A *restart strategy* is defined as an infinite sequence of time intervals $\tau_1, \tau_2, \tau_3, \ldots$ which define epochs $\tau_1, \tau_1 + \tau_2, \tau_1 + \tau_2 + \tau_3, \ldots$ when the algorithm is restarted from scratch. It can be shown that the optimal restart strategy uses $\tau_1 = \tau_2 = \cdots = \tau^*$, where $\tau^*$ is some (unknown) constant.

Implementing the optimal strategy may be difficult in practice because it requires inputting the constant value $\tau^*$. Runtimes can vary greatly for different combinations of algorithm, instance, and solution quality sought. Since usually one has no prior information about the runtime distribution of the stochastic search algorithm for the optimization problem under consideration, one runs the risk of choosing a value of $\tau^*$ that is either too small or too large. On the one hand, a value that is too small can cause the restart variant of the algorithm to take much longer to converge than a no-restart variant. On the other hand, a value that is too large may never lead to a restart, causing the restart-variant of the algorithm to take as long to converge as the no-restart variant. Figure 9.10 illustrates the restart strategies with time-to-target plots for the maximum cut instance *G12* on an 800-node graph with edge density of 0.63% with target solution value 554 for $\tau = 6, 9, 12, 18, 24, 30,$ and 42 seconds. For each value of $\tau$, 100 independent runs of a GRASP with path-relinking heuristic with restarts were performed. The variant with $\tau = \infty$ corresponds to the heuristic without restart. The figure shows that, for some values of $\tau$, the resulting heuristic outperformed its counterpart with no restart by a large margin.

In GRASP with path-relinking, the number of iterations between improvements of the incumbent (or best so far) solution tends to vary less than the runtimes for different combinations of instance and solution quality sought. If one takes this into account, a simple and effective restart strategy for GRASP with path-relinking is

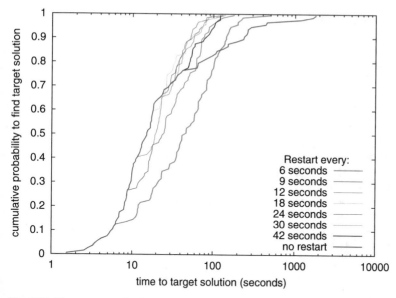

**Fig. 9.10** Time-to-target plot for target solution value of 554 for maximum cut instance *G12* using different values of τ.

to keep track of the last iteration when the incumbent solution was improved and restart the GRASP with path-relinking heuristic if κ iterations have gone by without improvement. We shall call such a strategy restart(κ). A restart consists in saving the incumbent and emptying out the elite set.

The pseudo-code shown in Figure 9.11 summarizes the steps of a GRASP with path-relinking heuristic using the restart(κ) strategy for a minimization problem. The algorithm keeps track of the current iteration (CurrentIter), as well as of the last iteration when an improving solution was found (LastImprov). If an improving solution is detected in line 16, then this solution and its cost are saved in lines 17 and 18, respectively, and the iteration of last improvement is set to the current iteration in line 19. If, in line 21, it is determined that more than κ iterations have gone by since the last improvement of the incumbent, then a restart is triggered, emptying out the elite set in line 22 and resetting the iteration of last improvement to the current iteration in line 23. If restart is not triggered, then in line 25 the current solution S is tested for inclusion in the elite set and the set is updated if S is accepted. The best overall solution found S* is returned in line 28 after the stopping criterion is satisfied.

As an illustration of the use of the restart(κ) strategy within a GRASP with path-relinking heuristic, consider the maximum cut instance *G12*. For the values κ = 50, 100, 200, 300, 500, 1000, 2000, and 5000, the heuristic was run independently 100 times, stopping when a cut of weight 554 or higher was found. A strategy without restarts was also implemented. Figures 9.12 and 9.13, as well as Table 9.1, summarize these runs, showing the average time to target solution as a function of the value

```
begin GRASP+PR+RESTARTS;
1   𝓔 ← ∅;
2   f* ← ∞;
3   LastImprov ← 0;
4   CurrentIter ← 0;
5   while stopping criterion not satisfied do
6       CurrentIter ← CurrentIter + 1;
7       S ← SEMI-GREEDY;
8       if S is not feasible then
9           S ← Repair(S);
10      end-if;
11      S ← LOCAL-SEARCH(S);
12      if |𝓔| > 0 then
13          Select an elite solution S' at random from 𝓔;
14          S ← FORWARD-PR(S, S');
15      end-if;
16      if f(S) < f* then
17          S* ← S;
18          f* ← f(S);
19          LastImprov ← CurrentIter;
20      end-if;
21      if CurrentIter − LastImprov > κ then
22          𝓔 ← ∅;
23          LastImprov ← CurrentIter;
24      else
25          UPDATE-ELITE-SET(S, 𝓔);
26      end-if;
27  end-while;
28  return S*;
end GRASP+PR+RESTARTS.
```

**Fig. 9.11** Pseudo-code of a template of a GRASP with path-relinking heuristic with restarts for a minimization problem.

of $\kappa$ and the time-to-target plots for different values of $\kappa$. These figures illustrate well the effect on running time of selecting a value of $\kappa$ that is either too small ($\kappa = 50, 100$) or too large ($\kappa = 2000, 5000$). They further show that there is a wide range of $\kappa$ values ($\kappa = 200, 300, 500, 1000$) that result in lower runtimes when compared to the strategy without restarts.

Figure 9.14 further illustrates the behavior of the restart(100), restart(500), and restart(1000) strategies for the previous example, when compared with the strategy without restarts on the same maximum cut instance *G12*. However, in this figure, for each strategy, we plot the number of iterations to the target solution value. It is interesting to note that, as expected, each strategy restart($\kappa$) behaves exactly like the strategy without restarts for the $\kappa$ first iterations, for $\kappa = 100, 500, 1000$. After this point, each trajectory deviates from that of the strategy without restarts. Among these strategies, restart(500) is the one with the best performance.

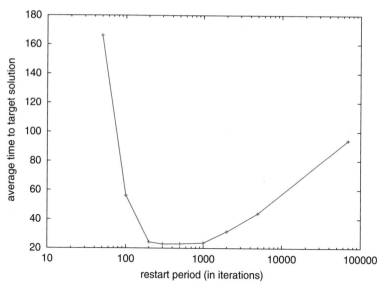

**Fig. 9.12** Average time-to-target solution for maximum cut instance *G12* using different values of κ. All runs of all strategies have found a solution at least as good as the target value of 554.

We conclude this chapter with some observations about these experiments. The effect of the restart strategies can be mainly observed in the column corresponding to the fourth quartile of Table 9.1. Entries in this quartile correspond to those in the heavy tails of the distributions. The restart strategies in general did not affect the other quartiles of the distributions, which is a desirable characteristic. Compared to the no-restart strategy, at least one restart strategy was always able to reduce the maximum number of iterations, the average number of iterations, and the standard deviation of the number of iterations. Compared to the no-restart strategy, restart strategies restart(500) and restart(1000) were able to reduce the maximum number of iterations, as well as the average and the standard deviation. Strategy restart(100) did so, too, but not as much as restart(500) and restart(1000). Restart strategies restart(500) and restart(1000) were clearly the best strategies of those tested.

**Table 9.1** Summary of computational results on maximum cut instance *G12* with four strategies. For each strategy, 100 independent runs were executed, each stopped when a solution as good as the target solution value 554 was found. For each strategy, the table shows the distribution of the number of iterations by quartile. For each quartile, the table gives the maximum number of iterations taken by all runs in that quartile, i.e., the slowest of the fastest 25% (1st), 50% (2nd), 75% (3rd), and 100% (4th) of the runs. The average number of iterations over the 100 runs and the standard deviation (st.dev.) are also given for each strategy.

| Strategy | Iterations in quartile | | | | Average | st.dev. |
|---|---|---|---|---|---|---|
| | 1st | 2nd | 3rd | 4th | | |
| Without restarts | 326 | 550 | 1596 | 68813 | 4525.1 | 11927.0 |
| restart(1000) | 326 | 550 | 1423 | 5014 | 953.2 | 942.1 |
| restart(500) | 326 | 550 | 1152 | 4178 | 835.0 | 746.1 |
| restart(100) | 509 | 1243 | 3247 | 8382 | 2055.0 | 2005.9 |

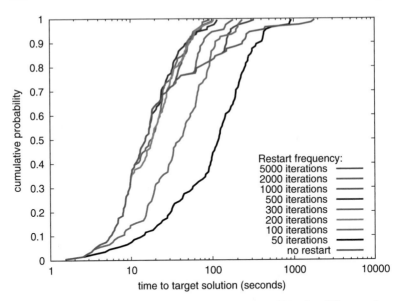

**Fig. 9.13** Time-to-target plots for maximum cut instance *G12* using different values of κ. The figure also shows the time-to-target plot for the strategy without restarts. All runs of all strategies found a solution at least as good as the target value of 554.

## 9.6 Bibliographical notes

GRASP with path-relinking as proposed in Section 9.3 was first introduced by Laguna and Martí (1999), where a forward path-relinking operator from the solution found by local search to a randomly selected elite solution was applied. This was followed by a number of applications of GRASP with path-relinking, e.g., to maximum cut (Festa et al., 2002), 2-path network design (Ribeiro and Rosseti, 2002), Steiner problem in graphs (Ribeiro et al., 2002), job-shop scheduling (Aiex et al., 2003), private virtual circuit routing (Resende and Ribeiro, 2003a), *p*-median (Resende and Werneck, 2004), quadratic assignment (Oliveira et al., 2004), set packing (Delorme et al., 2004), three-index assignment (Aiex et al., 2005), *p*-hub median (Pérez et al., 2005), uncapacitated facility location (Resende and Werneck, 2006), project scheduling (Alvarez-Valdes et al., 2008a), maximum weighted satisfiability (Festa et al., 2006), maximum diversity (Silva et al., 2007), network migration scheduling (Andrade and Resende, 2007a), capacitated arc routing (Labadi et al., 2008; Usberti et al., 2013), disassembly sequencing (Adenso-Díaz et al., 2008), flowshop scheduling (Ronconi and Henriques, 2009), multi-plant capacitated lot sizing (Nascimento et al., 2010), workover rig scheduling (Pacheco et al., 2010), max-min diversity (Resende et al., 2010a), biobjective orienteering (Martí et al., 2015), biobjective path dissimilarity (Martí et al., 2015), generalized quadratic assignment (Mateus et al., 2011), antibandwidth (Duarte et al., 2011), capacitated clustering (Deng and Bard, 2011), linear ordering (Chaovalitwongse et al., 2011), data

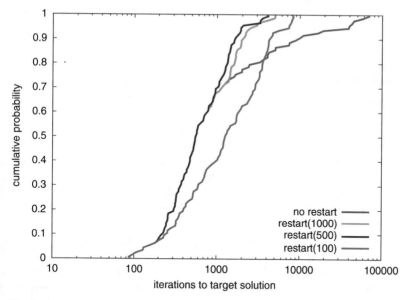

**Fig. 9.14** Comparison of the iterations-to-target plots for maximum cut instance *G12* using strategies restart(100), restart(500), and restart(1000). The figure also shows the iterations-to-target plot for the strategy without restarts. All runs of all strategies found a solution at least as good as the target value of 554.

clustering (Frinhani et al., 2011), two-echelon location routing (Nguyen et al., 2012), image registration (Santamaría et al., 2012), drawing proportional symbols in maps (Cano et al., 2013), family traveling salesperson (Morán-Mirabal et al., 2014), handover minimization in mobility networks (Morán-Mirabal et al., 2013b), facility layout (Silva et al., 2013b), survivable network design (Pedrola et al., 2013), equitable dispersion (Martí and Sandoya, 2013), 2D and 3D bin packing (Alvarez-Valdes et al., 2013), microarray data analysis (Cordone and Lulli, 2013), community detection (Nascimento and Pitsoulis, 2013), set *k*-covering (Pessoa et al., 2013), network load balancing (Santos et al., 2013), power optimization in ad hoc networks (Moraes and Ribeiro, 2013), capacitated vehicle routing (Sörensen and Schittekat, 2013), and symmetric Euclidean clustered traveling salesman (Mestria et al., 2013).

Surveys on GRASP with path-relinking can be found in Resende and Ribeiro (2005a), Aiex and Resende (2005), Resende (2008), Resende et al. (2010b), Resende and Ribeiro (2010), Ribeiro and Resende (2012), and Festa and Resende (2013). A special issue of *Computers & Operations Research* (Martí et al., 2013b) was dedicated to GRASP with path-relinking.

Section 9.4 discussed evolutionary path-relinking that was originally proposed by Resende and Werneck (2004), where it was used as a post-processing phase for a GRASP with path-relinking for the *p*-median problem. Andrade and Resende (2007a) were the first to apply evolutionary path-relinking periodically during the search. The term evolutionary path-relinking was introduced by Andrade and Resende (2007b). This was followed by a number of applications of GRASP with evolutionary path-relinking, e.g., to uncapacitated facility location (Resende and Werneck, 2006), max-min diversity (Resende et al., 2010a), image registration (Santamaría et al., 2010; Santamaría et al., 2012), power transmission network expansion planning (Rahmani et al., 2010), vehicle routing with trailers (Villegas, 2010), antibandwidth minimization (Duarte et al., 2011), truck and trailer routing (Villegas et al., 2011), parallel machine scheduling (Rodriguez et al., 2012), linear ordering (Duarte et al., 2012), family traveling salesperson (Morán-Mirabal et al., 2014), handover minimization in mobility networks (Morán-Mirabal et al., 2013b), set covering (Morán-Mirabal et al., 2013a), maximum cut (Morán-Mirabal et al., 2013a), node capacitated graph partitioning (Morán-Mirabal et al., 2013a), capacitated arc routing (Usberti et al., 2013), and 2D and 3D bin packing (Alvarez-Valdes et al., 2013),

Figures 9.3 and 9.4 show time-to-target plots comparing pure GRASP and GRASP with path-relinking implementations on instances of the three-index assignment problem (Aiex et al., 2005), maximum satisfiability (Festa et al., 2006), bandwidth packing (Resende and Ribeiro, 2003a), and the quadratic assignment problem (Oliveira et al., 2004).

Figure 9.8 shows results from Resende et al. (2010a), where a GRASP and GRASP with evolutionary path-relinking for max-min diversity were proposed. The simulated annealing and multistart algorithms were the ones described in Kincaid (1992) and Ghosh (1996), respectively.

The restart($\kappa$) strategy for GRASP with path-relinking discussed in Section 9.5 was proposed by Resende and Ribeiro (2011). Besides the experiments presented in this chapter for the maximum cut instance *G12*, that paper also considered five other instances of maximum cut, maximum weighted satisfiability, and bandwidth packing. Strategies for speeding up stochastic local search algorithms using restarts were first proposed by Luby et al. (1993), where they proved the result for an optimal restart strategy. Restart strategies in metaheuristics have been addressed in D'Apuzzo et al. (2006), Kautz et al. (2002), Nowicki and Smutnicki (2005), Palubeckis (2004), and Sergienko et al. (2004). Further work on restart strategies can be found in Shylo et al. (2011a) and Shylo et al. (2011b).

# Chapter 10
# Parallel GRASP heuristics

Parallel computers and parallel algorithms have increasingly found their way into metaheuristics. Most parallel implementations of GRASP found in the literature consist in either partitioning the search space or the GRASP iterations and assigning each partition to a processor. GRASP is applied to each partition in parallel. These implementations can be categorized as multiple-walk independent-thread, with the communication among processors during GRASP iterations being limited to the detection of program termination and gathering the best solution found over all processors. Approaches for the parallelization of GRASP with path-relinking can be categorized as either multiple-walk independent-thread or multiple-walk cooperative-thread, with processors sharing and exchanging information about elite solutions visited during the GRASP iterations. This chapter is an introduction to parallel GRASP heuristics, covering multiple-walk independent-thread strategies, multiple-walk cooperative-thread strategies, and some applications of parallel GRASP and parallel GRASP with path-relinking.

## 10.1 Multiple-walk independent-thread strategies

Most parallel implementations of GRASP follow the *multiple-walk independent-thread* strategy, based on the distribution of the iterations among the processors. In general, each search thread has to perform $MaxIterations/p$ iterations, where $p$ and $MaxIterations$ are, respectively, the number of processors and the total number of iterations. Each processor has a copy of the sequential algorithm, a copy of the problem data, and an independent seed to generate its own pseudo-random number sequence. To avoid that the processors find the same solutions, each processor uses a different sequence of pseudo-random numbers. A single global variable is required to store the best solution found over all processors. One of the processors acts as the master, reading and distributing problem data, generating the seeds which will be used by the pseudo-random number generator at each processor, distributing the iterations, and collecting the best solution found by each processor. Since the

© Springer Science+Business Media New York 2016
M.G.C. Resende, C.C. Ribeiro, *Optimization by GRASP*,
DOI 10.1007/978-1-4939-6530-4_10

iterations are independent and very little information is exchanged, linear speedups are easily obtained provided that no major load imbalance occurs. The *speedup* of a parallel GRASP heuristic running on $p$ processors is the ratio of the time taken by the sequential GRASP heuristic and the time taken by the parallel heuristic running on $p$ processors. To improve load balancing, the iterations can be evenly distributed over the processors or according to their demands.

Implementations of this strategy in machines with different architectures and using different software platforms have shown linear or almost-linear speedups for a number of applications. We illustrate the case for independent-thread strategies with two examples of parallel implementations.

The first example is of a parallel GRASP for the MAX-SAT problem running on a cluster of SUN-SPARC 10 workstations, sharing the same file system, with communication done using Parallel Virtual Machine (PVM). The parallel GRASP was applied to each test instance using 1, 5, 10, and 15 processors, and the maximum number of iterations was set at 1000, 200, 100, and 66, respectively. The computation time required to perform the specified number of iterations and the best solution found were recorded. Since communication was kept to a minimum, linear speedups were expected. Figure 10.1 shows individual speedups and average speedups for these runs. Figure 10.2 shows that the average quality of the solutions found was not greatly affected by the number of processors used.

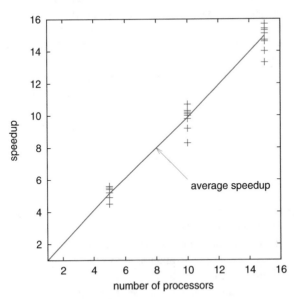

**Fig. 10.1** Average speedups on 5, 10, and 15 processors for the maximum satisfiability problem.

The second example is an implementation of a parallel GRASP for the Steiner problem in graphs. Parallelization was achieved by the distribution of 512 iterations over the processors, with the value of the restricted candidate list parameter $\alpha$

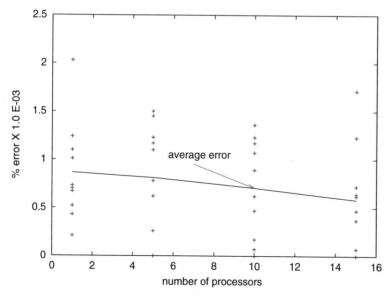

**Fig. 10.2** Error on 1, 5, 10, and 15 processors for the maximum satisfiability problem.

randomly chosen in the interval $[0.0, 0.3]$ at each iteration. The algorithm was tested on an IBM SP-2 machine with 32 processors, using the Message Passing Interface (MPI) library for communication. The 60 problems from series C, D, and E of the OR-Library were used for the computational experiments. The parallel implementation obtained 45 optimal solutions over the 60 test instances. The relative deviation with respect to the optimal value was never larger than 4%. Almost-linear speedups were observed for 2, 4, 8, and 16 processors with respect to the sequential implementation and are illustrated in Figure 10.3.

Path-relinking may also be used in conjunction with independent-thread parallel implementations of GRASP. An independent-thread implementation for the job shop scheduling problem keeps local sets (or pools) of elite solutions in each processor and path-relinking is applied to pairs of elite solutions stored in each local pool. Computational results using MPI on an SGI Challenge computer with 28 R10000 processors showed linear speedups for the 3-index assignment problem.

Multiple-walk independent-thread approaches for the parallelization of GRASP may benefit from load balancing techniques, whenever heterogeneous processors are used or if the parallel machine is simultaneously shared by several users. In this case, almost-linear speedups can be obtained with a heterogeneous distribution of the iterations among the $p$ processors in $q \geq p$ packets. Each processor starts performing one packet of $\lceil MaxIterations/q \rceil$ iterations and informs the master when it finishes its packet of iterations. The master stops the execution of each worker processor when there are no more iterations to be performed and collects the best solution found. Faster or less loaded processors will perform more iterations than the others. In the case of the parallel GRASP implemented for the problem of

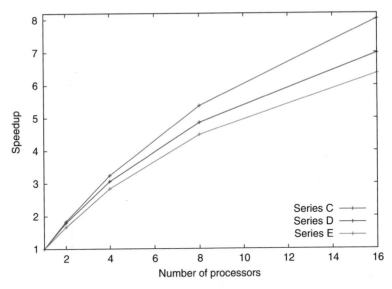

**Fig. 10.3** Average speedups on 2, 4, 8, and 16 processors on Steiner tree problem in graphs.

traffic assignment discussed in Chapter 7, this dynamic load balancing strategy allows reductions in the elapsed times of up to 15% with respect to the times observed for the static strategy, in which the iterations are uniformly distributed over the processors.

The efficiency of multiple-walk independent-thread parallel implementations of metaheuristics (based on running multiple copies of the same sequential algorithm) has been addressed in the literature. The *efficiency* of the parallel heuristic running on $p$ processors is given by its speedup divided by $p$. We have seen in Section 6.2 of this book that the time taken by a pure GRASP heuristic to find a solution with cost at least as good as a certain target value has been shown experimentally to behave as a random variable that fits an exponential distribution. In the case where the setup times are not negligible, the runtimes fit a two-parameter shifted exponential distribution. Therefore, the probability density function of the time-to-target random variable is given by $f(t) = (1/\lambda) \cdot e^{-t/\lambda}$ in the first case or by $f(t) = (1/\lambda) \cdot e^{-(t-\mu)/\lambda}$ in the second, with the parameters $\lambda \in \mathbb{R}^+$ and $\mu \in \mathbb{R}^+$ being associated with the shape and the shift of the exponential function, respectively.

Recall that the speedup of a parallel GRASP heuristic running on $p$ processors measures the ratio between the time needed to find a solution with value at least as good as the target value using a sequential algorithm and that taken by a parallel implementation with $p$ processors. The linear speedups of parallel GRASP implementations with negligible setup times naturally follow from the expression of the probability density function $f(t) = (1/\lambda) \cdot e^{-t/\lambda}$ of the exponentially distributed time-to-target random variable, as illustrated with the previous examples. For example, suppose $t_1, t_2, \ldots, t_p$ are $p$ independent exponentially distributed runtimes, each with parameter $\lambda$. Suppose $t_i$ is the runtime on processor $i = 1, 2, \ldots, p$. By definition, the expected value of $t_i$ is $E(t_i) = \lambda$, for $i = 1, 2, \ldots, p$. Define $\tau$ to be the runtime of the parallel process, i.e., the time taken by a parallel implementation with $p$ processors:

$$\tau = \min\{t_1, t_2, \ldots, t_p\},$$

which is the runtime of the fastest of the $p$ processes. Then,

$$
\begin{aligned}
P(\tau > a) &= P(\min\{t_1, t_2, \ldots, t_p\} > a) \\
&= P(t_1 > a, t_2 > a, \ldots, t_p > a) \\
&= P(t_1 > a) \cdot P(t_2 > a) \cdots P(t_p > a) \\
&= e^{-a/\lambda} \cdot e^{-a/\lambda} \cdots e^{-a/\lambda} \\
&= e^{-a/(\lambda/p)}.
\end{aligned}
$$

Therefore, the cumulative distribution function of $\tau$ is given by $F_\tau(a) = 1 - P(\tau > a) = 1 - e^{-a/(\lambda/p)}$. Hence, the random variable $\tau$ is also exponentially distributed with expected value $E(\tau) = \lambda/p$, showing that with $p$ processors there is an expected linear speedup of $p$.

Let $P_p(t)$ be the probability of not having found a given (target) solution value in $t$ time units with $p$ independent processors. If $P_1(t) = e^{-(t-\mu)/\lambda}$, with $\lambda \in \mathbb{R}^+$ and $\mu \in \mathbb{R}$, i.e., $P_1$ corresponds to a two-parameter exponential distribution, then $P_p(t) = e^{-p(t-\mu)/\lambda}$. This follows from the definition of the two-parameter exponential distribution. It implies that the probability of finding a solution of a given value in time $pt$ with one processor is equal to $1 - e^{-(pt-\mu)/\lambda}$, while the probability of finding a solution at least as good as that given target value in time $t$ with $p$ independent parallel processors is $1 - e^{-p(t-\mu)/\lambda}$. Note that if $\mu = 0$, corresponding to the case of nonshifted exponential distributions, then both probabilities are equal. Furthermore, since $p \geq 1$, then the two probabilities are approximately equal if $p|\mu| \ll \lambda$ and it is possible to approximately achieve linear speedup in solution time-to-target value using multiple independent processors.

The observation above suggests a test using a one-processor, sequential implementation to determine whether it is likely that a parallel implementation using multiple independent processors will be efficient. We say a parallel implementation is efficient if it achieves linear speedup (with respect to wall, or elapsed, time) to find a solution at least as good as a given target value. The test consists in performing a large number of independent runs of the sequential program to build a Q-Q plot and estimate the parameters $\mu$ and $\lambda$ of the shifted exponential distribution. If $p|\mu| \ll \lambda$, then we can predict that the parallel implementation will be efficient. Later in this chapter (on page 218), we illustrate this test.

## 10.2 Multiple-walk cooperative-thread strategies

In this section, we focus on the use of path-relinking as a mechanism for implementing GRASP as a *multiple-walk cooperative-thread* strategy, in which processors share and exchange information (in this case, about elite solutions previously visited) collected during previous GRASP iterations.

Path-relinking and its hybridization with GRASP heuristics have been extensively discussed in Chapters 8 and 9 of this book. The algorithm in Figure 10.4 recalls the pseudo-code of a hybrid GRASP with path-relinking for minimization, as already presented in Section 9.3.

```
begin GRASP+PR;
1   𝓔 ← ∅;
2   f* ← ∞;
3   while stopping criterion not satisfied do
4       S ← SEMI-GREEDY;
5       if S is not feasible then
6           S ← Repair(S);
7       end-if;
8       S ← LOCAL-SEARCH(S);
9       if |𝓔| > 0 then
10          Select an elite solution S' at random from 𝓔;
11          S ← FORWARD-PR(S, S');
12      end-if;
13      UPDATE-ELITE-SET(S, 𝓔);
14  end-while;
15  return S* = argmin{f(S) : S ∈ 𝓔};
end GRASP+PR.
```

**Fig. 10.4** Pseudo-code of a template of a basic GRASP with path-relinking heuristic for minimization (revisited).

Two basic mechanisms may be used to implement a multiple-walk cooperative-thread GRASP with path-relinking heuristic. In *distributed strategies*, each thread maintains its own pool of elite solutions. Each iteration of each thread consists initially of a GRASP construction, followed by local search. Then, the local optimum is combined with a randomly selected element of the thread's pool using path-relinking. The output of path-relinking is finally tested for insertion into the pool. If accepted for insertion, the solution is sent to the other threads, where it is tested for insertion into the other pools. Collaboration takes place at this point. Though there may be some communication overhead in the early iterations, this tends to ease up as pool insertions become less frequent.

The second mechanism corresponds to *centralized strategies* based on a single pool of elite solutions. As before, each GRASP iteration performed by each thread starts with the construction and local search phases. Next, an elite solution is requested and received from the centralized pool. Once path-relinking has been performed, the solution obtained as the output is sent to the pool and tested for insertion. Collaboration takes place when an elite solution is sent from a pool to a processor distinct from the one in which the solution was originally computed.

We note that, in both the distributed and the centralized strategies, each processor has a copy of the sequential algorithm and a copy of the data. One processor acts as the master, reading and distributing the problem data, generating the seeds used by the pseudo-random number generators at each processor, distributing the iterations, and collecting the best solution found by each processor. In the case of a distributed strategy, each processor has its own pool of elite solutions and all available processors perform GRASP iterations. Contrary to the case of a centralized strategy, one particular processor does not perform GRASP iterations and is used exclusively to store the pool and handle all operations involving communication requests between the pool and the workers. In the next section, we describe three examples of parallel implementations of GRASP with path-relinking.

## 10.3 Some parallel GRASP implementations

In this section, we report comparisons of multiple-walk independent-thread and multiple-walk cooperative-thread strategies for GRASP with path-relinking for the three-index assignment problem, the job shop scheduling problem, and the 2-path network design problem. For each problem, we first state the problem and describe the construction, local search, and path-relinking procedures. Next, we show numerical results comparing the different parallel implementations.

The experiments described in Sections 10.3.1 and 10.3.2 were done on an SGI Challenge computer (16 196-MHz MIPS R10000 processors and 12 194-MHz MIPS R10000 processors) with 7.6 Gb of memory. The algorithms were coded in Fortran and were compiled with the SGI MIPSpro F77 compiler using flags -O3 -static -u. The parallel codes used SGI's Message Passing Toolkit 1.4, which contains a fully compliant implementation of version 1.2 of the Message Passing Interface (MPI) specification. In the parallel experiments, wall clock times were measured with the MPI function MPI_WT. This was also the case for runs with a single processor that are compared to multiple-processor runs. Timing in the parallel runs excludes the time to read the problem data, to initialize the random number generator seeds, and to output the solution.

In the experiments described in Section 10.3.3, both variants of the parallel GRASP with path-relinking heuristic were coded in C and were compiled with version egcs-2.91.66 of the gcc compiler. MPI LAM 6.3.2 was used in the implementation. Computational experiments were performed on a cluster of 32 Pentium II 400MHz processors with 32 Mbytes of RAM memory each, running under the Red Hat 6.2 implementation of Linux. Processors were connected by a 10 Mbits/s IBM 8274 switch.

## *10.3.1 Three-index assignment*

### 10.3.1.1 Problem formulation

The three-index assignment problem (AP3) is a straightforward extension of the classical two-dimensional assignment problem and can be formulated as follows. Given three disjoint sets $I$, $J$, and $K$, with $|I| = |J| = |K| = n$, and a weight $c_{ijk}$ associated with each ordered triplet $(i, j, k) \in I \times J \times K$, find a minimum weight collection of $n$ disjoint triplets $(i, j, k) \in I \times J \times K$. Another way to formulate the AP3 is with permutations. There are $n^3$ cost elements. The optimal solution consists of the $n$ triplets with the smallest total cost, such that the constraints are not violated. The constraints are enforced if one assigns to each set $I$, $J$, and $K$, the numbers $1, 2, \ldots, n$ and none of the chosen triplets $(i, j, k)$ is allowed to have the same value for indices $i$, $j$, and $k$ as another. The permutation-based formulation for the AP3 is

$$\min_{p,q \in \pi_N} \sum_{i=1}^{n} c_{ip(i)q(i)},$$

where $\pi_N$ denotes the set of all permutations of the set of integers $N = \{1, 2, \ldots, n\}$.

### 10.3.1.2 GRASP construction

The construction phase selects $n$ triplets, one at a time, to form a three-index assignment $S$. A random choice in the interval $[0, 1]$ for the restricted candidate list parameter $\alpha$ is made at each iteration. The value remains constant during the entire construction phase. Construction begins with an empty solution $S$. The initial set $C$ of candidate triplets consists of the set of all triplets. Let $\underline{c}$ and $\overline{c}$ denote, respectively, the values of the smallest and largest cost triplets in $C$. All triplets $(i, j, k)$ in the candidate set $C$ having cost $c_{ijk} \leq \underline{c} + \alpha \cdot (\overline{c} - \underline{c})$ are placed in the restricted candidate list. Triplet $(i_p, j_p, k_p) \in C'$ is chosen at random and is added to the solution, i.e., $S = S \cup \{(i_p, j_p, k_p)\}$. Once $(i_p, j_p, k_p)$ is selected, any triplet $(i, j, k) \in C$ such that $i = i_p$ or $j = j_p$ or $k = k_p$ is removed from $C$. After $n - 1$ triplets have been selected, the set $C$ of candidate triplets contains one last triplet which is added to $S$, thus completing the construction phase.

### 10.3.1.3 Local search

If the solution of AP3 is represented by a pair of permutations $(p, q)$, then the solution space consists of all $(n!)^2$ possible combinations of permutations. If $p$ is a permutation vector, then a 2-exchange permutation of $p$ is a permutation vector that results from swapping two elements in $p$. In the 2-exchange neighborhood scheme used in this local search, the neighborhood of a solution $(p, q)$ consists of all 2-exchange permutations of $p$ plus all 2-exchange permutations of $q$. In the local search, the cost of each neighbor solution is compared with the cost of the current solution. If the cost of the neighbor is lower, then the solution is updated, the search is halted, and a search in the new neighborhood is initialized. The local search ends when no neighbor of the current solution has a lower cost than the current solution.

### 10.3.1.4 Path-relinking

A solution of AP3 can be represented by two permutation arrays of numbers $1, 2, \ldots, n$ in sets $J$ and $K$, respectively. Path-relinking is done between an initial solution

$$S = \{(p_1^S, p_2^S, \ldots, p_n^S), (q_1^S, q_2^S, \ldots, q_n^S)\}$$

and a guiding solution

$$T = \{(p_1^T, p_2^T, \ldots, p_n^T), (q_1^T, q_2^T, \ldots, q_n^T)\}.$$

Let the difference between $S$ and $T$ be defined by the two sets of indices

$$\delta_p^{S,T} = \{i = 1, \ldots, n \mid p_i^S \neq p_i^T\},$$
$$\delta_q^{S,T} = \{i = 1, \ldots, n \mid q_i^S \neq q_i^T\}.$$

During a path-relinking move, a permutation $\pi$ (for either $p$ or $q$) array in $S$, given by

$$(\ldots, \pi_i^S, \pi_{i+1}^S, \ldots, \pi_{j-1}^S, \pi_j^S, \ldots),$$

is replaced by a permutation array

$$(\ldots, \pi_j^S, \pi_{i+1}^S, \ldots, \pi_{j-1}^S, \pi_i^S, \ldots),$$

by exchanging permutation elements $\pi_i^S$ and $\pi_j^S$, where $i \in \delta_\pi^{S,T}$ and $j \in \{1, 2, \ldots, n\}$ are such that $\pi_j^T = \pi_i^S$.

### 10.3.1.5 Parallel independent-thread GRASP with path-relinking for AP3

We study the parallel efficiency of the multiple-walk independent-thread GRASP with path-relinking on AP3 instances B-S 20.1, B-S 22.1, B-S 24.1, and B-S 26.1, using 7, 8, 7, and 8 as target solution values, respectively. Table 10.1 shows the estimated shifted exponential distribution parameters for the multiple-walk independent-thread GRASP with path-relinking strategy, obtained from 200 independent runs of a sequential variant of the algorithm. In addition to the sequential variant, 60 independent runs of 2-, 4-, 8-, and 16-thread variants were run on the four test problems. Average speedups were computed dividing the sum of the execution times of the independent parallel program executing on one processor by the sum of the execution times of the parallel program on 2, 4, 8, and 16 processors, for

60 runs. The execution times of the independent parallel implementation executing on one processor and the execution times of the sequential program are approximately the same. The average speedups can be seen in Table 10.2 and Figure 10.5.

**Table 10.1** Estimated shifted exponential distribution parameters $\mu$ and $\lambda$ obtained with 200 independent runs of a sequential GRASP with path-relinking on AP3 instances B-S 20.1, B-S 22.1, B-S 24.1, and B-S 26.1, with target values 7, 8, 7, and 8, respectively.

|          | Estimated parameter |         |              |
|----------|---------|---------|--------------|
| Problem  | $\mu$   | $\lambda$ | $|\mu|/\lambda$ |
| B-S 20.1 | -26.46  | 1223.80 | 0.021        |
| B-S 22.1 | -135.12 | 3085.32 | 0.043        |
| B-S 24.1 | -16.76  | 4004.11 | 0.004        |
| B-S 26.1 | 32.12   | 2255.55 | 0.014        |
|          |         | average | 0.020        |

**Table 10.2** Speedups for multiple-walk independent-thread implementations of GRASP with path-relinking on instances B-S 20.1, B-S 22.1, B-S 24.1, and B-S 26.1, with target values 7, 8, 7, and 8, respectively. Speedups are computed with the average of 60 runs.

|          | Number of processors | | | | | | | |
|----------|---------|------------|---------|------------|---------|------------|---------|------------|
|          | 2       |            | 4       |            | 8       |            | 16      |            |
| Problem  | speedup | efficiency | speedup | efficiency | speedup | efficiency | speedup | efficiency |
| B-S 20.1 | 1.67    | 0.84       | 3.34    | 0.84       | 6.22    | 0.78       | 10.82   | 0.68       |
| B-S 22.1 | 2.25    | 1.13       | 4.57    | 1.14       | 9.01    | 1.13       | 14.37   | 0.90       |
| B-S 24.1 | 1.71    | 0.86       | 4.00    | 1.00       | 7.87    | 0.98       | 12.19   | 0.76       |
| B-S 26.1 | 2.11    | 1.06       | 3.89    | 0.97       | 6.10    | 0.76       | 11.49   | 0.72       |
| average  | 1.94    | 0.97       | 3.95    | 0.99       | 7.30    | 0.91       | **12.21** | 0.77     |

### 10.3.1.6 Parallel cooperative-thread GRASP with path-relinking for AP3

We now study the multiple-walk cooperative-thread strategy for GRASP with path-relinking on AP3. As with the independent-thread GRASP with path-relinking strategy, the target solution values 7, 8, 7, and 8 were used for instances B-S 20.1, B-S 22.1, B-S 24.1, and B-S 26.1, respectively. Table 10.3 and Figure 10.6 show super-linear speedups on instances B-S 22.1, B-S 24.1, and B-S 26.1 and about 90% efficiency for B-S 20.1. Super-linear speedups are possible
because good elite solutions are shared among the threads and are combined with GRASP solutions, whereas they would not be combined in an independent-thread implementation, making the parallel cooperative-thread GRASP with path-relinking converge faster to the target.

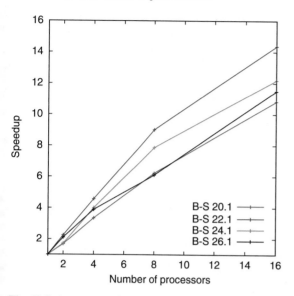

**Fig. 10.5** Average speedups on 2, 4, 8, and 16 processors for multiple-walk independent-thread parallel GRASP with path-relinking on AP3 instances B-S 20.1, B-S 22.1, B-S 24.1, and B-S 26.1.

**Table 10.3** Speedups for multiple-walk cooperative-thread implementations of GRASP with path-relinking on instances B-S 20.1, B-S 22.1, B-S 24.1, and B-S 26.1, with target values 7, 8, 7, and 8, respectively. Average speedups were computed over 60 runs.

| | Number of processors | | | | | | | |
| --- | --- | --- | --- | --- | --- | --- | --- | --- |
| | 2 | | 4 | | 8 | | 16 | |
| Problem | speedup | efficiency | speedup | efficiency | speedup | efficiency | speedup | efficiency |
| B-S 20.1 | 1.56 | 0.78 | 3.47 | 0.88 | 7.37 | 0.92 | 14.36 | 0.90 |
| B-S 22.1 | 1.64 | 0.82 | 4.22 | 1.06 | 8.83 | 1.10 | 18.78 | 1.04 |
| B-S 24.1 | 2.16 | 1.10 | 4.00 | 1.00 | 9.38 | 1.17 | 19.29 | 1.21 |
| B-S 26.1 | 2.16 | 1.08 | 5.30 | 1.33 | 9.55 | 1.19 | 16.00 | 1.00 |
| average | 1.88 | 0.95 | 4.24 | 1.07 | 8.78 | 1.10 | 17.10 | 1.04 |

Figure 10.7 compares the average speedup of the two implementations tested in this section, namely the multiple-walk independent-thread and the multiple-walk cooperative-thread GRASP with path-relinking implementations using target solution values 7, 8, 7, and 8, on the same instances. The figure shows that the cooperative variant of GRASP with path-relinking achieves the best parallelization, since the largest speedups are observed for that variant.

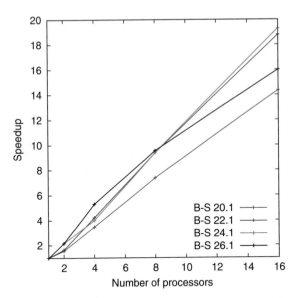

**Fig. 10.6** Average speedups on 2, 4, 8, and 16 processors for multiple-walk cooperative-thread parallel GRASP with path-relinking on AP3 instances B-S 20.1, B-S 22.1, B-S 24.1, and B-S 26.1.

## 10.3.2 Job shop scheduling

### 10.3.2.1 Problem formulation

The job shop scheduling problem (JSP) has long challenged researchers. It consists in processing a finite set of jobs on a finite set of machines. Each job is required to complete a set of operations in a fixed order. Each operation is processed on a specific machine for a fixed duration. Each machine can process at most one job at a time. Once a job initiates processing on a given machine, it must complete processing on that machine without interruption. A schedule is a mapping of the operations to time slots on the machines. The makespan is the maximum completion time of the jobs. The objective of the JSP is to find a schedule that minimizes the makespan.

A feasible solution of the JSP can be built from a permutation of the set of jobs $\mathcal{J}$ on each of the machines in the set $\mathcal{M}$, observing the precedence constraints, the restriction that a machine can process only one operation at a time, and requiring that once started, processing of an operation cannot be interrupted until its completion. Since each set of feasible permutations has a corresponding schedule, the objective of the JSP is to find, among the feasible permutations, the one with the smallest makespan.

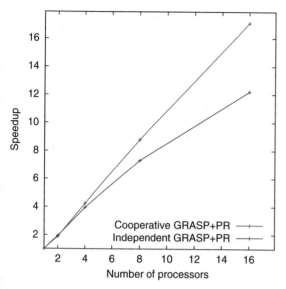

**Fig. 10.7** Average speedups on 2, 4, 8, and 16 processors for the parallel algorithms tested on instances of AP3: multiple-walk independent-thread GRASP with path-relinking and multiple-walk cooperative-thread GRASP with path-relinking.

### 10.3.2.2 GRASP construction

Each single operation is a building block of the GRASP construction phase for the JSP. A feasible schedule is built by scheduling individual operations, one at a time, until all operations have been scheduled.

While constructing a feasible schedule, not all operations can be selected at a given stage of the construction. An operation $\sigma_k^j$ can only be scheduled if all prior operations of job $j$ have already been scheduled. Therefore, at each construction phase iteration, at most $|\mathscr{J}|$ operations are candidates to be scheduled. Let this set of candidate operations be denoted by $\mathscr{O}_c$ and the set of already scheduled operations by $\mathscr{O}_s$. Denote the value of the greedy function for candidate operation $\sigma_k^j$ by $h(\sigma_k^j)$.

The greedy choice is to next schedule operation $\underline{\sigma}_k^j = \text{argmin}\{h(\sigma_k^j) \mid \sigma_k^j \in \mathscr{O}_c\}$. Let $\overline{\sigma}_k^j = \text{argmax}\{h(\sigma_k^j) \mid \sigma_k^j \in \mathscr{O}_c\}$, $\underline{h} = h(\underline{\sigma}_k^j)$, and $\overline{h} = h(\overline{\sigma}_k^j)$. Then, the GRASP restricted candidate list (RCL) is defined as

$$\text{RCL} = \{\sigma_k^j \in \mathscr{O}_c \mid \underline{h} \le h(\sigma_k^j) \le \underline{h} + \alpha(\overline{h} - \underline{h})\},$$

where $\alpha$ is a parameter such that $0 \le \alpha \le 1$.

A typical iteration of the GRASP construction is summarized as follows: a partial schedule (which is initially empty) is on hand, the next operation to be scheduled is selected from the RCL and is added to the partial schedule, resulting in a new partial schedule. The selected operation is inserted into the earliest available feasible time slot on machine $\mathcal{M}_{\sigma_k^j}$. Construction ends when the partial schedule is complete, i.e., all operations have been scheduled.

The algorithm uses two greedy functions. Even numbered iterations use a greedy function based on the makespan resulting from the inclusion of operation $\sigma_k^j$ to the already-scheduled operations, i.e., $h(\sigma_k^j) = \mathcal{C}_{max}$ for $\mathcal{O} = \{\mathcal{O}_s \cup \sigma_k^j\}$. On odd numbered iterations, solutions are constructed by favoring operations from jobs having long remaining processing times. The greedy function used is given by $h(\sigma_k^j) = -\sum_{\sigma_l^j \notin \mathcal{O}_s} p_l^j$, which measures the remaining processing time for job $j$. The use of two different greedy functions produce a greater diversity of initial solutions to be used by the local search.

### 10.3.2.3 Local search

As an attempt to decrease the makespan of the solution produced in the construction phase, we employ a 2-exchange local search procedure based on a disjunctive graph model.

### 10.3.2.4 Path-relinking

Path-relinking for job shop scheduling is similar to path-relinking for three-index assignment. Where in the case of three-index assignment each solution is represented by two permutation arrays, in the job shop scheduling problem, each solution is made up of $|\mathcal{M}|$ permutation arrays of numbers $1, 2, \ldots, |\mathcal{J}|$.

### 10.3.2.5 Parallel independent-thread GRASP with path-relinking for JSP

We study the efficiency of the multiple-walk independent-thread GRASP with path-relinking on JSP instances abz6, mt10, orb5, and la21 of ORLib using 943, 938, 895, and 1100 as target solution values, respectively. Table 10.4 shows the estimated shifted exponential distribution parameters for the multiple-walk independent-thread GRASP with path-relinking strategy obtained from 200 independent runs of a sequential variant of the algorithm. In addition to the sequential variant, 60 independent runs of 2-, 4-, 8-, and 16-thread variants were run on the four test problems. As before, the average speedups were computed dividing the sum of the execution times of the independent parallel program executing on one processor by the sum of the execution times of the parallel program on 2, 4, 8, and 16 processors, over 60 runs. The average speedups can be seen in Table 10.5 and Figure 10.8.

Compared to the efficiencies observed on the AP3 instances, those for these instances of the JSP were much worse. While with 16 processors average speedups of 12.2 were observed for AP3, average speedups of only 5.9 occurred for JSP. This is

**Table 10.4** Estimated shifted exponential distribution parameters $\mu$ and $\lambda$ obtained with 200 independent runs of a sequential GRASP with path-relinking on JSP instances `abz6`, `mt10`, `orb5`, and `la21`, with target values 943, 938, 895, and 1100, respectively.

| Problem | Estimated parameter | | |
|---|---|---|---|
| | $\mu$ | $\lambda$ | $|\mu|/\lambda$ |
| abz6 | 47.67 | 756.56 | 0.06 |
| mt10 | 305.27 | 524.23 | 0.58 |
| orb5 | 130.12 | 395.41 | 0.32 |
| la21 | 175.20 | 407.73 | 0.42 |
| | | average | 0.34 |

**Table 10.5** Speedups for multiple-walk independent-thread implementations of GRASP with path-relinking on instances `abz6`, `mt10`, `orb5`, and `la21`, with target values 943, 938, 895, and 1100, respectively. Speedups are computed with the average of 60 runs.

| | Number of processors | | | | | | | |
|---|---|---|---|---|---|---|---|---|
| | 2 | | 4 | | 8 | | 16 | |
| Problem | speedup | efficiency | speedup | efficiency | speedup | efficiency | speedup | efficiency |
| bz6 | 2.00 | 1.00 | 3.36 | 0.84 | 6.44 | 0.81 | 10.51 | 0.66 |
| mt10 | 1.57 | 0.79 | 2.12 | 0.53 | 3.03 | 0.39 | 4.05 | 0.25 |
| orb5 | 1.95 | 0.98 | 2.97 | 0.74 | 3.99 | 0.50 | 5.36 | 0.34 |
| la21 | 1.64 | 0.82 | 2.25 | 0.56 | 3.14 | 0.39 | 3.72 | 0.23 |
| average | 1.79 | 0.90 | 2.67 | 0.67 | 4.15 | 0.52 | 5.91 | 0.37 |

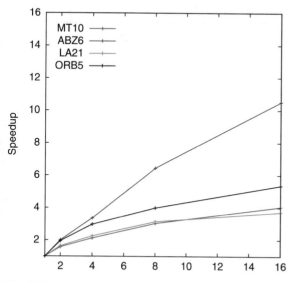

**Fig. 10.8** Average speedups on 2, 4, 8, and 16 processors for multiple-walk independent-thread parallel GRASP with path-relinking on JSP instances `abz6`, `mt10`, `orb5`, and `la21`.

consistent with the test proposed earlier in this chapter (page 209), since the average $|\mu|/\lambda$ values for AP3 and JSP are equal to 0.02 and 0.34, respectively.

### 10.3.2.6 Parallel cooperative-thread GRASP with path-relinking for JSP

We now study the multiple-walk cooperative-thread strategy for GRASP with path-relinking on the JSP. As with the independent-thread GRASP with path-relinking strategy, the target solution values 943, 938, 895, and 1100 were used for instances abz6, mt10, orb5, and la21, respectively. Table 10.6 and Figure 10.9 show super-linear speedups on instances abz6 and mt10, linear speedup on orb5 and about 70% efficiency for la21. As before, super-linear speedups are possible because good elite solutions are shared among the threads and these elite solutions are combined with GRASP solutions whereas they would not be combined in an independent-thread implementation.

**Table 10.6** Speedups for multiple-walk cooperative-thread implementations of GRASP with path-relinking on instances abz6, mt10, orb5, and la21, with target values 943, 938, 895, and 1100, respectively. Average speedups were computed over 60 runs.

| | Number of processors | | | | | | | |
|---|---|---|---|---|---|---|---|---|
| | 2 | | 4 | | 8 | | 16 | |
| Problem | speedup | efficiency | speedup | efficiency | speedup | efficiency | speedup | efficiency |
| abz6 | 2.40 | 1.20 | 4.21 | 1.05 | 11.43 | 1.43 | 23.58 | 1.47 |
| mt10 | 1.75 | 0.88 | 4.58 | 1.15 | 8.36 | 1.05 | 16.97 | 1.06 |
| orb5 | 2.10 | 1.05 | 4.91 | 1.23 | 8.89 | 1.11 | 15.76 | 0.99 |
| la21 | 2.23 | 1.12 | 4.47 | 1.12 | 7.54 | 0.94 | 11.41 | 0.71 |
| average | 2.12 | 1.06 | 4.54 | 1.14 | 9.05 | 1.13 | 16.93 | 1.06 |

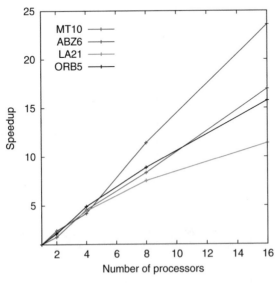

**Fig. 10.9** Average speedups on 2, 4, 8, and 16 processors for multiple-walk cooperative-thread parallel GRASP with path-relinking on JSP instances abz6, mt10, orb5, and la21.

Figure 10.10 compares the average speedup of the two implementations tested in this section, namely the multiple-walk independent-thread and the multiple-walk cooperative-thread GRASP with path-relinking implementations using target solution values 943, 938, 895, and 1100, on instances abz6, mt10, orb5, and la21, respectively. The figure shows that the cooperative variant of GRASP with path-relinking achieves the best parallelization.

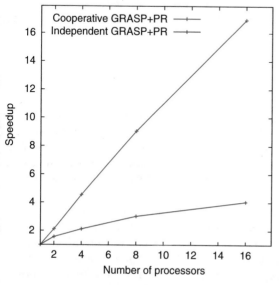

**Fig. 10.10** Average speedups on 2, 4, 8, and 16 processors for the parallel algorithms tested on instances of JSP: multiple-walk independent-thread GRASP with path-relinking and multiple-walk cooperative-thread GRASP with path-relinking.

## 10.3.3 2-path network design problem

### 10.3.3.1 Problem formulation

Let $G = (V, U)$ be a connected graph, where $V$ is the set of nodes and $U$ is the set of edges. A $k$-path between nodes $s, t \in V$ is a sequence of at most $k$ edges connecting $s$ and $t$. Given a non-negative weight function $w : U \to R_+$ associated with the edges of $G$ and a set $D$ of pairs of origin-destination nodes, the *2-path network design problem* (2PNDP) consists in finding a minimum weighted subset of edges $U' \subseteq U$ containing a 2-path between every origin-destination pair.

Applications of 2PNDP can be found in the design of communications networks, in which paths with few edges are sought to enforce high reliability and small delays.

**10.3.3.2 GRASP construction**

The construction of a new solution begins by the initialization of modified edge weights with the original edge weights. Each iteration of the construction phase starts by the random selection of an origin-destination pair still in $D$. A shortest 2-path between the extremities of this pair is computed, using the modified edge weights. The weights of the edges in this 2-path are set to zero until the end of the construction procedure, the origin-destination pair is removed from $D$, and a new iteration resumes. The construction phase stops when 2-paths have been computed for all origin-destination pairs.

**10.3.3.3 Local search**

The local search phase seeks to improve each solution built in the construction phase. Each solution may be viewed as a set of 2-paths, one for each origin-destination pair in $D$. To introduce diversity to drive different applications of the local search to different local optima, the origin-destination pairs are investigated at each GRASP iteration in a circular order, defined by a different random permutation of their original indices.

Each 2-path in the current solution is tentatively eliminated. The weights of the edges used by other 2-paths are temporarily set to zero, while those which are not used by other 2-paths in the current solution are restored to their original values. A new shortest 2-path between the extremities of the origin-destination pair under investigation is computed, using the modified weights. If the new 2-path improves the current solution, then the current solution is updated with the new 2-path; otherwise the previous 2-path is restored. The search stops if the current solution is not improved after a sequence of $|D|$ iterations along which all 2-paths are investigated. Otherwise, the next 2-path in the current solution is investigated for substitution and a new iteration resumes.

**10.3.3.4 Path-relinking**

A solution to 2PNDP is represented as a set of 2-paths connecting each origin-destination pair. Path-relinking starts by determining all origin-destination pairs whose associated 2-paths are different in the starting and guiding solutions. These computations amount to determining a set of moves which should be applied to the initial solution to reach the guiding solution. Each move is characterized by a pair of 2-paths, one to be inserted and the other to be eliminated from the current solution.

### 10.3.3.5 Parallel implementations of GRASP with path-relinking for 2PNDP

As for AP3 and JSP, in the case of an independent-thread parallel implementation of GRASP with path-relinking for 2PNDP, each processor has a copy of the sequential algorithm, a copy of the data, and its own pool of elite solutions. One processor acts as the master, reading and distributing the problem data, generating the seeds used by the pseudo-random number generators at each processor, distributing the iterations, and collecting the best solution found by each processor. All the $p$ available processors perform GRASP iterations.

On the other hand, in the case of a cooperative-thread parallel implementation of GRASP with path-relinking for 2PNDP, the master handles a centralized pool of elite solutions, collecting and distributing elite solutions upon request (recall that in the case of AP3 and JSP each processor had its own pool of elite solutions). The $p - 1$ workers exchange elite solutions found along their search trajectories. In this implementation for 2PNDP, each worker can send up to three different solutions to the master at each iteration: the solution obtained by local search, and solutions $w^1$ and $w^2$ obtained by forward and backward path-relinking between the same pair of starting and guiding solutions, respectively.

### 10.3.3.6 Computational results

The results illustrated in this section are for an instance with 100 nodes, 4950 edges, and 1000 origin-destination pairs. We use the methodology based on time-to-target plots showing empirical runtime distributions of the random variable *time to target solution value*. To plot the empirical distribution, we fix a solution target value and run each algorithm 200 times, recording the running time when a solution with cost at least as good as the target value is found. For each algorithm, we associate with the $i$-th sorted running time $t_i$ a probability $p_i = (i - \frac{1}{2})/200$ and plot the points $z_i = (t_i, p_i)$, for $i = 1, \ldots, 200$.

Results obtained for both the independent-thread and the cooperative-thread parallel implementations of GRASP with path-relinking on the above instance with the target value set at 683 are reported in Figure 10.11. The cooperative implementation is already faster than the independent implementation for eight processors. For fewer processors the independent implementation is naturally faster, since it employs all $p$ processors in the search (while only $p - 1$ worker processors effectively take part in the computations performed by the cooperative implementation).

Three strategies further improve the performance of the cooperative-thread implementation, by reducing the cost of the communication between the master and the workers when the number of processors increases:

- Strategy 1: Each send operation is broken into two parts. First, the worker only sends only the cost of the solution to the master. If this solution is better than the worst solution in the pool, then the full solution is sent. The number of messages increases, but most of them will be very small, with light memory requirements.

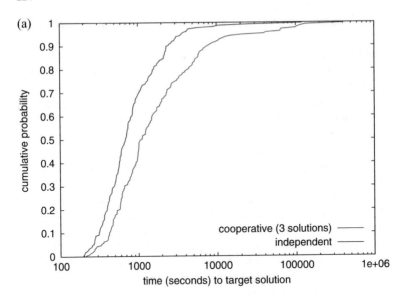

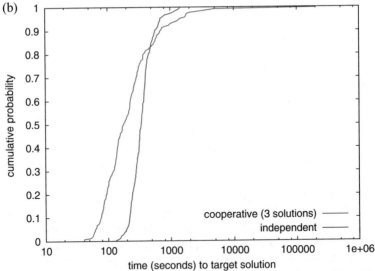

**Fig. 10.11** Running times for 200 runs of the multiple-walk independent-thread and the multiple-walk cooperative-thread implementations of GRASP with path-relinking using (a) two processors and (b) eight processors, with the target solution value set at 683.

- Strategy 2: Only one solution is sent to the pool at each GRASP iteration.
- Strategy 3: A distributed implementation, in which each worker handles its own pool of elite solutions. Every time a processor finds a new elite solution, the newly found elite solution is broadcast to the other processors.

Comparative results for these three strategies on the same problem instance are plotted in Figure 10.12. The first strategy outperforms the others.

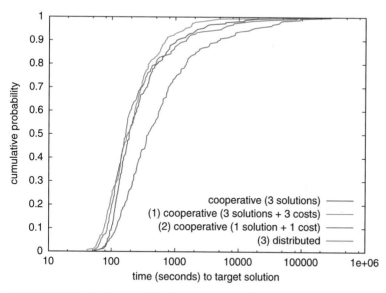

**Fig. 10.12** Strategies for improving the performance of the centralized multiple-walk cooperative-thread implementation on eight processors.

Table 10.7 lists the average computation times and the best solutions found over ten runs of each strategy when the total number of GRASP iterations is set at 3200. There is a clear degradation in solution quality for the independent-thread strategy when the number of processors increases, despite the fact that speedups are high, of the same order as the number of processors used in the computations. As fewer iterations are performed by each processor, the pool of elite solutions gets poorer with the increase in the number of processors. Since the processors do not communicate, the overall solution quality is worse. In the case of the cooperative strategy, the information shared by the processors guarantees the high quality of the solutions in the pool. The cooperative implementation is more robust: solution quality does not deteriorate and very good solutions are obtained as the number of processors increases. Smaller speedups than those obtained with the independent-thread strategy are observed. However, the efficiency remains close to one for up to 16 processors.

**Table 10.7**  Average times and best solutions over ten runs of 2PNDP.

| | Independent | | | Cooperative | | | |
|---|---|---|---|---|---|---|---|
| Processors | best value | avg. time (s) | speedup | best value | avg. time (s) | speedup | efficiency |
| 1 | 673 | 1310.1 | — | — | — | — | — |
| 2 | 676 | 686.8 | 1.91 | 676 | 1380.9 | 0.95 | 0.48 |
| 4 | 680 | 332.7 | 3.94 | 673 | 464.1 | 2.82 | 0.71 |
| 8 | 687 | 164.1 | 7.98 | 676 | 200.9 | 6.52 | 0.82 |
| 16 | 692 | 81.7 | 16.04 | 674 | 97.5 | 13.44 | 0.84 |
| 32 | 702 | 41.3 | 31.72 | 678 | 74.6 | 17.56 | 0.55 |

## 10.4  Bibliographical notes

Metaheuristics, such as GRASP, have found their way into the standard toolkit of combinatorial optimization methods. Parallel computers have increasingly found their way into metaheuristics.Verhoeven and Aarts (1995), Cung et al. (2002), Duni Ekşoğlu et al. (2002), Alba (2005), and Talbi (2009) presented good accounts of parallel implementations of metaheuristics.

Most multiple-walk independent-thread parallel implementations of GRASP (with or without path-relinking) described in Section 10.1 are based on partitioning the search space or the iterations among a number of processors and appeared in Alvim and Ribeiro (1998), Canuto et al. (2001), Feo et al. (1994), Drummond et al. (2002), Li et al. (1994), Martins et al. (1998), Martins et al. (1999), Martins et al. (2000), Martins et al. (2004), Murphey et al. (1998), Pardalos et al. (1995), Pardalos et al. (1996), Resende et al. (1998), and Ribeiro and Rosseti (2002), among other references. Linear speedups can be expected in parallel multiple-walk independent-thread implementations. This was illustrated with applications to the maximum satisfiability problem and to the Steiner problem in graphs. Pardalos et al. (1996) implemented a parallel GRASP for the MAX-SAT problem using PVM (Geist et al., 1994). Martins et al. (1998) implemented a parallel GRASP for the Steiner problem in graphs using MPI (Snir et al., 1998) on test problems taken from the OR-Library (Beasley, 1990a).

In the case of the multiple-walk independent-thread implementation described by Aiex et al. (2005) for the 3-index assignment problem and by Aiex et al. (2003) for the job shop scheduling problem, each processor applies path-relinking to pairs of elite solutions stored in a local pool. The test for predicting whether a parallel implementation using multiple independent processors will be efficient was proposed in Aiex and Resende (2005).

Path-relinking has been increasingly used to introduce memory in the otherwise memoryless original GRASP procedure and was also used in conjunction with parallel implementations of GRASP. The hybridization of GRASP and path-relinking led to some effective multiple-walk cooperative-thread implementations. Collaboration between the threads is usually achieved by sharing elite solutions, either in a single centralized pool or in distributed pools. In some of these implementations, super-linear speedups were achieved even for cases where small speedups occurred in multiple-walk independent-thread variants.

Section 10.2 dealt with multiple-walk cooperative-thread implementations of GRASP with path-relinking using distributed strategies that appeared in Aiex et al. (2003) and Aiex and Resende (2005), in which each thread maintains its own pool of elite solutions. Centralized strategies appeared in Martins et al. (2004) and Ribeiro and Rosseti (2002), in which only a single pool of elite solutions was used.

The three-index assignment problem (AP3) (Pierskalla, 1967) is a straightforward $NP$-hard (Frieze, 1983; Garey and Johnson, 1979) extension of the classical two-dimensional assignment problem. The parallel implementations and the computational experiments reported in Section 10.3.1 appeared originally in Aiex et al. (2005). Exact and heuristic algorithms exist for this problem in the literature (Balas and Saltzman, 1991; Burkard and Fröhlich, 1980; Burkard and Rudolf, 1993; Burkard et al., 1996; Crama and Spieksma, 1992; Hansen and Kaufman, 1973; Leue, 1972; Pardalos and Pitsoulis, 2000; Pierskalla, 1967; 1968; Vlach, 1967; Voss, 2000). Test instances were described in Balas and Saltzman (1991), Crama and Spieksma (1992), and Burkard et al. (1996).

The job shop scheduling problem (JSP) considered in Section 10.3.2 was proved to be $NP$-hard by Lenstra and Rinnooy Kan (1979). The GRASP construction phase is the one proposed in Binato et al. (2002) and Aiex et al. (2003). The 2-exchange local search is used in Aiex et al. (2003), Binato et al. (2002), and Taillard (1991), and is based on the disjunctive graph model of Roy and Sussmann (1964). We also refer to Aiex et al. (2003) and Binato et al. (2002) for a description of the implementation of the local search procedure. Test instances are available from the OR-Library (Beasley, 1990a).

Applications of the 2-path network design problem (2PNDP) introduced in Section 10.3.3 can be found in the design of communications networks, in which paths with few edges are sought to enforce high reliability and small delays. The problem was shown to be $NP$-hard by Dahl and Johannessen (2004). Ribeiro and Rosseti (2002; 2007) developed parallel GRASP heuristics for 2PNDP.

# Chapter 11
# GRASP for continuous optimization

Continuous GRASP, or C-GRASP, extends GRASP to the domain of continuous box-constrained global optimization. The algorithm searches the solution space over a dynamic grid. Each iteration of C-GRASP consists of two phases. In the construction (or diversification) phase, a greedy randomized solution is constructed. In the local search (or intensification) phase, a local search algorithm starts from the first phase solution and produces an approximate locally optimal solution. A deterministic rule triggers a restart after each C-GRASP iteration. This chapter addresses the construction phase and the restart strategy, and presents a local search procedure for continuous GRASP.

## 11.1 Box-constrained global optimization

Continuous global optimization seeks a minimum or maximum of a multimodal function over a continuous domain. In its minimization form, *global optimization* can be stated as finding a global minimum $S^* \in F \subseteq \mathbb{R}^n$ such that $f(S^*) \leq f(S)$, $\forall S \in F$, where $F$ is some region of $\mathbb{R}^n$ and the multimodal objective function is defined by $f : F \to \mathbb{R}$. In this chapter, we limit ourselves to box constraints: the domain is a hyper-rectangle $F = \{S = (S_1, \ldots, S_n) \in \mathbb{R}^n : \ell_i \leq S_i \leq u_i\}$, where $\ell_i, u_i \in \mathbb{R}$ such that $\ell_i \leq u_i$, for $i = 1, \ldots, n$. Therefore, the minimization problem considered here consists in finding $S^* = \operatorname{argmin}\{f(S) : \ell \leq S \leq u\}$, where $f : \mathbb{R}^n \to \mathbb{R}$, and $\ell, S, u \in \mathbb{R}^n$.

© Springer Science+Business Media New York 2016
M.G.C. Resende, C.C. Ribeiro, *Optimization by GRASP*,
DOI 10.1007/978-1-4939-6530-4_11

Five examples of classical box-constrained continuous global optimization problems are

- *Ackley function:*

$$\min A_n(x) = -20e^{-0.2\sqrt{\frac{1}{n}\sum_{i=1}^{n} x_i^2}} - e^{\frac{1}{n}\sum_{i=1}^{n}\cos(2\pi x_i)} + 20 + e,$$

where $(x_1,\ldots,S_n) \in [-15,30]^n$.

- *Bohachevsky function:*

$$\min B_2(x) = x_1^2 + 2x_2^2 - 0.3\cos(3\pi x_1) - 0.4\cos(4\pi x_2) + 0.7,$$

where $(x_1,x_2) \in [-50,100]^2$.

- *Schwefel function:*

$$\min SC_n(x) = 418.9829n - \sum_{i=1}^{n} x_i \sin(\sqrt{|x_i|}),$$

where $(x_1,\ldots,x_n) \in [-500,500]^n$.

- *Shekel function:*

$$\min S_{4,m}(x) = -\sum_{i=1}^{m}[(x-a_i)^T(x-a_i)+c_i]^{-1},$$

where $(x_1,x_2,x_3,x_4) \in [0,10]^4$,

$$a = \begin{bmatrix} 4.0 & 4.0 & 4.0 & 4.0 \\ 1.0 & 1.0 & 1.0 & 1.0 \\ 8.0 & 8.0 & 8.0 & 8.0 \\ 6.0 & 6.0 & 6.0 & 6.0 \\ 7.0 & 3.0 & 7.0 & 3.0 \\ 2.0 & 9.0 & 2.0 & 9.0 \\ 5.0 & 5.0 & 3.0 & 3.0 \\ 8.0 & 1.0 & 8.0 & 1.0 \\ 6.0 & 2.0 & 6.0 & 2.0 \\ 7.0 & 2.6 & 7.0 & 3.6 \end{bmatrix},$$

and $c = (0.1,0.2,0.2,0.4,0.4,0.6,0.3,0.7,0.5,0.5)$.

- *Shubert function:*

$$\min SH(x) = \left[\sum_{i=1}^{5} i\cos[(i+1)x_1+i]\right]\left[\sum_{i=1}^{5} i\cos[(i+1)x_2+i]\right],$$

where $(x_1,x_2) \in [-10,10]^2$.

## 11.2  C-GRASP for continuous box-constrained global optimization

*Continuous GRASP*, or C-GRASP, extends GRASP to the domain of continuous box-constrained global optimization. The algorithm searches the solution space over a dynamic grid with hypercubed cells of side $h$ each, fully contained in the domain. The initial grid is formed by hypercubes of sides of size $h_s$. As each approximate local minimum is found during local search, the grid density is increased by halving the side of the current hypercube. When the size of the grid side becomes very small, i.e., when $h < h_e$, for some given minimum grid size $h_e$, a restart of the search is triggered.

Figure 11.1 shows a hyper-rectangle approximated by three grids: a sparse grid on top (red grid); a medium-density grid in the middle (green grid); and a dense grid in the bottom (blue grid). In all hyper-rectangles, an optimal solution is represented as a point in the upper right-hand corner of the feasible domain. As the grid density increases, more grid points are placed in the hyper-rectangle and the upper right point on the grid becomes an increasingly better approximation of the solution.

The initial solution $S$ of the algorithm, as well as the sequence of initial solutions right after each restart, are randomly generated points in the interior of the hyper-rectangle, i.e., $\ell_i < S_i < u_i$ for $i = 1, \ldots, n$. Each iteration of C-GRASP consists of two phases. In the construction (or diversification) phase, a greedy randomized solution is constructed, while in the local search (or intensification) phase, a local search algorithm is applied, starting from the first phase solution and producing an approximate locally optimal solution. A deterministic rule can trigger a restart after each C-GRASP iteration.

The pseudo-code in Figure 11.2 shows the template of a continuous GRASP heuristic for the minimization of $f(S)$, with $\ell \leq S \leq u$, where $\ell, S, u \in \mathbb{R}^n$. The value $f^*$ of the best solution found is initialized in line 1. The loop from line 2 to 15 is repeated until some predefined stopping criterion is satisfied. In line 3, the current iterate $S$ is initialized (or reinitialized) with a point from the interior of the hyper-rectangle defined by the $n$-vectors $\ell$ and $u$, drawn randomly by procedure RANDOM-IN-BOX($\ell, u$). The grid size $h$ is initialized in line 4. The loop from line 5 to 14 is repeated until an approximate global minimum is found, i.e., while the grid size is not too small. The current solution is saved in line 6. In lines 7 and 8, the construction and local search phases of C-GRASP are applied, always starting from the current solution $S$ and using the grid size value $h$. If the current iterate $S$ obtained by local search is better than the best solution found so far, then the best found solution and its objective function value are updated in lines 10 and 11, respectively. Otherwise, if no improvement was found by either the construction or the local search phases, then the grid size is halved in line 13. If a stopping criterion is satisfied in line 2, then the best solution found $S^*$ (together with its objective function value $f(S^*)$) is returned as an approximate globally optimal solution in line 16.

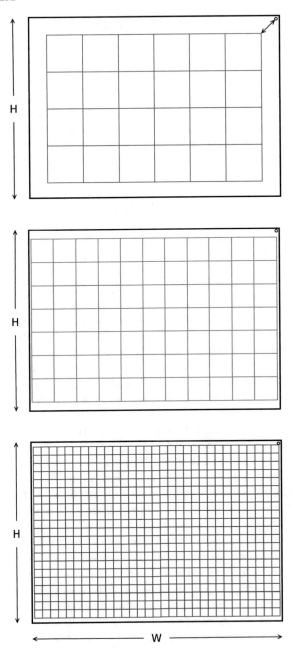

**Fig. 11.1** This figure illustrates the effect of changing the resolution of the dynamic grid. From top (red) to bottom (blue), both the grid density and the search resolution increase. As the grid density increases, the closest distance from a point on the grid to the solution represented by a point in the upper-right corner of the domain decreases and a point on the grid better approximates this solution.

```
begin C-GRASP(h_s, h_e);
1   f* ← ∞;
2   while stopping criterion not satisfied do
3       S ← RANDOM-IN-BOX(ℓ, u);
4       h ← h_s;
5       while h > h_e do
6           Ŝ ← S;
7           S ← CONTINUOUS-RANDOMIZED-GREEDY(S, h, ℓ, u);
8           S ← CONTINUOUS-LOCAL-SEARCH(S, h, ℓ, u);
9           if f(S) < f* then
10              S* ← S;
11              f* ← f(S);
12          end-if;
13          if f(S) = f(Ŝ) then h ← h/2;
14      end-while;
15  end-while;
16  return S*, f(S*);
end C-GRASP.
```

**Fig. 11.2** Pseudo-code of the basic C-GRASP heuristic for box-constrained continuous global minimization.

## 11.3 C-GRASP construction phase

The *construction phase* of C-GRASP mimics the construction phase of GRASP. The main difference is that while the GRASP construction starts from scratch, the construction in C-GRASP starts from a given initial solution $S$.

At each of its iterations, the construction modifies one of the $n$ components of $S$. It does so by performing a discrete line search in each yet unmodified canonical basis direction to build a restricted candidate list (RCL) of canonical components. A *discrete line search* is an approximate search that evaluates the objective function for a discrete set of points, all laying on a line, defined by a given canonical basis direction, that passes through the current iterate. The line search returns the point with the best objective function value. The restriction for the RCL is by value and is based on a parameter $\alpha$. A component is selected at random from the RCL, its value is set to the value found in the discrete line search, and its index is removed from further consideration. This is repeated until all components are examined and possibly modified.

The pseudo-code in Figure 11.3 summarizes the steps of the construction procedure of C-GRASP. In line 1, the set of yet unconsidered indices of search directions (which corresponds to the set of all still unfixed components) is initialized to correspond to all directions. In line 2, the RCL parameter $\alpha$ is assigned to a random real value in the interval $[0, 1]$. Each iteration of the loop from lines 3 to 21 potentially modifies the value of one component of $S$. In lines 4 and 5, the best and worst values that will be obtained by line search over all possible directions are initialized. The for loop in lines 6 to 12 evaluates the objective function, for all still unfixed

```
begin CONTINUOUS-RANDOMIZED-GREEDY(S, h, ℓ, u);
1      UnFixed ← {1, ..., n};
2      Set the RCL parameter α to a random real value in the interval [0, 1];
3      while UnFixed ≠ ∅ do
4          g ← +∞;
5          ḡ ← −∞;
6          forall i ∈ UnFixed do
7              S*ᵢ ← DISCRETE-LINE-SEARCH(S, h, i, ℓ, u);
8              Let Šⁱ denote S with its i-th component set to S*ᵢ;
9              gᵢ ← f(Šⁱ);
10             if g > gᵢ then g ← gᵢ;
11             if ḡ < gᵢ then ḡ ← gᵢ;
12         end-forall;
13         RCL ← ∅;
14         threshold ← g + α · (ḡ − g);
15         forall i ∈ UnFixed do
16             if gᵢ ≤ threshold then RCL ← RCL ∪ {i};
17         end-forall;
18         Let j be a randomly selected element from the restricted candidate list RCL;
19         Sⱼ ← S*ⱼ;
20         UnFixed ← UnFixed \ {j};
21     end-while;
22     return S, f(S);
end CONTINUOUS-RANDOMIZED-GREEDY.
```

**Fig. 11.3** Pseudo-code of the C-GRASP construction phase.

components of the solution being constructed. Since the line search is always per-formed along one of the directions of the canonical basis, it can modify at most one component of the solution. Line 7 invokes DISCRETE-LINE-SEARCH and re-turns the potentially modified component $S_i^*$ that minimizes $f(S)$ along the canon-ical direction $e_i$. In line 8, the current iterate is tentatively modified with its $i$-th component taking on the value $S_i^*$. The tentative solution is evaluated in line 9 and the lower and upper bounds on the solutions obtained by line search are updated in lines 10 and 11, if necessary. These values, along with $\alpha$ and the objective function values produced by the line searches, are used in lines 13 to 17 to set up the RCL. In line 18, an index $j$ is selected at random from the RCL and the $j$-th component of $S$ is set, in line 19, to the value $S_j^*$ found in the line search corresponding to the $j$-th canonical basis direction. In line 20, index $j$ is removed from the set of unfixed components. Finally, in line 22 the constructed solution $S$ and its objective function value $f(S)$ are returned.

The randomized greedy procedure CONTINUOUS-RANDOMIZED-GREEDY of Figure 11.3 can be made more efficient by noting that, right after line 18, the values of $S_j$ and $S_j^*$ may be identical. In this case, there will be no change, or move-ment, of solution $S$ with the assignment made in line 19. Consequently, in the next iteration of the while loop from line 3 to line 21, the values returned by the line

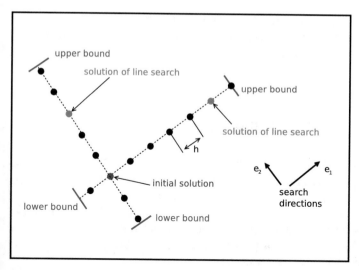

**Fig. 11.4** First iteration of a bi-dimensional randomized greedy construction: discrete line searches start from a solution represented by the blue point and are performed in both directions to populate the RCL with solutions represented by the green points. Each line search starts at the blue point (initial solution) and evaluates the blue point and the black points determined by the initial solution, each search direction, the grid size $h$, and the upper and lower bounds. Suppose the green point furthest to the right is selected at random from the RCL. It will be represented as the blue point in Figure 11.5 and will act as the new initial solution for the discrete line search of the second iteration of the bi-dimensional randomized greedy construction.

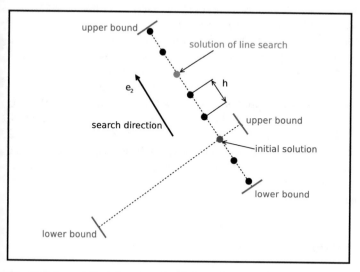

**Fig. 11.5** Second (last) iteration of a bi-dimensional randomized greedy construction: the blue point in this figure is the green point chosen at random from the RCL in Figure 11.4. It is the starting point for the discrete line search in the last direction of the randomized construction. As for the line search of Figure 11.4, the blue point and black points are evaluated. The green point is the best solution among the evaluated points and is the final solution produced by the randomized greedy construction procedure.

search procedure DISCRETE-LINE-SEARCH in line 7 will be identical to those produced in the current iteration. The pair $(S_i^*, g_i)$ computed in the current iteration can therefore be reused in the next iteration, where there will be no need to compute lines 7 to 9.

## 11.4 Approximate discrete line search

At each iteration of algorithm CONTINUOUS-RANDOMIZED-GREEDY in Figure 11.3, the approximate discrete line search DISCRETE-LINE-SEARCH procedure is applied several times along different directions (line 7 of the pseudo-code of Figure 11.3). This line search evaluates a discrete set of points on the line determined by the starting solution $S$ and one of the canonical search directions $e_i, i = 1, \ldots, n$. The set of visited points depends on the current iterate $S$, the grid size $h$, and the upper and lower bounds defined by the box constraints.

Figures 11.4 and 11.5 show two iterations of the randomized construction procedure, where the discrete line search is applied to a two-dimensional function. In the first iteration (shown in Figure 11.4), the line search is applied in two directions, each defined by the current iterate (or starting solution) and a canonical basis direction ($e_1 = (1,0)$ or $e_2 = (0,1)$). Note that the points in which the function is evaluated are determined by the initial solution, by the grid size $h$, and by the upper and lower bounds, as well as by the canonical basis directions. Once a solution is chosen in one of the line searches, this point becomes the new initial solution and another line search is performed (see Figure 11.5).

Figure 11.6 shows the pseudo-code of algorithm DISCRETE-LINE-SEARCH to perform a discrete line search for minimization. Lines 1 and 2 initialize, respectively, the best objective function value $f^*$ and the distance $\Delta$ to the next point to be visited. The loop in lines 3 to 9 perform the search from the initial solution $S$ along the canonical direction $e_i$ while the upper bound $u_i$ is not violated, using $h$ as the step size. If a new visited point improves the best solution along this direction in line 4, then $S^*$ and its cost $f^*$ are updated in lines 5 and 6, respectively. The step size is updated in line 8 and a new iteration resumes. Lines 10 to 17 perform the same search from the initial solution $S$ along the opposite direction, while the lower bound $\ell_i$ is not violated, once again using $h$ as the step size. Line 18 returns $S_i^*$, i.e., the $i$-th component of the best solution found $S^*$.

**Example of approximate discrete line search**

Suppose we want to minimize $f(x_1, x_2) = \sqrt{x_1 + x_2}$, with $0 \le x_1 \le 1$ and $0 \le x_2 \le 2$. We use C-GRASP and consider that at some iteration we wish to perform an approximate discrete line search starting from $(0.25, 0.25)$ along the canonical direction $e_1 = (1,0)$, with the grid parameter $h = 0.15$. The points to be evaluated along the line search are all defined by $(0.25, 0.25) + k \cdot h \cdot e_1 = (0.25, 0.25) + k \cdot h \cdot (1,0)$, for $k = -1, 0, 1, 2, 3, 4, 5$, i.e., $(0.10, 0.25)$, $(0.25, 0.25)$, $(0.40, 0.25)$, $(0.55, 0.25)$, $(0.70,$

```
begin DISCRETE-LINE-SEARCH(S, h, i, ℓ, u);
1      f* ← f(S);
2      Δ ← h;
3      while S + Δ · e_i ≤ u_i do
4          if f(S + Δ · e_i) < f* then
5              f* ← f(S + Δ · e_i);
6              S* ← S + Δ · e_i;
7          end-if;
8          Δ ← Δ + h;
9      end-while;
10     Δ ← h;
11     while S − Δ · e_i ≥ l_i do
12         if f(S − Δ · e_i) < f* then
13             f* ← f(S − Δ · e_i);
14             S* ← S − Δ · e_i;
15         end-if;
16         Δ ← Δ + h;
17     end-while;
18     return S_i*;
end DISCRETE-LINE-SEARCH.
```

**Fig. 11.6** Pseudo-code of the approximate discrete line search algorithm.

0.25), (0.85, 0.25), and (1.00, 0.25). Note that any other point along this direction will violate constraint $0 \le x_1 \le 1$. Of the seven trial points, $(x_1^*, x_2^*) = (0.10, 0.25)$ is the one minimizing $\sqrt{x_1 + x_2}$. Then, algorithm DISCRETE-LINE-SEARCH will return 0.10 as the best value for the first component.     ∎

## 11.5  C-GRASP local search

The local search procedure CONTINUOUS-LOCAL-SEARCH is called from line 8 of the pseudo-code of algorithm CONTINUOUS-GRASP in Figure 11.2 as an attempt to improve the constructed solution $S$ with a search on the largest grid of size $h$ that fits in the domain $F$ and for which one of its grid points coincides with the current iterate $S$.

The local search described in this section makes no use of derivatives. Though derivatives can be easily computed for many functions, there are some for which they cannot be computed or are computationally difficult to compute. The approach described in this section can be seen as approximating the role of the gradient of the objective function $f$.

From a given input point $S \in F$, the local improvement algorithm generates a neighborhood and determines at which points in the neighborhood, if any, the objective function improves. If an improving point is found, then it is made the current point and the local search continues from this new solution.

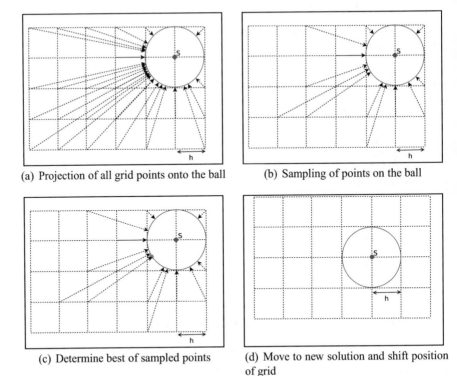

(a) Projection of all grid points onto the ball  (b) Sampling of points on the ball

(c) Determine best of sampled points   (d) Move to new solution and shift position
                                       of grid

**Fig. 11.7** One step of procedure CONTINUOUS-LOCAL-SEARCH: (a) All gridpoints in $F$ are projected onto the ball of radius $h$ centered at the current solution $S$; (b) A subset of the projected points on the ball is sampled; (c) Best sampled point is determined (in green); and (d) Solution $S$ is moved to best point, the grid is shifted to coincide with the new solution, and new ball of radius $h$ is centered at the new current solution.

Let $S$ be the current solution and $h$ be the current grid discretization parameter. Define

$$F_h(S) = \{S' \; : \; \ell \leq S' \leq u, \; S' = S + h \cdot \tau, \; \forall \, \tau \in \mathbb{Z}^n\}$$

to be a lattice of points in $F$ whose coordinates are integer steps (of size $h$) away from those of $S$ (see Figure 11.7(a)). Let

$$B_h(S) = \{S'' \; : \; S'' = S + h \cdot (S' - S)/\|S' - S\|, \; \forall \, S' \in F_h(S) \setminus \{S\}\}$$

be the projection of the points in $F_h(S) \setminus \{S\}$ onto the ball of radius $h$ centered at $S$ (see Figure 11.7(a)). The $h$-*neighborhood* of solution $S$ is defined as the set of points in $B_h(S)$. The size of this neighborhood is bounded from above by $\prod_{i=1}^n \lceil (u_i - \ell_i)/h + 1 \rceil$. If all of these points are examined and no improving solution is found, then the current solution $S^*$ is called an $h$-*local minimum*. Since the number of points in $B_h(S)$ can be huge, it may be only feasible to evaluate the objective function on a subset of them. If a subset of these points is examined and no improving point is found, then the current solution $S^*$ is considered an *approximate $h$-local minimum*.

```
begin CONTINUOUS-LOCAL-SEARCH(S, h, k^max, ℓ, u);
1       S* ← S;
2       f* ← f(S);
3       k ← 0;
4       while k < k^max do
5              S ← RandomlySelectElement (B_h(S*));
6              k ← k + 1;
7              if ℓ ≤ S ≤ u and f(S) < f* then
8                     S* ← S;
9                     f* ← f(S);
10                    k ← 0;
11             end-if;
12      end-while;
13      return S*;
end CONTINUOUS-LOCAL-SEARCH.
```

**Fig. 11.8** Pseudo-code of the C-GRASP local search phase.

The pseudo-code of the algorithm that performs the local search phase is shown in Figure 11.8. It takes as input the current solution $S \in F$, the grid size $h$, and the maximum number of points to be sampled in each neighborhood $B_h(S)$. The current best solution $S^*$ is initialized to $S$ in line 1. The cost of the best known solution is set in line 2. The number of points sampled in the neighborhood is initialized in line 3. Starting from $S^*$, the loop in lines 4 to 12 investigates at most $k^{max}$ neighbors of the current solution. A new neighbor $S \in B_h(S^*)$ is selected in line 5 and the number of solutions sampled in this neighborhood is incremented by one in line 6. If line 7 detects that the newly selected neighbor $S$ is feasible and better than $S^*$, then a move is performed: solution $S^*$ is set to $S$ in line 8, the cost of the best solution is updated in line 9, and the process restarts with $S^*$ as the new best solution after the counter $k$ is reset to 0 in line 10. Local improvement terminates if an approximate $h$-local minimum solution $S^*$ is found. At that point, $S^*$ is returned in line 13 as the solution produced by local search.

## 11.6 Computing global optima with C-GRASP

We conclude this chapter by showing the results of running an implementation of C-GRASP on the five functions listed in Section 11.1 of this chapter: Ackley (for $n = 10$), Bohachevsky, Schwefel (for $n = 10$), Shekel, and Shubert. These functions have global optimal objective function values of, respectively, 0, 0, 0, $-10.5364$, and $-186.7309$. Since all global optima are known, the inner loop of the algorithm is made to stop when either the grid size $h \le h_e$ (as in the case when the global optimum is unknown) or when the gap

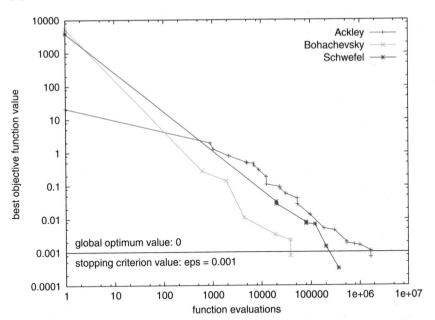

**Fig. 11.9** Best objective function value as a function of the number of function evaluations for C-GRASP runs on three functions: Ackley, Bohachevsky, and Schwefel. All three functions have global optima of value zero.

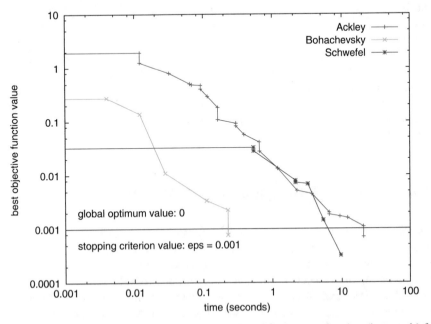

**Fig. 11.10** Best objective function value as a function of the computation time (in seconds) for C-GRASP runs on three functions: Ackley, Bohachevsky, and Schwefel. All three functions have global optima of value zero.

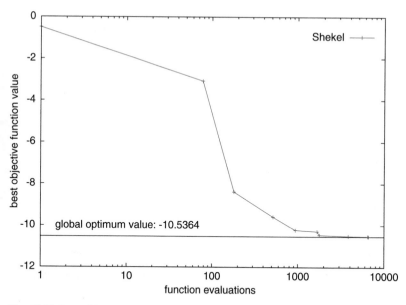

**Fig. 11.11** Best objective function value $f(S)$ as a function of the number of function evaluations for one C-GRASP run on function Shekel whose global optimum objective function value is $f(S^*) = -10.5364$.

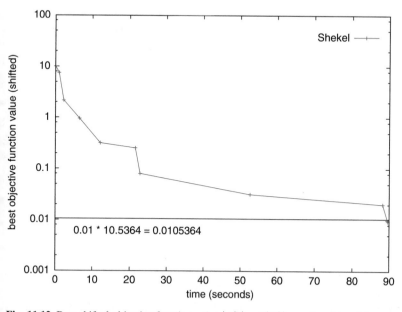

**Fig. 11.12** Best shifted objective function value $|f(S) - f(S^*)|$ as a function of the computation time (in seconds) for one C-GRASP run on function Shekel whose global optimum objective function value is $f(S^*) = -10.5364$. Optimization ends when $|f(S) - f(S^*)| < \varepsilon \cdot f(S^*) = 0.0105364$.

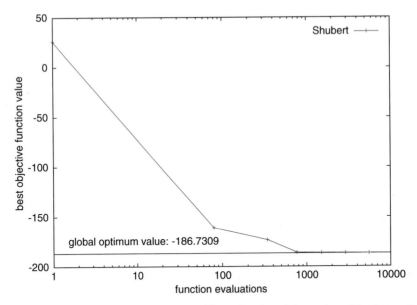

**Fig. 11.13** Best objective function value $f(S)$ as a function of the number of function evaluations for one C-GRASP run on function Shubert whose global optimum objective function value is $f(S^*) = -186.7309$.

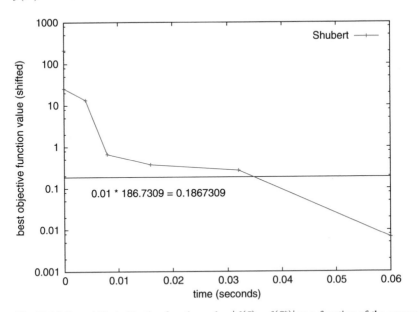

**Fig. 11.14** Best shifted objective function value $|f(S) - f(S^*)|$ as a function of the computation time (in seconds) for one C-GRASP run on function Shubert whose global optimum objective function value is $f(S^*) = -186.7309$. Optimization ends when $|f(S) - f(S^*)| < \varepsilon \cdot f(S^*) = 0.1867309$.

$$|f(S) - f(S^*)| \leq \begin{cases} \varepsilon, & \text{if } f(S^*) = 0, \\ \varepsilon \cdot |f(S^*)|, & \text{if } f(S^*) \neq 0, \end{cases} \tag{11.1}$$

where $S$ is the current best solution found by the heuristic, $S^*$ is the known global minimum solution, and $\varepsilon = 0.001$.

Each run used the same parameter values: initial grid size $h_s = 0.5$, final grid size $h_e = 0.0001$, and local search maximum sampling parameter $k^{max} = 100$.

For all runs, the algorithm stopped because of stopping rule (11.1), before the size of the grid $h$ became less than 0.0001. Therefore, only a single outer iteration was carried out. Figures 11.9 to 11.14 show the convergence of C-GRASP on the five test functions. Convergence is shown both as a function of the number of function evaluations and of the computation time (in seconds).

Figures 11.9 and 11.10 show, respectively, the convergence of the algorithm with respect to the number of function evaluations and the computation time for the three functions whose global minima are zero, i.e., for Ackley, Bohachevsky, and Schwefel. The true optimum for Ackley is $S^* = (0, \ldots, 0)$ with $f(S^*) = 0$, while C-GRASP stopped with

$$S = (0.000189, 0.000277, 0.000212, 0.000083, 0.000120,$$
$$0.000160, -0.000051, 0.000150, 0.000187, -0.000217)$$

and $f(S) = 0.000708$. The true optimum for Bohachevsky is $S^* = (0,0)$ with $f(S^*) = 0$, while C-GRASP stopped with $S = (-0.004350, -0.003859)$ and $f(S) = 0.000771$. The true optimum for Schwefel is $S^* = (420.9687, \ldots, 420.9687)$ with $f(S^*) = 0$, while C-GRASP stopped with

$$S = (420.970126, 420.962594, 420.981758, 420.974012, 420.945996.$$
$$420.963734, 420.957748, 420.952840, 420.986939, 420.975983)$$

and $f(S) = 0.000321$. Of those three functions, Ackley was the most difficult to optimize, requiring 1,681,424 function evaluations and 20.9 seconds of running time, while Bohachevsky required only 38,415 evaluations and 0.22 seconds. The final grid sizes for Ackley, Bohachevsky, and Schwefel were, respectively, 0.000977, 0.015625, and 0.062500.

Figures 11.11 and 11.12 illustrate, respectively, the convergence of the algorithm with respect to the number of function evaluations and the computation time for the function Shekel with $m = 10$. The true optimum for Shekel is $S^* = (4,4,4,4)$ with $f(S^*) = -10.5365$, while C-GRASP stopped with

$$S = (4.002639, 3.998661, 3.995356, 3.999527)$$

and $f(S) = -10.529192$. C-GRASP required 6565 function evaluations and 89.6 seconds to find this solution. The final grid size was 0.015625.

Finally, Figures 11.13 and 11.14 show, respectively, the convergence of the algorithm with respect to the number of function evaluations and computation time for the function Shubert. The function Shubert has many global minima, all with $f(S^*) = -186.7309$. C-GRASP stopped with $S = (4.859558, 5.483684)$ and $f(S) = -186.724170$. C-GRASP required 5550 function evaluations and 0.06 seconds to find this solution. The final grid size was 0.015625.

In each of the five runs of C-GRASP, procedure RANDOM-IN-BOX$(\ell, u)$ was never called more than once, since the grid size $h$ was never smaller or equal than $h_e = 0.0001$.

## 11.7 Bibliographical notes

C-GRASP was first introduced by Hirsch et al. (2007b) and in the Ph.D. thesis of Hirsch (2006). Hirsch et al. (2010) made several observations to speed up the computations of C-GRASP, including the reuse of line search results. The local search algorithm GENCAN (Birgin and Martínez, 2002), an active-set method for bound-constrained local minimization, was used by Birgin et al. (2010) to play the role of local search in a C-GRASP for minimization of functions for which gradients can be computed. Martin et al. (2013) proposed improvements to C-GRASP, including the use of direct searches in the local search phase. Araújo et al. (2015) presented several direct search procedures that are used in a C-GRASP heuristic. They named their algorithm DC-GRASP. Silva et al. (2013a) described libcgrpp, a GNU-style dynamic shared Python/C library for quick implementation of C-GRASP heuristics.

C-GRASP has been applied to a range of problems. These include sensor registration in a sensor network (Hirsch et al., 2006), finding correspondence of projected 3D points and lines (Hirsch et al., 2011), solving systems of nonlinear equations (Hirsch et al., 2009), determining the relationship between drug combinations and adverse reactions (Hirsch et al., 2007a), economic dispatch of thermal units (Vianna Neto et al., 2010), robot path planning (Macharet et al., 2011), thermodynamics (Guedes et al., 2011), target tracking (Hirsch et al., 2012), and finding the largest ellipse, with prescribed eccentricity, inscribed in a nonconvex polygon (da Silva et al., 2012).

The global minima for the five test functions used in Section 11.6 were computed with the Python/C library of Silva et al. (2013a). Andrade et al. (2014) presented a parallel implementation of C-GRASP construction using a GPU. Speedups of up to 1.56 were measured, even though construction only accounts for 10 to 40% of the execution time in C-GRASP.

# Chapter 12
# Case studies

In this final chapter of the book, we consider four case studies to illustrate the application and implementation of GRASP heuristics. These heuristics are for 2-path network design, graph planarization, unsplittable multicommodity flows, and maximum cut in a graph. The key point here is not to show numerical results or compare these GRASP heuristics with other approaches, but instead simply show how to customize the GRASP metaheuristic for each particular problem.

## 12.1 2-path network design problem

Let $G = (V, U)$ be a connected undirected graph, where $V$ is the set of nodes and $U$ is the set of edges. A $k$-path between nodes $s, t \in V$ is a sequence of at most $k$ edges connecting $s$ and $t$. Given a non-negative weight function $w : U \to R_+$ associated with the edges of $G$ and a set $D$ of pairs of origin-destination nodes, the *2-path network design problem* (2PNDP) consists in finding a minimum weighted subset of edges $U' \subseteq U$ containing a 2-path between every origin-destination pair.

## 12.1.1 GRASP with path-relinking for 2-path network design

In the remainder of this section, we customize a parallel GRASP heuristic for the 2-path network design problem. We describe the construction and local search procedures, as well as a path-relinking intensification strategy.

© Springer Science+Business Media New York 2016
M.G.C. Resende, C.C. Ribeiro, *Optimization by GRASP*,
DOI 10.1007/978-1-4939-6530-4_12

### 12.1.1.1 Solution construction

The construction of a new solution begins with the initialization of modified edge weights with their original weights. Each iteration of the construction phase starts by the random selection of an origin-destination pair still in $D$. A shortest 2-path between the extremities of this pair is computed, using the modified edge weights. The weights of the edges in this 2-path are set to zero until the end of the construction procedure, the origin-destination pair is removed from $D$, and a new iteration begins. The construction phase stops when 2-paths have been computed for all origin-destination pairs.

### 12.1.1.2 Local search

The local search phase seeks to improve each solution built in the construction phase. Each solution can be viewed as a set of 2-paths, one for each origin-destination pair in $D$. To introduce diversity by driving different applications of the local search to different local optima, the origin-destination pairs are investigated at each GRASP iteration in a circular order defined by a different random permutation of their original indices.

Each 2-path in the current solution is tentatively eliminated. The weights of the edges used by other 2-paths are temporarily set to zero, while those that are not used by other 2-paths in the current solution are restored to their original values. A new shortest 2-path between the extremities of the origin-destination pair under investigation is computed, using the modified weights. If the new 2-path improves the current solution, then the current solution is modified; otherwise the previous 2-path is restored. The search stops if the current solution is not improved after a sequence of $|D|$ iterations along which all 2-paths are investigated. Otherwise, the next 2-path in the current solution is investigated for substitution and a new iteration begins.

### 12.1.1.3 Path-relinking

Path-relinking is applied to solution pairs composed of an initial solution, chosen at random from a pool formed by a limited number of previously found elite solutions, and by the solution produced by local search, which we call the guiding solution. The pool is initially empty. Each locally optimal solution is considered a candidate to be inserted into the pool if it is different from every other solution currently in the pool. If the pool is full and the candidate is better than the worst elite solution, then the candidate replaces the worst elite solution. If the pool is not full, then the candidate is simply inserted.

The algorithm starts by determining all origin-destination pairs whose associated 2-paths are different in the initial and guiding solutions. These computations amount to determining a set of moves that should be applied to the initial solution to reach the guiding solution. Each move is characterized by a pair of 2-paths, one to be inserted and the other to be eliminated from the current solution. The best solution is initialized with the initial solution. At each path-relinking iteration, the best yet unselected move is applied to the current solution, the incumbent solution is updated, and the selected move is removed from the set of candidate moves, until the guiding solution is reached. The incumbent is returned as the best solution found by path-relinking and is inserted into the pool if it satisfies the membership conditions.

### 12.1.1.4 Parallel GRASP implementation and numerical results

Parallel implementations of metaheuristics such as GRASP are more robust than their sequential versions. We describe a parallel implementation of the GRASP sequential heuristic described in the previous sections, corresponding to a typical multiple-walk independent-thread strategy introduced in Chapter 10. The iterations are evenly distributed over the processors. However, to improve load balancing, the iterations could also be distributed by demand, when faster processors perform more iterations than slower processors.

The processors perform MaxIterations/$p$ iterations each, where $p$ is the number of processors and MaxIterations is the total number of iterations. Each processor has a copy of the sequential GRASP algorithm, a copy of the problem data, and its own pool of elite solutions. One of the processors acts as the master, reading and distributing the problem data, generating the seeds that are used by the pseudo-random number generators at each processor, distributing the iterations, and collecting the best solution found by each processor.

The results of the parallel GRASP algorithm are compared with those obtained by a greedy heuristic, using two samples of solution values and Student's $t$-test for unpaired observations. The main statistics are summarized in Table 12.1. These results show with 40% confidence level that GRASP finds better solutions than the greedy heuristic. The average value of the solutions obtained by GRASP was 2.2% smaller than that of the solutions obtained by the greedy heuristic. The dominance of GRASP is even stronger when harder or larger instances are considered. The parallel GRASP was applied to problems with up to 400 nodes, 79,800 edges, and 4,000 origin-destination pairs, while the greedy heuristic solved problems with no more than 120 nodes, 7,140 edges, and 60 origin-destination pairs.

**Table 12.1** Statistics for GRASP (sample A) and the greedy heuristic (sample B).

|                    | Parallel GRASP (sample A) | Greedy (sample B)   |
| ------------------ | ------------------------- | ------------------- |
| Size               | $n_A = 100$               | $n_B = 30$          |
| Mean               | $\mu_A = 443.73$          | $\mu_B = 453.67$    |
| Standard deviation | $S_A = 40.64$             | $S_B = 61.56$       |

## 12.2 Graph planarization

A graph is said to be *planar* if it can be drawn on the plane in such a way that no two of its edges cross. Given a graph $G = (V, E)$ with vertex set $V$ and edge set $E$, the objective of *graph planarization* is to find a minimum cardinality subset of edges $F \subseteq E$ such that the graph $G' = (V, E \setminus F)$ resulting from the removal of the edges $F$ from $G$, is planar. This problem is also known as the maximum planar subgraph problem. A *maximal planar subgraph* is a planar subgraph $G' = (V', E')$ of $G = (V, E)$, such that the addition of any edge $e \in E \setminus E'$ to $G'$ destroys the planarity of the subgraph. Applications of graph planarization include graph drawing and numerous layout problems. Graph planarization is known to be *NP*-hard.

We begin with a review of a two-phase heuristic used as part of the GRASP heuristic for graph planarization. Then, the GRASP heuristic itself is described. Finally, we describe a post-optimization algorithm to further improve the solution obtained by GRASP.

### 12.2.1 Two-phase heuristic

In this section, we review the main components of the GT two-phase heuristic for graph planarization. The first phase of this heuristic is depicted in Figure 12.1 and consists in devising a sequence $\Pi$ of the set of vertices $V$ of the input graph $G$. Next, the vertices of $G$ are placed on a line according to the sequence $\Pi$. Let $\pi(v)$ denote the relative position of vertex $v \in V$ within vertex sequence $\Pi$. Furthermore, let $e_1 = (a, b)$ and $e_2 = (c, d)$ be two edges of $G$, such that, without loss of generality, $\pi(a) < \pi(b)$ and $\pi(c) < \pi(d)$. These edges are said to cross with respect to sequence $\Pi$ if $\pi(a) < \pi(c) < \pi(b) < \pi(d)$ or $\pi(c) < \pi(a) < \pi(d) < \pi(b)$. Basically, the second phase of GT partitions the edge set $E$ of $G$ into subsets $\mathscr{B}$, $\mathscr{R}$, and $\mathscr{P}$ in such a way that $|\mathscr{B} + \mathscr{R}|$ is large (or ideally maximum) and no two edges both in $\mathscr{B}$ or both in $\mathscr{R}$ cross with respect to the sequence $\Pi$ devised in the first phase.

Let $H = (E, I)$ be a graph where each of its vertices corresponds to an edge of the input graph $G$. Vertices $e_1$ and $e_2$ of $H$ are connected by an edge if the corresponding edges of $G$ cross with respect to sequence $\Pi$. A graph is called an *overlap graph* if its vertices can be placed in one-to-one correspondence with a family of intervals on a line. Two intervals are said to overlap if they cross and none is contained in the other. Two vertices of the overlap graph are connected by an edge if and only if their corresponding intervals overlap. Hence, the graph $H$ as constructed above is the overlap graph associated with the representation of $G$ defined by sequence $\Pi$.

The second phase of the GT two-phase heuristic consists in two-coloring a maximum number of vertices of the overlap graph $H$ such that each of the two color classes $\mathscr{B}$ (blue) and $\mathscr{R}$ (red) forms an independent set. Equivalently, the second phase seeks a *maximum induced bipartite subgraph* of the overlap graph $H$, i.e., a bipartite subgraph having the largest number of vertices. This problem is equivalent to drawing the edges of the input graph $G$ above or below the line where its

```
begin FirstPhaseGT;
1   d ← min_{v∈V}{deg_G(v)};
2   RCL ← {v ∈ V : deg_G(v) = d};
3   Select v_1 from RCL;
4   𝒱 ← V \ {v_1};
5   G_1 ← graph induced on G by 𝒱;
6   for k = 2,...,|V| do
7       d ← min_{v∈𝒱}{deg_{G_{k-1}}(v)};
8       if ADJ_{G_{k-1}}(v_{k-1}) ≠ ∅ then
9           RCL ← {v ∈ ADJ_{G_{k-1}}(v_{k-1}) : deg_{G_{k-1}}(v) = d};
10      else
11          RCL ← {v ∈ 𝒱 : deg_{G_{k-1}}(v) = d};
12      end-if
13      Select v_k from RCL;
14      𝒱 ← 𝒱 \ {v_k};
15      G_k ← graph induced on G by 𝒱;
16  end-for;
17  return Π = (v_1, v_2,..., v_{|V|});
end FirstPhaseGT.
```

**Fig. 12.1** Pseudo-code of the first phase of the GT heuristic.

vertices have been placed according to sequence $\Pi$. Since the decision version of the problem of finding a maximum induced bipartite subgraph of an overlap graph is *NP*-complete, a greedy algorithm is used in the GT heuristic to construct a maximal induced bipartite subgraph of the overlap graph. This algorithm finds a maximum independent set $\mathscr{B} \subseteq E$ of the overlap graph $H = (E, I)$, reduces this overlap graph by removing from the vertex set $E$ all vertices in $\mathscr{B}$ and from the edge set $I$ all edges incident to vertices in $\mathscr{B}$, and then finds a maximum independent set $\mathscr{R} \subseteq E \setminus \mathscr{B}$ in the resulting overlap graph $H' = (E \setminus \mathscr{B}, I')$. The two independent sets obtained induce a bipartite subgraph of the original overlap graph, not necessarily with a maximum number of vertices. This procedure has polynomial-time complexity, since finding a maximum independent set of an overlap graph is polynomially solvable in time $O(|E|^3)$, where $|E|$ is the number of vertices of the overlap graph $H = (E, I)$. The pseudo-code of the second phase of heuristic GT is given in Figure 12.2. The set $\mathscr{B} \cup \mathscr{R}$ corresponds to the edges that can be drawn without crossings.

```
begin SecondPhaseGT(Π);
1   Build overlap graph H = (E, I) using sequence Π;
2   Find a maximum independent set ℬ in the overlap graph H = (E, I);
3   Reduce the overlap graph H = (E, I);
4   Find a maximum independent set ℛ in the reduced overlap graph H' = (E \ ℬ, I');
5   return ℬ, ℛ;
end SecondPhaseGT.
```

**Fig. 12.2** Pseudo-code of the second phase of the GT heuristic.

**begin** ConstructGreedyRandomizedSolution-GP($\alpha$, seed);
1   $\underline{d} \leftarrow \min_{v \in V}\{\deg_G(v)\}$;
2   $\bar{d} \leftarrow \max_{v \in V}\{\deg_G(v)\}$;
3   RCL $\leftarrow \{v \in V : \underline{d} \le \deg_G(v) \le \alpha(\bar{d}-\underline{d})+\underline{d}\}$;
4   $v_1 \leftarrow$ random(seed, RCL);
5   $\mathcal{V} \leftarrow V \setminus \{v_1\}$;
6   $G_1 \leftarrow$ graph induced on $G$ by $\mathcal{V}$;
7   **for** $k = 2,\ldots,|V|$ **do**
8       $\underline{d} \leftarrow \min_{v \in \mathcal{V}}\{\deg_{G_{k-1}}(v)\}$;
9       $\bar{d} \leftarrow \max_{v \in \mathcal{V}}\{\deg_{G_{k-1}}(v)\}$;
10      **if** $\text{ADJ}_{G_{k-1}}(v_{k-1}) \ne \varnothing$ **then**
11          RCL $\leftarrow \{v \in \text{ADJ}_{G_{k-1}}(v_{k-1}) : \underline{d} \le \deg_{G_{k-1}}(v) \le \alpha(\bar{d}-\underline{d})+\underline{d}\}$;
12      **else**
13          RCL $\leftarrow \{v \in \mathcal{V} : \underline{d} \le \deg_{G_{k-1}}(v) \le \alpha(\bar{d}-\underline{d})+\underline{d}\}$;
14      **end-if**;
15      $v_k \leftarrow$ random(seed, RCL);
16      $\mathcal{V} \leftarrow \mathcal{V} \setminus \{v_k\}$;
17      $G_k \leftarrow$ graph induced on $G$ by $\mathcal{V}$;
18  **end-for**;
19  **return** $\Pi = (v_1,\ldots,v_{|V|})$;
**end** ConstructGreedyRandomizedSolution-GP.

**Fig. 12.3** Pseudo-code of the GRASP construction phase (vertex sequencing).

This two-phase algorithm is not guaranteed to produce an optimal (i.e., maximum) planar subgraph. Furthermore, even under a simple neighborhood definition, it does not necessarily produce a locally optimal solution. The first phase of GT is based on an adaptive greedy algorithm to produce a vertex sequence. This vertex sequence appears to affect the size of the planar subgraph found in the second phase of GT. However, it is not clear that the sequence produced by the adaptive greedy algorithm is the best. To produce other, possibly better, sequences, randomization and local search can be introduced in the adaptive greedy algorithm. We next explore these ideas and describe a GRASP heuristic for graph planarization that finds a locally optimal planar subgraph, often improving on the solution found by GT.

### 12.2.2 GRASP for graph planarization

The two-phase heuristic presented in the previous section uses an adaptive greedy algorithm to produce the vertex sequencing of its first phase. In the following, we show an alternative to the adaptive greedy algorithm: a GRASP for the first phase vertex sequencing problem. The construction phase of this GRASP heuristic is described in the pseudo-code of Figure 12.3.

The procedure takes as input the graph $G = (V,E)$ to be planarized, the restricted candidate list (RCL) parameter $0 \le \alpha \le 1$, and a seed for the pseudo-random number generator. Let $\deg_G(v)$ be the degree of vertex $v$ with respect

```
begin LocalSearch-GP(Π);
1    while Π is not locally optimal do
2        Find Π' ∈ 𝒩(Π) such that χ(Π') < χ(Π);
3        Π ← Π';
4    end-while;
5    return Π = (v₁,...,v|V|);
end LocalSearch-GP.
```

**Fig. 12.4** Pseudo-code of the GRASP local search phase for graph planarization.

to $G$, $\underline{d} = \min_{v \in V}\{\deg_G(v)\}$ and $\bar{d} = \max_{v \in V}\{\deg_G(v)\}$. The first vertex in the sequence is determined in lines 1 to 4, where all vertices having degree in the range $[\underline{d}, \alpha(\bar{d} - \underline{d}) + \underline{d}]$ are placed in the RCL and a single vertex is selected at random from the list. The working vertex set $\mathcal{V}$ and graph $G_1$ are defined in lines 5 and 6.

The loop from lines 7 to 18 determines the sequence of the remaining $|V| - 1$ vertices. To assign the $k$-th vertex (iteration $k$ of the loop), two cases can occur. Define $G_k$ to be the graph induced on $G$ by $V \setminus \{v_1, v_2, \ldots, v_k\}$. Let $\text{ADJ}_{G_{k-1}}(v_{k-1})$ be the set of vertices of $G_{k-1}$ adjacent to $v_{k-1}$ in $G$. The RCL is made up of all vertices in $\text{ADJ}_{G_{k-1}}(v_{k-1})$ having degree in the range $[\underline{d}, \alpha(\bar{d} - \underline{d}) + \underline{d}]$ in $G_k$. Otherwise, if $\text{ADJ}_{G_{k-1}}(v_{k-1}) = \varnothing$, the RCL is made up of all unselected vertices having degree in the range $[\underline{d}, \alpha(\bar{d} - \underline{d}) + \underline{d}]$ in $G_k$. In line 15, the $k$-th vertex in the sequence is determined by selecting a vertex, at random, from the RCL. The working vertex set and the working graph are updated in lines 16 and 17. The vertex sequence $\Pi = (v_1, \ldots, v_{|V|})$ is returned in line 19.

The first phase of the GT heuristic seeks a sequence of the vertices, followed by a second phase minimizing the number of edges that need to be removed to eliminate all edge crossings with respect to the first phase sequence. One possible strategy (not taken in GT) is to attempt to reduce the number of crossing edges by locally searching a neighborhood of the current vertex sequence prior to the second phase. The local search procedure makes use of a neighborhood $\mathcal{N}(\Pi)$ of the vertex sequence $\Pi$ that is formed by all vertex sequences $\Pi'$ differing from $\Pi$ in exactly two positions, i.e.,

$$\mathcal{N}(\Pi) = \{\Pi' = (v'_1, v'_2, \ldots, v'_{|V|}) : v'_i = v_i, \forall i \neq j, k, \; v'_j = v_k, \; v'_k = v_j \; j \neq k\}.$$

Let $\chi(\Pi)$ be the number of pairs of edges that cross if the vertex sequence $\Pi$ is adopted. The pseudo-code in Figure 12.4 describes the local search procedure used in the GRASP heuristic, based on a slightly more restricted neighborhood that only considers the exchange of consecutive vertices.

Putting together the randomized vertex sequencing procedure displayed in Figure 12.3, the local search algorithm displayed in Figure 12.4, and the second phase of the GT heuristic provided in Figure 12.2 we obtain a GRASP for graph planarization, whose pseudo-code is given in Figure 12.5.

The number of edges in the maximal planar subgraph corresponding to the best solution found is initialized in line 1. The iterative GRASP procedure in lines 2 to 11

```
begin GRASP-GP(α, seed, MaxIter);
1   BR* ← −∞;
2   for k = 1,..., MaxIter do
3       Π ← ConstructGreedyRandomizedSolution-GP(α, seed);
4       Π ← LocalSearch-GP(Π);
5       ℬ, ℛ ← SecondPhaseGT(Π);
6       if |ℬ| + |ℛ| > BR* then
7           Π* ← Π;
8           ℬ* ← ℬ;
9           ℛ* ← ℛ;
10      end-if;
11  end-for;
12  return Π*, ℬ*, ℛ*;
end GRASP-GP.
```

**Fig. 12.5** Pseudo-code of the GRASP heuristic for graph planarization.

is repeated MaxIter times. In each iteration, a greedy randomized solution (vertex sequence $\Pi$) is constructed in line 3. In line 4, the local search phase attempts to produce a vertex sequence that has fewer crossings of pairs of edges than the one generated in line 3. The vertex sequence $\Pi$ is given as input to the second phase heuristic of GT in line 5 to produce a planar subgraph of $G$. If the new solution improves the number of the edges in the planar subgraph, then the best solution found is updated in lines 7 to 9. The best solution found is returned in line 12.

### 12.2.3 Enlarging the planar subgraph

As already observed, there is no guarantee that the planar subgraph produced by SecondPhaseGT is optimal. Three edge sets are output: $\mathcal{B}$ (blue edges), $\mathcal{R}$ (red edges), and $\mathcal{P}$ (the remaining edges, which we refer to as the pale edges). By construction, $\mathcal{B}$, $\mathcal{R}$, and $\mathcal{P}$ are such that no red or pale edge can be colored blue. Likewise, pale edges cannot be colored red. However, if there exists a pale edge $p$ such that all blue edges that cross with $p$ (let $\mathcal{B}_p \subseteq \mathcal{B}$ be the set of such blue edges) do not cross with any red edge, then all blue edges in $\mathcal{B}_p$ can be colored red and $p$ can be colored blue. Consequently, this reassignment of color classes increases the size of the planar subgraph by one edge.

Figure 12.6 shows the pseudo-code of procedure EnlargePlanarGraph that seeks pale and blue edges allowing the above color class reassignment and enlarges the planar subgraph whenever such edges are encountered. The pale edges are scanned in the loop in lines 1 to 17. Set $\mathcal{B}_p$ is initialized in line 2 and the pale edge $p$ is temporarily made a candidate to be recolored by setting variable enlarge to .TRUE. in line 3. The loop in lines 4 to 11 scans the blue edges to construct the set $\mathcal{B}_p$ for each pale edge $p \in \mathcal{P}$. Any blue edge that crosses with the pale edge $p$ is added to the candidate set $\mathcal{B}_p$ in line 6. Red edges are scanned in the loop in lines 7 to 9. If a blue edge $b \in \mathscr{B}_p$ crosses any red edge, then the pale edge $p$ will be discarded by setting the variable `enlarge` to .FALSE. in line 8. If none of the blue edges in $\mathscr{B}_p$ crosses a red edge, then all blue edges in $\mathscr{B}_p$ will be recolored as red and the pale edge $p$ will be colored as blue in lines 13 to 15. The possibly enlarged solution is returned in line 18.

```
begin EnlargePlanarGraph(ℬ,ℛ,𝒫);
1   forall p ∈ 𝒫 do
2       ℬ_p ← ∅;
3       enlarge ← .TRUE.;
4       forall b ∈ ℬ do
5           if edges p and b cross then
6               ℬ_p ← ℬ_p ∪ {b};
7               forall r ∈ ℛ do
8                   if edges r and b cross then enlarge ← .FALSE.;
9               end-forall;
10          end-if;
11      end-forall;
12      if enlarge = .TRUE. then
13          ℬ ← ℬ ∪ {p} \ ℬ_p;
14          ℛ ← ℛ ∪ ℬ_p;
15          𝒫 ← 𝒫 \ {p};
16      end-if;
17  end-forall;
18  return ℬ,ℛ;
end EnlargePlanarGraph.
```

**Fig. 12.6** Pseudo-code of the improvement procedure to enlarge a planar subgraph.

The improvement procedure EnlargePlanarGraph can be applied to each solution obtained in line 5 of the GRASP heuristic displayed in Figure 12.5 or, alternatively, exclusively to the best solution found returned by GRASP-GP in line 12.

## 12.3 Unsplittable multicommodity network flow: Application to bandwidth packing

Telecommunication service providers offer virtual private networks to customers by provisioning a set of permanent (long-term) private virtual circuits (PVCs) between endpoints on a large backbone network. During the provisioning of a PVC, routing decisions are made either automatically by the routing equipment (the router) or by the network operator, through the use of preferred routing assignments and without any knowledge of future requests. Over time, these decisions usually cause inefficiencies in the network and occasional rerouting of the PVCs is needed. The new routing scheme is then implemented on the network through preferred routing assignments. Given a preferred routing assignment, the switch will move the PVC from its current route to the new preferred route as soon as this move becomes feasible.

One possible way to create preferred routing assignments is to appropriately order the set of PVCs currently in the network and apply an algorithm that mimics the routing algorithm used by the router to each PVC in that order. However, more elaborate routing algorithms, which take into account factors not considered by the router, could further improve the efficiency of network resource utilization.

Typically, the routing scheme used by the routers to automatically provision PVCs is also used to reroute the PVCs in the case of trunk or card failures. Therefore, this routing algorithm should be efficient in terms of running time, a requirement that can be traded off for improved network resource utilization when building preferred routing assignments offline.

We discuss variants of a GRASP with path-relinking algorithm for the problem of routing offline a set of PVC demands over a backbone network, such that a combination of the delays due to propagation and congestion is minimized. This problem and its variants are also known in the literature as bandwidth packing problems. The set of PVCs to be routed can include not only all or a subset of the PVCs currently in the network, but also possibly a set of forecast PVCs. The explicit handling of propagation delays, as opposed to just handling the number of hops, is particularly important in international networks, where distances between backbone nodes vary considerably. The minimization of network congestion is important for providing the maximum flexibility to handle overbooking (which is typically used by network operators to account for noncoincidence of traffic), rerouting (due to link or card failures), and bursting above the committed rate (which is not only allowed, but sold to customers as one of the attractive features of the service).

We next formulate the offline PVC routing problem as an integer multicommodity flow problem with additional constraints and a hybrid objective function, which takes into account delays due to propagation as well as delays due to network congestion. Minimum cost multicommodity network flow problems are characterized by a set of commodities flowing through an underlying network, each commodity having an associated integral demand that must flow from its source to its destination. The flows are simultaneous and the commodities share network resources. We conclude this section by describing variants of a GRASP with path-relinking heuristic for this problem.

### 12.3.1 Problem formulation

Let $G = (V, E)$ be an undirected graph representing a backbone network. Denote by $V = \{1, \dots, n\}$ the set of backbone nodes where routers reside, while $E$ is the set of trunks (or edges) that connect the backbone nodes, with $|E| = m$. Parallel trunks are allowed. Since $G$ is an undirected graph, flows through each trunk $(i, j) \in E$ have two components to be summed up, one in each direction. However, for modeling purposes, costs and capacities are associated only with ordered pairs $(i, j) \in E$ satisfying $i < j$. For each trunk $(i, j) \in E$, we denote by $b_{ij}$ its maximum allowed bandwidth (in kbits/second), while $c_{ij}$ denotes the maximum number of PVCs that can be routed through it and $d_{ij}$ is the propagation (or hopping) delay associated with the trunk. Each commodity $k \in K = \{1, \ldots, p\}$ is a PVC to be routed, associated with an origin-destination pair and with a bandwidth requirement $r_k$ (or demand, also known as its effective bandwidth). It takes into account the actual bandwidth required by the customer in the forward and reverse directions, as well as an overbooking factor.

The ultimate objective of the offline PVC routing problem is to minimize propagation delays or network congestion, subject to several technological constraints. Queuing delays are often associated with network congestion and in some networks account for a large part of the total delay. In other networks, distances can be long and loads low, causing the propagation delay to account for a large part of the total delay. Two common measures of network congestion are the load on the most utilized trunk, and the average delay in a network of independent $M/M/1$ queues. Another measure, which we use here, is a cost function that penalizes heavily loaded trunks. This function resembles the average delay function, except that it allows loads to exceed trunk capacities. Routing assignments with minimum propagation delays may not achieve the least network congestion. Likewise, routing assignments having the least congestion may not minimize propagation delays. A compromising objective is to route the PVCs such that a desired point in the trade-off curve between propagation delays and network congestion is achieved.

The upper bound on the number of PVCs allowed on a trunk depends on the technology used to implement it. A set of routing assignments is feasible if and only if, for every trunk $(i, j) \in E$, the total PVC effective bandwidth requirements routed through it does not exceed its maximum bandwidth $b_{ij}$ and the number of PVCs routed through it is not greater than $c_{ij}$.

Let $x_{ij}^k$ be a 0-1 variable such that $x_{ij}^k = 1$ if and only if trunk $(i, j) \in E$ is used to route commodity $k \in K$ from node $i$ to node $j$. The following linear integer program models the problem:

$$\min \phi(x) = \sum_{(i,j) \in E, i<j} \phi_{ij}(x_{ij}^1, \cdots, x_{ij}^p, x_{ji}^1, \cdots, x_{ji}^p) \tag{12.1}$$

subject to

$$\sum_{k \in K} r_k(x_{ij}^k + x_{ji}^k) \le b_{ij}, \ \forall (i, j) \in E, i < j, \tag{12.2}$$

$$\sum_{k \in K} (x_{ij}^k + x_{ji}^k) \le c_{ij}, \ \forall (i, j) \in E, i < j, \tag{12.3}$$

$$\sum_{(i,j) \in E} x_{ij}^k - \sum_{(i,j) \in E} x_{ji}^k = a_i^k, \ \forall i \in V, \forall k \in K, \tag{12.4}$$

$$x_{ij}^k \in \{0, 1\}, \ \forall (i, j) \in E, \forall k \in K. \tag{12.5}$$

Constraints of type (12.2) limit the total flow on each trunk to at most its capacity. Constraints of type (12.3) enforce the limit on the number of PVCs routed through each trunk. Constraints of type (12.4) are flow conservation equations, which together with constraints (12.5) state that the flow associated with each PVC cannot be split, where $a_i^k = 1$ if node $i$ is the source for commodity $k$, $a_i^k = -1$ if node $i$ is the destination for commodity $k$, and $a_i^k = 0$ otherwise.

The cost function $\phi_{ij}(x_{ij}^1, \cdots, x_{ij}^p, x_{ji}^1, \cdots, x_{ji}^p)$ associated with each trunk $(i, j) \in E$ with $i < j$ is the linear combination of a trunk propagation delay component and a trunk congestion component. The propagation delay component is defined as

$$\phi_{ij}^d(x_{ij}^1, \cdots, x_{ij}^p, x_{ji}^1, \cdots, x_{ji}^p) = d_{ij} \cdot \sum_{k \in K} \rho_k(x_{ij}^k + x_{ji}^k), \qquad (12.6)$$

where coefficients $\rho_k$ are used to model two plausible delay functions:

- If $\rho_k = 1$, then this component leads to the minimization of the number of hops weighted by the propagation delay on each trunk.
- If $\rho_k = r_k$, then the minimization takes into account the effective bandwidth routed through each trunk weighted by its propagation delay.

Let $y_{ij} = \sum_{k \in K} r_k(x_{ij}^k + x_{ji}^k)$ be the total flow through trunk $(i, j) \in E$ with $i < j$. The trunk congestion component depends on the utilization rates $u_{ij} = y_{ij}/b_{ij}$ of each trunk $(i, j) \in E$ with $i < j$. This piecewise linear function,

$$\phi_{ij}^b(x_{ij}^1, \cdots, x_{ij}^p, x_{ji}^1, \cdots, x_{ji}^p) = b_{ij} \cdot \begin{cases} u_{ij}, & u_{ij} \in [0, 1/3) \\ 3 \cdot u_{ij} - 2/3, & u_{ij} \in [1/3, 2/3), \\ 10 \cdot u_{ij} - 16/3, & u_{ij} \in [2/3, 9/10), \\ 70 \cdot u_{ij} - 178/3, & u_{ij} \in [9/10, 1), \\ 500 \cdot u_{ij} - 1468/3, & u_{ij} \in [1, 11/10), \\ 5000 \cdot u_{ij} - 16318/3, & u_{ij} \in [11/10, \infty), \end{cases}$$
$$(12.7)$$

depicted in Figure 12.7, increasingly penalizes flows approaching or violating the capacity limits. The value

$$\Omega = \max_{(i,j) \in E, i < j} \{u_{ij}\}$$

is a global measure of the maximum congestion in the network.

Let weights $(1 - \delta)$ and $\delta$ correspond, respectively, to the propagation delay and to the network congestion components, with $\delta \in [0, 1]$. The cost function

$$\phi_{ij}(x_{ij}^1, \cdots, x_{ij}^p, x_{ji}^1, \cdots, x_{ji}^p) =$$
$$(1 - \delta) \cdot \phi_{ij}^d(x_{ij}^1, \cdots, x_{ij}^p, x_{ji}^1, \cdots, x_{ji}^p) + \delta \cdot \phi_{ij}^b(x_{ij}^1, \cdots, x_{ij}^p, x_{ji}^1, \cdots, x_{ji}^p) \quad (12.8)$$

is associated with each trunk $(i, j) \in E$ with $i < j$. Note that if $\delta > 0$, then the network congestion component is present in the objective function, which allows us to relax capacity constraints (12.2). This will be assumed in the algorithms discussed in Section 12.3.2.

Model (12.1)–(12.5) proposed in this section has two distinctive features. First, it takes into account a two component objective function, which is able to handle

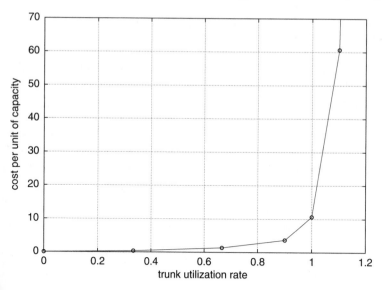

**Fig. 12.7** Piecewise linear load balance cost component associated with each trunk.

both delays and load balance. Second, it enforces constraints that limit the maximum number of PVCs that can be routed through any trunk. A GRASP with path-relinking heuristic for its solution is described in the next section.

## 12.3.2 GRASP with path-relinking for PVC routing

In the remainder of this section, we customize a GRASP heuristic for the offline PVC routing problem. We describe construction and local search procedures, as well as a path-relinking intensification strategy.

### 12.3.2.1 Construction phase

In the construction phase, the routes are determined, one at a time. In each iteration of construction a new PVC is selected to be routed. To reduce the computation times, we make use of a combination of the strategies usually employed by GRASP and heuristic-biased stochastic sampling. We create a restricted candidate list (RCL) with a fixed number of elements $n_c$. At each iteration, the RCL is formed by the $n_c$ unrouted PVC pairs with the largest demands. An element $\ell$ is selected at random from this list with probability $\pi(\ell) = r_\ell / \sum_{k \in \text{RCL}} r_k$.

Once a PVC $\ell \in K$ is selected, it is routed on a shortest path from its origin to its destination. The capacity constraints (12.2) are relaxed and handled via the penalty function introduced by the load balance component (12.7) of the edge weights. The constraints of type (12.3) are explicitly taken into account by forbidding routing through trunks already using its maximum number of PVCs. The weight $\Delta\phi_{ij}$ of each edge $(i,j) \in E$ is given by the increment of the cost function value $\phi_{ij}(x_{ij}^1, \cdots, x_{ij}^p, x_{ji}^1, \cdots, x_{ji}^p)$ associated with routing $r_\ell$ additional units of demand through edge $(i,j)$.

More precisely, let $\underline{K} \subseteq K$ be the set of previously routed PVCs and $\underline{K}_{ij} \subseteq \underline{K}$ be the subset of PVCs that are routed through trunk $(i,j) \in E$. Likewise, let $\bar{K} = \underline{K} \cup \{\ell\} \subseteq K$ be the new set of routed PVCs and $\bar{K}_{ij} = \underline{K}_{ij} \cup \{\ell\} \subseteq \bar{K}$ be the new subset of PVCs that are routed through trunk $(i,j)$. Then, we define $\underline{x}_{ij}^\ell = 1$ if PVC $\ell \in \underline{K}$ is routed through trunk $(i,j) \in E$ from $i$ to $j$, $\underline{x}_{ij}^\ell = 0$ otherwise. Similarly, we define $\bar{x}_{ij}^\ell = 1$ if PVC $\ell \in \bar{K}$ is routed through trunk $(i,j) \in E$ from $i$ to $j$, $\bar{x}_{ij}^\ell = 0$ otherwise. According with (12.8), the cost associated with each edge $(i,j) \in E$ in the current solution is given by $\phi_{ij}(\underline{x}_{ij}^1, \cdots, \underline{x}_{ij}^p, \underline{x}_{ji}^1, \cdots, \underline{x}_{ji}^p)$. In the same manner, the cost associated with each edge $(i,j) \in E$ after routing PVC $\ell$ will be $\phi_{ij}(\bar{x}_{ij}^1, \cdots, \bar{x}_{ij}^p, \bar{x}_{ji}^1, \cdots, \bar{x}_{ji}^p)$. Then, the incremental edge weight $\Delta\phi_{ij}$ associated with routing PVC $\ell \in K$ through edge $(i,j) \in E$, used in the shortest path computations, is given by

$$\Delta\phi_{ij} = \phi_{ij}(\bar{x}_{ij}^1, \cdots, \bar{x}_{ij}^p, \bar{x}_{ji}^1, \cdots, \bar{x}_{ji}^p) - \phi_{ij}(\underline{x}_{ij}^1, \cdots, \underline{x}_{ij}^p, \underline{x}_{ji}^1, \cdots, \underline{x}_{ji}^p). \qquad (12.9)$$

The enforcement of type (12.3) constraints may lead to unroutable demand pairs. In this case, the current solution is discarded and a new construction phase starts.

### 12.3.2.2 Local search

Each solution built in the first phase may be viewed as a set of routes, one for each PVC. The local search procedure seeks to improve each route in the current solution. For each PVC $k \in K$, we start by removing $r_k$ units of flow from each edge in its current route. Next, we compute incremental edge weights $\Delta\phi_{ij}$ associated with routing this demand through each trunk $(i,j) \in E$ according to equation (12.9), as described in Section 12.3.2.1. A tentative new shortest path route is computed using the incremental edge weights. If the new route improves the solution, then it replaces the current route of PVC $k$. This is continued until no improving route can be found.

### 12.3.2.3 Path-relinking

Path-relinking for bandwidth packing is applied to pairs $\{y,z\}$ of solutions, where one solution is the locally optimum obtained after local search and the other solution is randomly chosen from an elite set $\mathscr{E}$ formed by a limited number $n_\mathscr{E}$ of elite solutions found along the search. This elite pool is initially empty. Each locally optimal solution obtained by local search is considered as a candidate to be inserted into the pool if it differs by at least one trunk in one route from every other solution currently in the pool. If the pool already has $n_{\mathscr{E}}$ solutions and the candidate is better than the worst solution in the pool, then the candidate replaces the worst solution. If the pool is not full, the candidate is simply inserted in the pool.

```
begin GRASP+PR-BPP;
1    φ* ← ∞;
2    Pool ← ∅;
3    for k = 1,...,Max_Iterations do
4        Construct a greedy randomized solution x;
5        Obtain y by applying local search to x;
6        if y satisfies the membership conditions then update Pool with y;
7        Randomly select an elite solution z ∈ Pool with uniform probability;
8        Compute K_{y,z};
9        Let ȳ be the best solution found by applying path-relinking to y and z;
10       if ȳ satisfies the membership conditions then update Pool with ȳ;
11       if φ(ȳ) < φ* then
12           x* ← ȳ;
13           φ* ← φ(x*);
14       end-if;
15   end-for;
16   return x*;
end GRASP+PR-BPP.
```

**Fig. 12.8** Pseudo-code of the GRASP with path-relinking procedure for the bandwidth packing problem.

Either $y$ or $z$ is selected to be the initial solution, while the other will be the guiding solution. The algorithm starts by computing the set of moves that should be applied to the initial solution to reach the guiding solution. Starting from the initial solution, the best move still not performed is applied to the current solution, until the guiding solution is attained. The best solution found along this trajectory is also considered as a candidate for insertion in the pool and the incumbent is updated. Several alternatives have been considered and combined to explore trajectories connecting $y$ and $z$. All these alternatives involve the trade-offs between computation time and solution quality, as already discussed in Chapters 8 and 9.

In this application of path-relinking, the set of moves between any pair $\{y,z\}$ of solutions is the subset $K_{y,z} \subseteq K$ of PVCs routed through different routes in $y$ and $z$. Without loss of generality, let us suppose that path-relinking starts from any elite solution $z$ in the pool and uses the locally optimal solution $y$ as the guiding solution.

The best solution $\bar{y}$ along the new path to be constructed is initialized with $z$. For each PVC $k \in K_{y,z}$, the same shortest path computations described in Sections 12.3.2.1 and 12.3.2.2 are used to evaluate the cost of the new solution obtained by rerouting the demand associated with PVC $k$ through the route used in the guiding solution $y$ instead of the route used in the current solution originated from $z$. The best move is selected and removed from $K_{y,z}$. The new solution obtained by rerouting the above selected PVC is computed, the incumbent $\bar{y}$ is updated, and a new

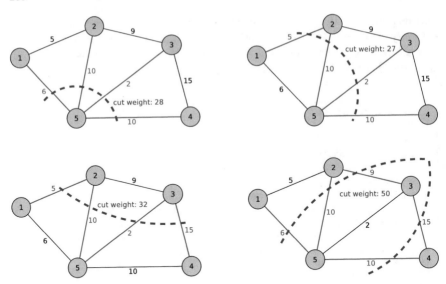

**Fig. 12.9** Example of the maximum cut problem on a graph with five vertices and seven edges. Four cuts are shown. The maximum cut is $(S, \bar{S}) = (\{1,2,4\}, \{3,5\})$ and has a weight $w(S, \bar{S}) = 50$.

iteration resumes. These steps are repeated, until the guiding solution $y$ is reached. The incumbent $\bar{y}$ is returned as the best solution found by path-relinking and inserted into the pool if it satisfies the membership conditions.

The pseudo-code with the complete description of procedure GRASP+PR-BPP for the bandwidth packing problem arising in the context of offline PVC rerouting is given in Figure 12.8. This description incorporates the construction, local search, and path-relinking phases.

## 12.4 Maximum cut in a graph

Given an undirected graph $G = (V, U)$, where $V$ is the set of vertices and $U$ is the set of edges, and weights $w_{uv}$ associated with each edge $(u, v) \in U$, the *maximum cut* (MAX-CUT) *problem* consists in finding a nonempty proper subset of vertices $S \subset V$ ($S \neq \varnothing$), such that the weight of the cut $(S, \bar{S})$, given by

$$w(S, \bar{S}) = \sum_{u \in S, v \in \bar{S}} w_{uv},$$

is maximized.

Figure 12.9 shows four cuts having different weights on a graph with five nodes and seven edges. The maximum cut has $S = \{1,2,4\}$ and $\bar{S} = \{3,5\}$, with weight $w(S, \bar{S}) = 50$.

```
begin GRASP+PR-MAXCUT;
1   w* ← −∞;
2   𝓔 ← ∅;
3   while stopping criterion is not satisfied do
4       (S,S̄) ← SEMI-GREEDY-MAXCUT;
5       (S,S̄) ← LOCAL-SEARCH-MAXCUT((S,S̄));
6       if 𝓔 ≠ ∅ then
7           Select a solution (Sᵍ,S̄ᵍ) from the pool 𝓔;
8           (S,S̄) ← PATH-RELINKING-MAXCUT((S,S̄), (Sᵍ,S̄ᵍ), (S̄ᵍ,Sᵍ));
9       end-if;
10      if (S,S̄) satisfies the membership conditions then
11          Update the pool 𝓔 with (S,S̄);
12      end-if;
13      w(S,S̄) = Σ_{i∈S,j∈S̄} w_{ij};
14      if w(S,S̄) > w* then
15          (S*,S̄*) ← (S,S̄);
16          w* ← w(S,S̄);
17      end-if;
18  end-while;
19  return (S*,S̄*),w*;
end GRASP+PR-MAXCUT.
```

**Fig. 12.10** Pseudo-code of a GRASP with path-relinking for the MAX-CUT problem.

## 12.4.1 GRASP with path-relinking for the maximum cut problem

A GRASP with path-relinking heuristic for the MAX-CUT problem consists in repeatedly constructing a cut $(S,\bar{S})$ with a semi-greedy algorithm, applying local search from $(S,\bar{S})$ to produce a locally maximal solution, and applying path-relinking from the local maximum to a solution $(S^g,\bar{S}^g)$ selected from a pool $\mathscr{E}$ of elite solutions. The best local maximum $(S^*,\bar{S}^*)$, found over all GRASP iterations, is returned as the GRASP solution.

The pseudo-code of a GRASP with path-relinking heuristic for the MAX-CUT problem is shown in Figure 12.10. In line 1, the value $w^*$ of the best cut found is initialized to $-\infty$ and in line 2 the pool of elite solutions $\mathscr{E}$ is initialized empty. The while loop from line 3 to line 18 carries out the GRASP with path-relinking iterations. The algorithm terminates when a stopping criterion is satisfied. In line 4, a semi-greedy solution $(S,\bar{S})$ is constructed and, in line 5, it is tentatively improved with local search. The local maximal produced by local search in line 5 is $(S,\bar{S})$. Path-relinking is applied if the pool has at least one elite solution. In that case, a guiding solution $(S^g,\bar{S}^g)$ is selected at random from the pool in line 7 and the path-relinking operator is applied from the locally maximal cut $(S,\bar{S})$ to the guiding solution in line 8. The solution obtained by path-relinking is saved in $(S,\bar{S})$. If solution $(S,\bar{S})$, obtained by either local search or path-relinking, satisfies the membership conditions, then the pool of elite solutions $\mathscr{E}$ is updated in line 11. Though the cut weight $w(S,\bar{S})$ is computed in the local search procedure and in the path-relinking procedure, its computation is shown in line 13 of the pseudo-code. If the weight of the local maximum is greater than the weight $w^*$ of the best cut found so far, then the best cut $(S^*, \bar{S}^*)$ and its weight are updated in lines 15 and 16, respectively. The best cut and the best solution found are returned in line 19.

In the remainder of this chapter we describe the components of this GRASP with path-relinking heuristic in more detail.

### 12.4.1.1 A greedy algorithm for the maximum cut problem

The construction phase of the GRASP for the MAX-CUT problem described here is a semi-greedy algorithm. Recall that we wish to build a proper subset $S \subset V$, such that $(S, \bar{S})$ forms a partition of $V$, i.e., $S \cup \bar{S} = V$ and $S \cap \bar{S} = \varnothing$. The ground set for the MAX-CUT problem is the set $V$ of vertices of graph $G = (V, U)$.

We first describe a greedy algorithm that is the basis for this GRASP. It builds a solution incrementally in sets $X$ and $Y$ by assigning vertices from the ground set $V$ to either $X$ or $Y$. Initially, sets $X$ and $Y$ each contain an endpoint of a largest-weight edge. At each other step of the construction, a new ground set element $v \in V$ is added to either set $X$ or set $Y$ of the partial solution. This is repeated until $X \cup Y = V$, at which point we set $S$ to $X$, $\bar{S}$ to $Y$, and a feasible solution $(S, \bar{S})$ is on hand.

At each iteration of this greedy construction, an element is selected from a candidate list whose elements are the yet-unassigned ground set elements, i.e., $V \setminus (X \cup Y)$, according to an adaptive greedy function described next.

The greedy function takes into account the contribution to the objective function (the weight of the partial cut) achieved by assigning a particular element to either set $X$ or set $Y$. Formally, let $(X, Y)$ be the partial solution under construction. Recall that, for any partial solution, $X \cup Y \subset V$. For each yet-unassigned vertex $v \in V \setminus (X \cup Y)$, define

$$\sigma_X(v) = \sum_{u \in Y} w_{vu} \tag{12.10}$$

and

$$\sigma_Y(v) = \sum_{u \in X} w_{vu} \tag{12.11}$$

to be, respectively, the incremental contributions to the cut weight resulting from the assignment of node $v$ to sets $X$ and $Y$ of the partial partition $(X, Y)$. The greedy function

$$g(v) = \max\{\sigma_X(v), \sigma_Y(v)\},$$

for $v \in V \setminus (X \cup Y)$, measures how much additional weight results from the assignment of vertex $v$ to $X$ or $Y$. The greedy choice is

$$v^* = \operatorname{argmax}\{g(v) : v \in V \setminus (X \cup Y)\}.$$

Vertex $v^*$ is assigned to set $X$ if $\sigma_X(v) > \sigma_Y(v)$ or to set $Y$, otherwise.

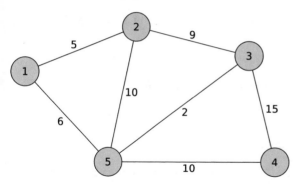

**Fig. 12.11** Five-node graph for maximum cut problem.

**Maximum cut problem – Adaptive greedy algorithm to find a large-weight cut**

Consider the following example on the five-node graph $G = (V,U)$ of Figure 12.11, for which we seek a large-weight cut. We build a partition $(X,Y)$ of the nodes of $G$ incrementally, with the greedy algorithm described above. Initially, sets $X$ and $Y$ are such that each contains an endpoint of a largest-weight edge of $G$. Since edge $(3,4)$ is the one with the largest weight, then $X = \{3\}$, $Y = \{4\}$, and the weight of the partial cut is $w(X,Y) = w_{3,4} = 15$.

To select the next node to be added to the partial cut, we consider only nodes 2 and 5, since node 1 is not adjacent to any node in $X \cup Y$ and, consequently, $\sigma_X(1) = \sigma_Y(1) = 0$. Consider first node 2. Its contribution to the cut if added to set $Y$ is $\sigma_Y(2) = \sum_{u \in X} w_{2,u} = w_{2,3} = 9$. Since it is not adjacent to any node in set $Y$, then it will not contribute to the partial cut if added to $X$, i.e., $\sigma_X(2) = \sum_{u \in Y} w_{2,u} = 0$. Now consider node 5. Its contribution to the cut if added to sets $X$ and $Y$ are, respectively, $\sigma_X(5) = \sum_{u \in Y} w_{5,u} = w_{5,4} = 10$ and $\sigma_Y(5) = \sum_{u \in X} w_{5,u} = w_{5,3} = 2$. Since the greedy function values are

$$g(1) = \max\{\sigma_X(1), \sigma_Y(1)\} = 0,$$
$$g(2) = \max\{\sigma_X(2), \sigma_Y(2)\} = 9,$$
$$g(5) = \max\{\sigma_X(5), \sigma_Y(5)\} = 10,$$

then the greedy choice is node 5, because

$$\text{argmax}\{g(1), g(2), g(5)\} = 5.$$

Furthermore, since $\sigma_X(5) > \sigma_Y(5)$, then node 5 is assigned to set $X$. The partial cut becomes $(X,Y) = (\{3,5\}, \{4\})$ with weight 25.

The remaining nodes are 1 and 2. Consider first node 1. Its contribution to the cut if added to set $Y$ is $\sigma_Y(1) = \sum_{u \in X} w_{1,u} = w_{1,5} = 6$. Since node 1 is not adjacent to node 4, the only node in $Y$, then $\sigma_X(1) = \sum_{u \in Y} w_{1,u} = 0$. Now, consider node 2. Its contribution to the cut if added to set $Y$ is $\sigma_Y(2) = \sum_{u \in X} w_{2,u} = w_{2,5} + w_{2,3} = 19$.

Since node 2 is not adjacent to node 4, the only node in $Y$, then $\sigma_X(2) = \sum_{u \in Y} w_{2,u} = 0$. Since the greedy function values are

$$g(1) = \max\{\sigma_X(1), \sigma_Y(1)\} = 6,$$
$$g(2) = \max\{\sigma_X(2), \sigma_Y(2)\} = 19,$$

then the greedy choice is node 2, because

$$\text{argmax}\{g(1), g(2)\} = 2.$$

Furthermore, since $\sigma_Y(2) > \sigma_X(2)$, then node 2 is assigned to set $Y$. The partial cut becomes $(X,Y) = (\{3,5\}, \{2,4\})$ with weight 44.

Finally, consider node 1 that is the last remaining. Its contribution to the cut if added to sets $X$ and $Y$ are, respectively, $\sigma_X(1) = \sum_{u \in Y} w_{1,u} = w_{1,2} = 5$ and $\sigma_Y(1) = \sum_{u \in X} w_{1,u} = w_{1,5} = 6$. Since $\sigma_Y(1) > \sigma_X(1)$, then node 1 is assigned to set $Y$. The final cut is $(S, \bar{S}) = (X,Y) = (\{3,5\}, \{1,2,4\})$ with weight 50. This cut is the best one of the four shown in Figure 12.9.  ∎

### 12.4.1.2 A semi-greedy algorithm for the maximum cut problem

Chapter 3 introduced the randomization of greedy algorithms to construct their semi-greedy variants. We next present a semi-greedy variant of the greedy algorithm for maximum cut described in Section 12.4.1.1.

To start the construction process, a large-weight edge $(i^*, j^*) \in U$ is selected and each of its endpoints is assigned to a different subset of the partial solution, e.g., $i^*$ is assigned to $X$ and $j^*$ to $Y$. To add variability to this choice process, one adopts a greedy randomized approach by building a restricted candidate list with all edges in $U$ having weights above the cutoff threshold $\mu = w_{min} + \alpha \cdot (w^{max} - w_{min})$, where $w_{min}$ and $w^{max}$ are, respectively, the smallest and largest edge weights of edges in $U$ and $\alpha$ is a real number in the interval $[0, 1]$. Edge $(i^*, j^*)$ is randomly selected from this initial restricted candidate list.

To define the construction mechanism for the restricted candidate list used at each iteration, let

$$w_{min} = \min\{\min_{v \in V'} \sigma_X(v), \min_{v \in V'} \sigma_Y(v)\}$$

and

$$w^{max} = \max\{\max_{v \in V'} \sigma_X(v), \max_{v \in V'} \sigma_Y v)\},$$

where $V' = V \setminus (X \cup Y)$ is the set of nodes that are not yet assigned to either subset $X$ or subset $Y$. Denoting by $\mu = w_{min} + \alpha \cdot (w^{max} - w_{min})$ the cut-off value, where $\alpha$ is a parameter such that $0 \leq \alpha \leq 1$, the restricted candidate list is made up by all

```
begin SEMI-GREEDY-MAXCUT;
1   Generate at random a real-valued parameter α ∈ [0,1];
2   w_min ← min{w_ij : (i,j) ∈ U};
3   w^max ← max{w_ij : (i,j) ∈ U};
4   μ ← w_min + α · (w^max − w_min);
5   RCL_e ← {(i,j) ∈ U : w_ij ≥ μ};
6   Select edge (i*, j*) at random from RCL_e;
7   X ← {i*};
8   Y ← {j*};
9   while X ∪ Y ≠ V do
10      V' ← V \ (X ∪ Y);
11      forall v ∈ V' do
12          σ_X(v) ← Σ_{u∈Y} w_vu;
13          σ_Y(v) ← Σ_{u∈X} w_vu;
14      end-forall;
15      w_min ← min{min_{v∈V'} σ_X(v), min_{v∈V'} σ_Y(v)};
16      w^max ← max{max_{v∈V'} σ_X(v), max_{v∈V'} σ_Y(v)};
17      μ ← w_min + α · (w^max − w_min);
18      RCL_v ← {v ∈ V' : max{σ_X(v), σ_Y(v)} ≥ μ};
19      Select vertex v* at random from RCL_v;
20      if σ_X(v*) > σ_Y(v*) then
21          X ← X ∪ {v*};
22      else
23          Y ← Y ∪ {v*};
24      end-if;
25  end-while;
26  S ← X;
27  S̄ ← Y;
28  return (S, S̄), w(S, S̄);
end SEMI-GREEDY-MAXCUT.
```

**Fig. 12.12** Pseudo-code of the semi-greedy GRASP construction phase algorithm for the MAX-CUT problem.

nodes whose value of the greedy function is greater than or equal to $\mu$. A node $v$ is randomly selected from the list. If $\sigma_X(v) > \sigma_Y(v)$, then node $v \in V'$ is placed in $X$; otherwise it is placed in $Y$.

The pseudo-code of the semi-greedy GRASP construction procedure for the maximum cut problem is shown in Figure 12.12. The restricted candidate list parameter $\alpha$ is generated at random in line 1. The initial edge of the cut is determined in lines 2 to 8. Lines 2 and 3 determine the smallest and largest edge weights $w_{min}$ and $w^{max}$, respectively. The cutoff value $\mu$ is computed in line 4 and the restricted candidate list $RCL_e$ is set up in line 5. Finally, in line 6, edge $(i^*, j^*)$ is randomly selected from $RCL_e$ and each endpoint of the selected edge is assigned in lines 7 and 8.

The while loop in lines 9 to 25 builds the remainder of the cut. It stops when a cut $(X, Y)$ is on hand, i.e., when $X \cup Y = V$. In line 10, the set $V'$ of candidate vertices still to be added to each side of the cut under construction is determined.

In lines 11 to 14, the incremental contributions $\sigma_X(v)$ and $\sigma_Y(v)$ associated with the addition of each vertex $v \in V'$ to subsets $X$ and $Y$, respectively, are computed. Lines 15 and 16 compute, respectively, the smallest and largest contributions of vertex $v \in V'$. Line 17 computes the cutoff value for membership in the restricted candidate list $RCL_v$, which is set up in line 18. The next vertex, $v^*$, to be added to $X$ or $Y$ is selected at random from $RCL_v$ in line 19. If it contributes more to the cut by being added to $X$, then it is added to that set in line 21. Otherwise, it is added to $Y$ in line 23. In lines 26 and 27, $X$ and $Y$ are assigned, respectively, to sets $S$ and $\bar{S}$ that form the cut $(S, \bar{S})$. Line 28 returns the constructed cut $(S, \bar{S})$ and its weight $w(S, \bar{S})$.

### 12.4.1.3 Local search for the maximum cut problem

Since a solution $(S, \bar{S})$ generated with the semi-greedy algorithm of Section 12.4.1.2 is not guaranteed to be locally optimum with respect to any neighborhood structure, a local search algorithm may improve its weight. We base the local search algorithm presented next on the following neighborhood structure. To each vertex $v \in V$, we associate either the neighbor $(S \setminus \{v\}, \bar{S} \cup \{v\})$ if $v \in S$, or the neighbor $(S \cup \{v\}, \bar{S} \setminus \{v\})$, otherwise. In other words, we move vertex $v$ from one side of the cut to the other. Let

$$\sigma_S(v) = \sum_{u \in \bar{S}} w_{vu} \qquad (12.12)$$

be the sum of the weights of the edges incident to $v$ that have their other endpoint in $\bar{S}$ and

$$\sigma_{\bar{S}}(v) = \sum_{u \in S} w_{vu}. \qquad (12.13)$$

be the sum of the weights of the edges incident to $v$ that have their other endpoint in $S$. The value

$$\delta(v) = \begin{cases} \sigma_{\bar{S}}(v) - \sigma_S(v), & \text{if } v \in S, \\ \sigma_S(v) - \sigma_{\bar{S}}(v), & \text{if } v \in \bar{S}, \end{cases}$$

represents the change in the objective function associated with moving vertex $v$ from one subset of the cut to the other. All possible moves are investigated. The current solution is replaced by the first improving neighbor found. The search stops after all possible moves have been evaluated and no improving neighbor is found.

The pseudo-code of the local search procedure is given in Figure 12.13. The loop from line 2 to line 16 is repeated until no improving move is possible. In line 3 the move indicator variable *change* is initialized to indicate that no move has been made. The for loop from line 4 to line 15 scans all vertices and attempts to move vertices from one set of the cut to the other. The loop concludes scanning the vertices either if all moves are tested and no move improves the total weight of the cut or if an improving move is found. Lines 5 to 7 moves node $v$ from $S$ to $\bar{S}$ if $\sigma_S(v) > \sigma_{\bar{S}}(v)$, where $\sigma_S(v)$ and $\sigma_{\bar{S}}(v)$ are defined, respectively, in equations (12.12) and (12.13). Line 8 sets the indicator variable *change* to indicate a move has been made. Line 10

```
begin LOCAL-SEARCH-MAXCUT((S,S̄));
1   change ← .TRUE.;
2   while change do
3       change ← .FALSE.;
4       for v = 1,...,|V| while .NOT.change do
5           if v ∈ S and σ_S̄(v) − σ_S(v) > 0 then
6               S ← S\{v};
7               S̄ ← S̄∪{v};
8               change ← .TRUE.;
9           else
10              if v ∈ S̄ and σ_S(v) − σ_S̄(v) > 0 then
11                  S̄ ← S̄\{v};
12                  S ← S∪{v};
13                  change ← .TRUE.;
13              end-if;
14          end-if;
15      end-for;
16  end-while;
17  return (S,S̄), w(S,S̄);
end LOCAL-SEARCH-MAXCUT.
```

**Fig. 12.13** Pseudo-code of the GRASP local search phase algorithm for the MAX-CUT problem.

to 12 moves node $v$ from $\bar{S}$ to $S$ if $\sigma_{\bar{S}}(v) > \sigma_S(v)$. Line 13 sets the indicator variable *change* to indicate a move has been made. A locally maximal cut $(S,\bar{S})$ and its weight $w(S,\bar{S})$ are returned by the local search procedure in line 17.

### 12.4.1.4 GRASP with path-relinking for maximum cut

As we saw in Chapter 9, during GRASP with path-relinking, trajectories connecting high-quality solutions in the search space graph $\mathscr{G} = (F,M)$ are explored in search of other high-quality solutions. In the GRASP with path-relinking algorithm described here, we apply the usual strategy of saving high-quality, or elite, solutions in an elite set $\mathscr{E}$ of size $n_{\mathscr{E}}$. This elite set is initially empty and it is constructed during the initial GRASP iterations. Then, after each GRASP iteration, one or more paths in $\mathscr{G}$ connecting the solution produced by the local search procedure and one of the elite solutions is explored in search for other high-quality solutions. In a forward path-relinking scheme, we specify the local search solution to be the initial solution and a randomly selected elite solution to be the guiding solution. A series of moves take the current solution from the initial solution to the guiding solution. Each move along the path introduces attributes contained in the guiding solution into the current solution. At each step, the chosen move to a solution in a restricted neighborhood of the current solution is usually one that maximizes some greedy criteria. As we saw in Chapter 8, choices other than a greedy choice are possible. However, in this discussion we assume that the move made is one that increases the weight of the current solution the most or, if no move increases its weight, then we choose one that decreases it the least.

We next describe a forward path-relinking procedure for the MAX-CUT problem, going from an initial solution $(S^i, \bar{S}^i)$ to a guiding elite solution $(S^g, \bar{S}^g)$. Note that other forms of path-relinking, such as backward and mixed path-relinking, can be applied in place of forward path-relinking. Also note that the guiding solution can be represented not only as $(S^g, \bar{S}^g)$, but also as $(\bar{S}^g, S^g)$. Since different solutions can be traversed in the path from an initial solution $(S^i, \bar{S}^i)$ to a guiding solution $(S^g, \bar{S}^g)$ and in the path from $(S^i, \bar{S}^i)$ to $(\bar{S}^g, S^g)$, then traversing both paths may enable the algorithm to find better solutions.

The path-relinking procedure starts by initializing the current solution with the initial solution, i.e., setting $(S, \bar{S}) = (S^i, \bar{S}^i)$ and computing the restricted neighborhood

$$N((S, \bar{S}) : (S^g, \bar{S}^g)) = N^+((S, \bar{S}) : (S^g, \bar{S}^g)) \cup N^-((S, \bar{S}) : (S^g, \bar{S}^g)),$$

where

$$N^+((S, \bar{S}) : (S^g, \bar{S}^g)) = \{(S \cup \{v\}, \bar{S} \setminus \{v\}) : v \in ((S \cup S^g) \setminus (S \cap S^g)) \cap S^g\}$$

and

$$N^-((S, \bar{S}) : (S^g, \bar{S}^g)) = \{(S \setminus \{v\}, \bar{S} \cup \{v\}) : v \in ((S \cup S^g) \setminus (S \cap S^g)) \cap S\}.$$

Set $N^+((S, \bar{S}) : (S^g, \bar{S}^g))$ is the neighborhood formed by solutions that result from moving from $\bar{S}$ to $S$ a vertex $v$ that belongs to $S^g$ (but not to $S$). Conversely, set $N^-((S, \bar{S}) : (S^g, \bar{S}^g))$ corresponds to the neighborhood formed by solutions that result from moving from $S$ to $\bar{S}$ a vertex $v$ that does not belong to $S^g$ (but belongs to $S$).

### Maximum cut problem – Path-relinking iteration

Consider the following example on the graph in Figure 12.11, where $V = \{1, 2, 3, 4, 5\}$. Let the current solution $(S, \bar{S})$ be such that $S = \{1, 2\}$ and $\bar{S} = \{3, 4, 5\}$ and let the guiding solution $(S^g, \bar{S}^g)$ be such that $S^g = \{1, 3, 5\}$ and $\bar{S}^g = \{2, 4\}$. Since $S \cup S^g = \{1, 2, 3, 5\}$ and $S \cap S^g = \{1\}$, then

$$((S \cup S^g) \setminus (S \cap S^g)) \cap S^g = \{2, 3, 5\} \cap \{1, 3, 5\} = \{3, 5\}$$

and

$$((S \cup S^g) \setminus (S \cap S^g)) \cap S = \{2, 3, 5\} \cap \{1, 2\} = \{2\}.$$

Consequently,

$$N((S,\bar{S}) : (S^g,\bar{S}^g)) = N^+((S,\bar{S}) : (S^g,\bar{S}^g)) \cup N^-((S,\bar{S}) : (S^g,\bar{S}^g))$$
$$= \{(S \cup \{v\}, \bar{S} \setminus \{v\}) : v \in ((S \cup S^g) \setminus (S \cap S^g)) \cap \bar{S}^g\} \cup$$
$$\{(S \setminus \{v\}, \bar{S} \cup \{v\}) : v \in ((S \cup S^g) \setminus (S \cap S^g)) \cap S\}$$
$$= \{(S \cup \{v\}, \bar{S} \setminus \{v\}) : v \in \{3,5\}\} \cup$$
$$\{(S \setminus \{v\}, \bar{S} \cup \{v\}) : v \in \{2\}\}$$
$$= \{(\{1,2,3\},\{4,5\}), (\{1,2,5\},\{3,4\}), (\{1\},\{2,3,4,5\})\}.$$

Since $w(\{1,2,3\},\{4,5\}) = 33$, $w(\{1,2,5\},\{3,4\}) = 21$, and $w(\{1\},\{2,3,4,5\}) = 11$, path-relinking would make the greedy choice and make the move to solution

$$(S,\bar{S}) = (\{1,2,3\},\{4,5\}).$$

∎

The best solution in the restricted neighborhood $N((S,\bar{S}) : (S^g,\bar{S}^g)) = N^+((S,\bar{S}) : (S^g,\bar{S}^g)) \cup N^-((S,\bar{S}) : (S^g,\bar{S}^g))$ is selected, the current solution is updated, and a new path-relinking iteration is performed, until the guiding solution is reached. The total number of iterations performed by path-relinking is

$$|N((S^i,\bar{S}^i) : (S^g,\bar{S}^g))| + |N((S^i,\bar{S}^i) : (\bar{S}^g,S^g))|,$$

where $(S^i,\bar{S}^i)$ is the starting solution and $(S^g,\bar{S}^g)$ and $(\bar{S}^g,S^g)$ are the two representations of the guiding solution.

Figure 12.14 shows the pseudo-code for a forward path-relinking algorithm for the maximum cut problem. The procedure takes as input an initial solution $(S^i,\bar{S}^i)$ and two representations of a guiding solution, $(S^g,\bar{S}^g)$ and $(\bar{S}^g,S^g)$, and returns a locally maximum cut $(S,\bar{S})$ in line 25. Lines 1 through 12 traverse a path from $(S^i,\bar{S}^i)$ to $(S^g,\bar{S}^g)$, while in lines 13 through 24 a path from $(S^i,\bar{S}^i)$ to $(\bar{S}^g,S^g)$ is traversed. Lines 1 and 2 initialize the current solution $(S,\bar{S})$ with the initial solution $(S^i,\bar{S}^i)$ and line 3 initializes the largest cut weight to $-\infty$. Traversal of the path from $(S^i,\bar{S}^i)$ to $(S^g,\bar{S}^g)$ in the solution space graph takes place in the while loop from line 4 to line 11. This loop is applied until the guiding solution is reached, i.e., until $N((S,\bar{S}) : (S^g,\bar{S}^g))$ becomes empty. In line 5, a best solution among all solutions in the restricted neighborhood $N((S,\bar{S}) : (S^g,\bar{S}^g))$ is assigned to $(S^+,\bar{S}^+)$ and, in line 6, the move to $(S^+,\bar{S}^+)$ is made. If the weight of the cut corresponding to the new solution is the largest seen so far in this path, then the cut and its weight are saved in $(S^*,\bar{S}^*)$ (line 8) and in $w^*$ (line 9), respectively. Since there is no guarantee that the solutions in the path traversed by path-relinking are local maxima, local search is applied to $(S^*,\bar{S}^*)$ in line 12 and the local maximum found is $(S^1,\bar{S}^1)$. Traversal of the second path is similar and takes place in lines 13 to 24. The local maximum found starting from the best solution in this path is $(S^2,\bar{S}^2)$ and is computed in line 24. The best solution $(S,\bar{S})$ returned in line 25 is the solution with the largest cut weight among $(S^1,\bar{S}^1)$ and $(S^2,\bar{S}^2)$.

Once a new solution $(S,\bar{S})$ is produced by path-relinking, the GRASP with path-relinking procedure verifies in lines 10 to 12 of the pseudo-code in Figure 12.10 if

```
begin PATH-RELINKING-MAXCUT((S^i, S̄^i), (S^g, S̄^g), (S̄^g, S̄^g));
1   S ← S^i;
2   S̄ ← S̄^i;
3   w* ← −∞;
4   while N((S, S̄) : (S^g, S̄^g)) ≠ ∅ do
5       (S^+, S^+) ← argmax{w(X,Y) : (X,Y) ∈ N((S, S̄) : (S^g, S̄^g))};
6       (S, S̄) ← (S^+, S^+);
7       if w(S, S̄) > w* then
8           (S*, S*) ← (S, S̄);
9           w* ← w(S, S̄);
10      end-if;
11  end-while;
12  (S^1, S̄^1) ← LOCAL-SEARCH-MAXCUT(S*, S̄*);
13  S ← S^i;
14  S̄ ← S̄^i;
15  w* ← −∞;
16  while N((S, S̄) : (S̄^g, S^g)) ≠ ∅ do
17      (S^+, S^+) ← argmax{w(X,Y) : (X,Y) ∈ N((S, S̄) : (S̄^g, S^g))};
18      (S, S̄) ← (S^+, S^+);
19      if w(S, S̄) > w* then
20          (S*, S*) ← (S, S̄);
21          w* ← w(S, S̄);
22      end-if;
23  end-while;
24  (S^2, S̄^2) ← LOCAL-SEARCH-MAXCUT(S*, S̄*);
25  return (S, S̄) = argmax{w(S^1, S̄^1), w(S^2, S̄^2)};
end PATH-RELINKING-MAXCUT.
```

**Fig. 12.14** Pseudo-code of the forward path-relinking procedure for the MAX-CUT problem.

this solution can be inserted into the elite set $\mathscr{E}$. Denote by $n_{\mathscr{E}}$ the maximum number of elite elements. If $|\mathscr{E}| < n_{\mathscr{E}}$, then $(S, \bar{S})$ is simply inserted into $\mathscr{E}$. Otherwise, if $|\mathscr{E}| = n_{\mathscr{E}}$, then two cases can arise. In the first case, if $w(S, \bar{S})$ is greater than the largest cut weight in $\mathscr{E}$, then $(S, \bar{S})$ is inserted into the pool $\mathscr{E}$. In the second case, it will be added to the pool $\mathscr{E}$ only if it is sufficiently different from all pool elements. In both cases, an elite solution $(S^-, \bar{S}^-)$ of less weight than $w(S, \bar{S})$ is replaced by $(S, \bar{S})$ in the pool. There can be one or more such lower-weight solutions in the pool. Among those, the chosen solution $(S^-, \bar{S}^-)$ is the one that is most similar to $(S, \bar{S})$.

## 12.5 Bibliographical notes

Applications of the 2-path network design problem can be found in the design of communication networks, in which paths with few edges are sought to enforce high reliability and small delays. Dahl and Johannessen (2004) proved that the decision version of the 2-path network design problem is *NP*-complete and proposed a greedy heuristic based on the linear relaxation of its integer programming formulation.

They also gave an exact cutting plane algorithm and presented computational results for randomly generated problems with up to 120 nodes, 7,140 edges, and 60 origin-destination pairs. The numerical results reported in Section 12.1 appeared originally in Ribeiro and Rosseti (2002), where 100 test instances with 70 nodes and 35 origin-destination pairs were randomly generated with the same parameters of those used by Dahl and Johannessen (2004), since neither the instances nor the code of the greedy heuristic were available for a straightforward comparison with the parallel GRASP algorithm. Student's $t$-test for unpaired observations was applied as in Jain (1991). Efficient parallel cooperative implementations of GRASP heuristics for the 2-path network design problem appeared in Ribeiro and Rosseti (2007), as did the implementation details of the algorithms used in the computational experiments.

The graph planarization problem discussed in Section 12.2 is NP-hard, see Liu and Geldmacher (1977). Applications of graph planarization include graph drawing, such as in CASE tools (Tamassia and Di Battista, 1988), automated graphical display systems, and numerous layout problems, such as circuit layout and the layout of industrial facilities (Hassan and Hogg, 1987). A survey of some of these applications appeared in Mutzel (1994). The two-phase GT heuristic for graph planarization was originally proposed by Goldschmidt and Takvorian (1994). Since the decision version of the problem of finding a maximum induced bipartite subgraph of an overlap graph is NP-complete (Sarrafzadeh and Lee, 1989), a greedy algorithm is used by the two-phase heuristic GT to construct a maximal induced bipartite subgraph of the overlap graph. Finding the maximum independent set of an overlap graph has been shown by Gavril (1973) to be polynomially solvable in time $O(|E|^3)$, where $|E|$ is the number of vertices of the overlap graph $H = (E, I)$ (see Golumbic (1980) for another description of this algorithm). The GRASP heuristic for graph planarization was developed by Resende and Ribeiro (1997), where extensive computational results have been reported. Its code is detailed and available from Ribeiro and Resende (1999).

The bandwidth packing problem introduced in Section 12.3 can be solved in polynomial time (Ouorou et al., 2000) if the cost function in each edge is convex. However, the problem is NP-hard if the flows are required to be integral (Even et al., 1976) or if each commodity is required to follow a single path from its source to its destination (Chlamtac et al., 1994). The cost function adopted in model (12.1)-(12.5) is the same piecewise linear function originally proposed by Fortz and Thorup (2000).

Several heuristics have been proposed for different variants of the bandwidth packing problem. One of the first algorithms for routing virtual circuits in communication networks was proposed by Yee and Lin (1992). Resende and Resende (1999) proposed a GRASP for frame relay permanent virtual circuit routing different from the one described in this chapter. Sung and Park (1995) developed a Lagrangean heuristic for a similar variant of this problem. Laguna and Glover (1993) considered a bandwidth packing problem in which they want to assign calls to paths in a capacitated graph, such that capacities are not violated and some measure of the total profit is maximized. They developed a tabu search algorithm which makes use of an efficient implementation of the $k$-shortest path algorithm.

Amiri et al. (1999) proposed another formulation for the bandwidth packing problem, considering both revenue losses and costs associated with communication delays as part of the objective. A heuristic procedure based on Lagrangean relaxation was applied for finding bounds and solutions. Shyur and Wen (2001) proposed a tabu search algorithm for optimizing the system of virtual paths. The objective function consisted in minimizing the maximum link load, by requiring that each route visits the minimum number of hubs. The load of a link is defined as the sum of the virtual path demands, summed over the virtual paths that traverse this link.

A number of exact approaches for solving variants of the bandwidth packing problem have also appeared in the literature. Parker and Ryan (1994) described a branch-and-bound procedure for optimally solving a bandwidth packing problem, in which the linear relaxation of the associated integer programming problem is solved by using column generation. LeBlanc et al. (1999) addressed packet switched telecommunication networks, considering restrictions on paths and flows: hop limits, node and link capacity constraints, and high- and low-priority flows. They minimize the expected queuing time and do not impose integrality constraints on the flows. Dahl et al. (1999) studied a network configuration problem in telecommunications, searching for paths in a capacitated network to accommodate a given traffic demand matrix. Their model also involves an intermediate pipe layer. The problem is formulated as an integer linear program. An associated integral polytope is studied and different classes of facets are described. Barnhart et al. (2000) proposed a branch-and-cut-and-price algorithm for origin-destination integer multicommodity flow problems. This problem is a constrained version of the linear multicommodity network flow problem, in which each flow can use only one path from its origin to its destination.

The GRASP algorithm for the bandwidth packing problem discussed in this chapter was originally proposed by Resende and Ribeiro (2003a), where computational results illustrating the trade-offs between different implementation strategies and their application in practice are reported in detail. The construction mechanism based on heuristic-biased stochastic sampling was introduced by Bresina (1996), in which the candidates are ranked according to the greedy function. Binato et al. (2002) also used this selection procedure, but restricted to elements of the RCL.

The GRASP with path-relinking heuristic for the maximum cut problem, discussed in this chapter, was originally proposed in Festa et al. (2002). In that paper, a variable neighborhood search (VNS) heuristic for MAX-CUT and its hybridization with path-relinking were also proposed.

The decision version of the MAX-CUT problem was proved to be NP-complete by Karp (1972). Applications of MAX-CUT are found in VLSI design and statistical physics, see, e.g., Barahona et al. (1988), Chang and Du (1987), Chen et al. (1983), and Pinter (1984), among others. The reader is referred to Poljak and Tuza (1995) for an introductory survey of MAX-CUT.

The idea that the MAX-CUT problem can be naturally relaxed to a semidefinite programming problem was first observed by Lovász (1979) and Shor (1987). Goemans and Williams (1995) proposed a randomized algorithm that uses semidefinite programming to achieve a performance guarantee of 0.87856 if the weights are non-negative. Algorithms for solving the semidefinite programming relaxation of MAX-CUT are particularly efficient because they explore the structure of the problem. One approach along this line is the use of interior point methods (Benson et al., 2000; Fujisawa et al., 1997; 2000).

Other nonlinear programming approaches have also been presented for the MAX-CUT semidefinite programming relaxation, solution see Helmberg and Rendl (2000) and Homer and Peinado (1997). Homer and Peinado (1997) reformulated the constrained problem as an unconstrained one and used the standard steepest-ascent method on the latter. A variant of the Homer and Peinado algorithm was proposed by Burer and Monteiro (2001). Their idea is based on the constrained nonlinear programming reformulation of the MAX-CUT semidefinite programming relaxation obtained by a change of variables.

Burer et al. (2001) proposed a rank-2 relaxation heuristic for MAX-CUT and described the circut computer code that produces better solutions in practice than the randomized algorithm of Goemans and Williamson.

Following the paper by Festa et al. (2002) that proposed GRASP, path-relinking, and variable neighborhood search for the MAX-CUT, other heuristics based on metaheuristic concepts were proposed, including a hierarchical social heuristic (Duarte et al., 2004), ant colony optimization (Gao et al., 2008), scatter search (Martí et al., 2009), memetic algorithms (Wu and Hao, 2012), genetic algorithms (Seo et al., 2012), tabu search (Kochenberger et al., 2013), and breakout local search (Benlic and Hao, 2013).

# References

E.H.L. Aarts and J. Korst. *Simulated annealing and Boltzmann machines: A stochastic approach to combinatorial optimization and neural computing.* Wiley, New York, 1989.

B. Adenso-Díaz, S. García-Carbajal, and S.M. Gupta. A path-relinking approach for a bi-criteria disassembly sequencing problem. *Computers & Operations Research*, 35:3989–3997, 2008.

R. Agrawal and R. Srikant. Fast algorithms for mining association rules. In *Proceedings of the 20th International Conference on Very Large Data Bases*, pages 487–499. Morgan Kaufmann Publishers, 1994.

R.M. Aiex and M.G.C. Resende. Parallel strategies for GRASP with path-relinking. In T. Ibaraki, K. Nonobe, and M. Yagiura, editors, *Metaheuristics: Progress as real problem solvers*, pages 301–331. Springer, New York, 2005.

R.M. Aiex, M.G.C. Resende, and C.C. Ribeiro. Probability distribution of solution time in GRASP: An experimental investigation. *Journal of Heuristics*, 8:343–373, 2002.

R.M. Aiex, S. Binato, and M.G.C. Resende. Parallel GRASP with path-relinking for job shop scheduling. *Parallel Computing*, 29:393–430, 2003.

R.M. Aiex, M.G.C. Resende, P.M. Pardalos, and G. Toraldo. GRASP with path relinking for three-index assignment. *INFORMS Journal on Computing*, 17: 224–247, 2005.

R.M. Aiex, M.G.C. Resende, and C.C. Ribeiro. TTTPLOTS: A perl program to create time-to-target plots. *Optimization Letters*, 1:355–366, 2007.

E. Alba. *Parallel metaheuristics: A new class of algorithms.* Wiley, New York, 2005.

D. Aloise and C.C. Ribeiro. Adaptive memory in multistart heuristics for multicommodity network design. *Journal of Heuristics*, 17:153–179, 2011.

G.A. Alvarez-Perez, J.L. González-Velarde, and J.W. Fowler. Crossdocking – Just in time scheduling: An alternative solution approach. *Journal of the Operational Research Society*, 60:554–564, 2008.

© Springer Science+Business Media New York 2016
M.G.C. Resende, C.C. Ribeiro, *Optimization by GRASP*,
DOI 10.1007/978-1-4939-6530-4

R. Alvarez-Valdes, F. Parreño, and J.M. Tamarit. A GRASP algorithm for constrained two-dimensional non-guillotine cutting problems. *Journal of the Operational Research Society*, 56:414–425, 2004.

R. Alvarez-Valdes, E. Crespo, J.M. Tamarit, and F. Villa. GRASP and path relinking for project scheduling under partially renewable resources. *European Journal of Operational Research*, 189:1153–1170, 2008a.

R. Alvarez-Valdes, F. Parreño, and J.M. Tamarit. Reactive GRASP for the strip-packing problem. *Computers & Operations Research*, 35:1065–1083, 2008b.

R. Alvarez-Valdes, F. Parreño, and J.M. Tamarit. A GRASP/path relinking algorithm for two- and three-dimensional multiple bin-size bin packing problems. *Computers & Operations Research*, 40:3081–3090, 2013.

A.C. Alvim and C.C. Ribeiro. Load balancing for the parallelization of the GRASP metaheuristic. In *Proceedings of the X Brazilian Symposium on Computer Architecture*, pages 279–282, Búzios, 1998.

A. Amiri, E. Rolland, and R. Barkhi. Bandwidth packing with queueing delay costs: Bounding and heuristic solution procedures. *European Journal of Operational Research*, 112:635–645, 1999.

K.P. Anagnostopoulos, P.D. Chatzoglou, and S. Katsavounis. A reactive greedy randomized adaptive search procedure for a mixed integer portfolio optimization problem. *Managerial Finance*, 36:1057–1065, 2010.

D.V. Andrade and M.G.C. Resende. GRASP with path-relinking for network migration scheduling. In *Proceedings of the International Network Optimization Conference*, Spa, 2007a. URL http://bit.ly/1NfaTK0. Last visited on April 16, 2016.

D.V. Andrade and M.G.C. Resende. GRASP with evolutionary path-relinking. In *Proceedings of the Seventh Metaheuristics International Conference*, Montreal, 2007b.

L.M.M.S. Andrade, R.B. Xavier, L.A.F. Cabral, and A.A. Formiga. Parallel construction for continuous GRASP optimization on GPUs. In *Anais do XLVI Simpósio Brasileiro de Pesquisa Operacional*, pages 2393–2404, Salvador, 2014. URL http://bit.ly/1SS3lte. Last visited on April 16, 2016.

A.A. Andreatta and C.C. Ribeiro. Heuristics for the phylogeny problem. *Journal of Heuristics*, 8:429–447, 2002.

C.H. Antunes, E. Oliveira, and P. Lima. A multi-objective GRASP procedure for reactive power compensation planning. *Optimization and Engineering*, 15:199–215, 2014.

D.L. Applegate, R.E. Bixby, V. Chvátal, and W.J. Cook. *The traveling salesman problem: A computational study*. Princeton University Press, Princeton, 2006.

T.M.U. Araújo, L.M.M.S. Andrade, C. Magno, L.A.F. Cabral, R.Q. Nascimento, and C.N. Meneses. DC-GRASP: Directing the search on continuous-GRASP. *Journal of Heuristics*, 2015. doi: 10.1007/s10732-014-9278-6. Published online on 6 January 2015.

V.A. Armentano and O.C.B. Araujo. GRASP with memory-based mechanisms for minimizing total tardiness in single machine scheduling with setup times. *Journal of Heuristics*, 12:427–446, 2006.

J.E.C. Arroyo, P.S. Vieira, and D.S. Vianna. A GRASP algorithm for the multi-criteria minimum spanning tree problem. *Annals of Operations Research*, 159: 125–133, 2008.

L. Bahiense, G.C. Oliveira, M. Pereira, and S. Granville. A mixed integer disjunctive model for transmission network expansion. *IEEE Transactions on Power Systems*, 16:560–565, 2001.

E. Balas and M.J. Saltzman. An algorithm for the three-index assignment problem. *Operations Research*, 39:150–161, 1991.

J. Bang-Jensen, G. Gutin, and A. Yeo. When the greedy algorithm fails. *Discrete Optimization*, 1:121–127, 2004.

F. Barahona, M. Grötschel, M. Jürgen, and G. Reinelt. An application of combinatorial optimization to statistical optimization and circuit layout design. *Operations Research*, 36:493–513, 1988.

H. Barbalho, I. Rosseti, S.L. Martins, and A. Plastino. A hybrid data mining GRASP with path-relinking. *Computers & Operations Research*, 40:3159–3173, 2013.

J.F. Bard, Y. Shao, and A.I. Jarrah. A sequential GRASP for the therapist routing and scheduling problem. *Journal of Scheduling*, 17:109–133, 2014.

C. Barnhart, C.A. Hane, and P.H. Vance. Using branch-and-price-and-cut to solve origin-destination integer multicommodity flow problems. *Operations Research*, 48:318–326, 2000.

V. Bartkutė and L. Sakalauskas. Statistical inferences for termination of Markov type random search algorithms. *Journal of Optimization Theory and Applications*, 141:475–493, 2009.

V. Bartkutė, G. Felinskas, and L. Sakalauskas. Optimality testing in stochastic and heuristic algorithms. Technical Report 12, Vilnius Gediminas Technical University, Vilnius, 2006.

R. Battiti and G. Tecchiolli. Parallel biased search for combinatorial optimization: Genetic algorithms and tabu. *Microprocessors and Microsystems*, 16:351–367, 1992.

J.E. Beasley. An algorithm for set-covering problems. *European Journal of Operational Research*, 31:85–93, 1987.

J.E. Beasley. OR-Library: Distributing test problems by electronic mail. *Journal of the Operational Research Society*, 41:1069–1072, 1990a.

J.E. Beasley. A Lagrangean heuristic for set-covering problems. *Naval Research Logistics*, 37:151–164, 1990b.

J.E. Beasley. Lagrangean relaxation. In C.R. Reeves, editor, *Modern heuristic techniques for combinatorial problems*, pages 243–303. Blackwell Scientific Publications, Oxford, 1993.

U. Benlic and J.-K. Hao. Breakout local search for the Max-Cut problem. *Engineering Applications of Artificial Intelligence*, 26:1162–1173, 2013.

S. Benson, Y. Ye, and X. Zhang. Solving large-scale sparse semidefinite programs for combinatorial optimization. *SIAM Journal on Optimization*, 10:443–461, 2000.

D. Berger, B. Gendron, J.-Y Potvin, S. Raghavan, and P. Soriano. Tabu search for a network loading problem with multiple facilities. *Journal of Heuristics*, 6:253–267., 2000.

D. Bertsimas and R. Weismantel. *Optimization over integers*. Dynamic Ideas, Belmont, 2005.

S. Binato and G.C. Oliveira. A reactive GRASP for transmission network expansion planning. In C.C. Ribeiro and P. Hansen, editors, *Essays and surveys in metaheuristics*, pages 81–100. Kluwer Academic Publishers, Boston, 2002.

S. Binato, W.J. Hery, D. Loewenstern, and M.G.C. Resende. A GRASP for job shop scheduling. In C.C. Ribeiro and P. Hansen, editors, *Essays and surveys in metaheuristics*, pages 59–79. Kluwer Academic Publishers, Boston, 2002.

E.G. Birgin and J.M. Martínez. Large-scale active-set box-constrained optimization method with spectral projected gradients. *Computational Optimization and Applications*, 23:101–125, 2002.

E.G. Birgin, E.M. Gozzi, M.G.C. Resende, and R.M.A. Silva. Continuous GRASP with a local active-set method for bound-constrained global optimization. *Journal of Global Optimization*, 48:289–310, 2010.

C.G.E. Boender and A.H.G. Rinnooy Kan. Bayesian stopping rules for multistart global optimization methods. *Mathematical Programming*, 37:59–80, 1987.

J.A. Bondy and U.S.R. Murty. *Graph theory with applications*. Elsevier, 1976.

M. Boudia, M.A.O. Louly, and C. Prins. A reactive GRASP and path relinking for a combined production–distribution problem. *Computers & Operations Research*, 34:3402–3419, 2007.

J.L. Bresina. Heuristic-biased stochastic sampling. In *Proceedings of the Thirteenth National Conference on Artificial Intelligence*, pages 271–278, Portland, 1996. Association for the Advancement of Artificial Intelligence.

S. Burer and R.D.C. Monteiro. A projected gradient algorithm for solving the Max-Cut SDP relaxation. *Optimization Methods and Software*, 15:175–200, 2001.

S. Burer, R.D.C. Monteiro, and Y. Zhang. Rank-two relaxation heuristics for MAX-CUT and other binary quadratic programs. *SIAM Journal on Optimization*, 12: 503–521, 2001.

R.E. Burkard and K. Fröhlich. Some remarks on 3-dimensional assignment problems. *Methods of Operations Research*, 36:31–36, 1980.

R.E. Burkard and R. Rudolf. Computational investigations on 3-dimensional axial assignment problems. *Belgian Journal of Operational Research, Statistics and Computer Science*, 32:85–98, 1993.

R.E. Burkard, R. Rudolf, and G.J. Woeginger. Three-dimensional axial assignment problems with decomposable cost coefficients. *Discrete Applied Mathematics*, 65:123–139, 1996.

E.K. Burke and G. Kendall, editors. *Search methodologies: Introductory tutorials in optimization and decision support techniques*. Springer, New York, 2005.

E.K. Burke and G. Kendall, editors. *Search methodologies: Introductory tutorials in optimization and decision support techniques*. Springer, New York, 2nd edition, 2014.

S.I. Butenko, C.W. Commander, and P.M. Pardalos. A GRASP for broadcast scheduling in ad-hoc TDMA networks. In *Proceedings of the International Conference on Computing, Communications, and Control Technologies*, volume 5, pages 322–328, Austin, 2004.

R.G. Cano, G. Kunigami, C.C. de Souza, and P.J. de Rezende. A hybrid GRASP heuristic to construct effective drawings of proportional symbol maps. *Computers & Operations Research*, 40:1435–1447, 2013.

S.A. Canuto, M.G.C. Resende, and C.C. Ribeiro. Local search with perturbations for the prize-collecting Steiner tree problem in graphs. *Networks*, 38:50–58, 2001.

B. Cao and F. Glover. Tabu search and ejection chains – Application to a node weighted version of the cardinality-constrained TSP. *Management Science*, 43: 908–921, 1997.

S. Casey and J. Thompson. GRASPing the examination scheduling problem. In E. Burke and P. De Causmaecker, editors, *Practice and theory of automated timetabling IV*, volume 2740 of *Lecture Notes in Computer Science*, pages 232–244. Springer, Berlin, 2003.

L. Cavique, C. Rego, and I. Themido. Subgraph ejection chains and tabu search for the crew scheduling problem. *Journal of the Operational Research Society*, 50: 608–616, 1999.

J.M. Chambers, W.S. Cleveland, B. Kleiner, and P.A. Tukey. *Graphical methods for data analysis*. Duxbury Press, Boston, 1983.

K.C. Chang and D.-Z. Du. Efficient algorithms for layer assignment problems. *IEEE Transactions on Computer-Aided Design*, CAD-6:67–78, 1987.

W.A. Chaovalitwongse, C.A.S Oliveira, B. Chiarini, P.M. Pardalos, and M.G.C. Resende. Revised GRASP with path-relinking for the linear ordering problem. *Journal of Combinatorial Optimization*, 22:572–593, 2011.

I. Charon and O. Hudry. The noising method: A new method for combinatorial optimization. *Operations Research Letters*, 14:133–137, 1993.

I. Charon and O. Hudry. The noising methods: A survey. In C.C. Ribeiro and P. Hansen, editors, *Essays and surveys in metaheuristics*, pages 245–261. Kluwer Academic Publishers, Boston, 2002.

R. Chen, Y. Kajitani, and S. Chan. A graph-theoretic via minimization algorithm for two-layer printed circuit boards. *IEEE Transactions on Circuits and Systems*, CAS-30:284–299, 1983.

M. Chica, O. Cordón, S. Damas, and J. Bautista. A multiobjective GRASP for the 1/3 variant of the time and space assembly line balancing problem. In N. García-Pedrajas, F. Herrera, C. Fyfe, J. Benítez, and M. Ali, editors, *Trends in applied intelligent systems*, volume 6098 of *Lecture Notes in Computer Science*, pages 656–665. Springer, Berlin, 2010.

I. Chlamtac, A. Faragó, and T. Zhang. Optimizing the system of virtual paths. *IEEE/ACM Transactions on Networking*, 2:581–587, 1994.

V. Chvátal. A greedy heuristic for the set-covering problem. *Mathematics of Operations Research*, 4:233–235, 1979.

A. Cobham. The intrinsic computational difficulty of functions. In Y. Bar-Hillel, editor, *Proceedings of the 1964 International Congress for Logical Methodology and Philosophy of Science*, pages 24–30, Amsterdam, 1964. North Holland.

C.W. Commander, S.I. Butenko, P.M. Pardalos, and C.A.S. Oliveira. Reactive GRASP with path relinking for broadcast scheduling. In *Proceedings of the 40th Annual International Telemetry Conference*, pages 792–800, San Diego, 2004.

S.A. Cook. The complexity of theorem-proving procedures. In M.A. Harrison, R.B. Banerji, and J.D. Ullman, editors, *Proceedings of the Third Annual ACM Symposium on Theory of Computing*, pages 151–158, New York, 1971. ACM.

R. Cordone and G. Lulli. A GRASP metaheuristic for microarray data analysis. *Computers & Operations Research*, 40:3108–3120, 2013.

T.H. Cormen, C.E. Leiserson, R.L. Rivest, and C. Stein. *Introduction to Algorithms*. MIT Press, Cambridge, 3rd edition, 2009.

C. Cotta and A.J. Fernández. A hybrid GRASP–evolutionary algorithm approach to Golomb ruler search. In X. Yao, E.K. Burke, J.A. Lozano, J. Smith, J.J. Merelo-Guervós, J.A. Bullinaria, J.E. Rowe, P. Tiňo, A. Kabán, and H.-P. Schwefel, editors, *Parallel Problem Solving from Nature*, volume 3242 of *Lecture Notes in Computer Science*, pages 481–490. Springer, Berlin, 2004.

Y. Crama and F.C.R. Spieksma. Approximation algorithms for three-dimensional assignment problems with triangle inequalities. *European Journal of Operational Research*, 60:273–279, 1992.

G.L. Cravo, G.M. Ribeiro, and L.A.N. Lorena. A greedy randomized adaptive search procedure for the point-feature cartographic label placement. *Computers & Geosciences*, 34:373–386, 2008.

G.A. Croes. A method for solving traveling-salesman problems. *Operations Research*, 6:791–812, 1958.

W.B. Crowston, F. Glover, G.L. Thompson, and J.D. Trawick. Probabilistic and parametric learning combinations of local job shop scheduling rules. Technical Report 117, Carnegie-Mellon University, Pittsburgh, 1963.

V.-D. Cung, S.L. Martins, C.C. Ribeiro, and C. Roucairol. Strategies for the parallel implementation of metaheuristics. In C.C. Ribeiro and P. Hansen, editors, *Essays and surveys in metaheuristics*, pages 263–308. Kluwer Academic Publishers, Boston, 2002.

V.B. da Silva, M. Ritt, J.B. da Paz Carvalho, M.J. Brusso, and J.T. da Silva. Identificação da maior elipse com excentricidade prescrita inscrita em um polígono não convexo através do Continuous GRASP. *Revista Brasileira de Computação Aplicada*, 4:61–70, 2012.

G. Dahl and B. Johannessen. The 2-path network design problem. *Networks*, 43: 190–199, 2004.

G. Dahl, A. Martin, and M. Stoer. Routing through virtual paths in layered telecommunication networks. *Operations Research*, 47:693–702, 1999.

G.B. Dantzig. *Linear programming and extensions*. Princeton University Press, Princeton, 1953.

M.M. D'Apuzzo, A. Migdalas, P.M. Pardalos, and G. Toraldo. Parallel computing in global optimization. In E. Kontoghiorghes, editor, *Handbook of parallel computing and statistics*. Chapman & Hall / CRC, Boca Raton, 2006.

S. Das and S.M. Idicula. Application of reactive GRASP to the biclustering of gene expression data. In *Proceedings of the International Symposium on Biocomputing*, page 14, Calicut, 2010. ACM.

H. Davoudpour and M. Ashrafi. Solving multi-objective SDST flexible flow shop using GRASP algorithm. *The International Journal of Advanced Manufacturing Technology*, 44:737–747, 2009.

H. Delmaire, J.A. Díaz, E. Fernández, and M. Ortega. Reactive GRASP and tabu search based heuristics for the single source capacitated plant location problem. *INFOR*, 37:194–225, 1999.

X. Delorme, X. Gandibleux, and J. Rodriguez. GRASP for set packing problems. *European Journal of Operational Research*, 153:564–580, 2004.

X. Delorme, X. Gandibleux, and F. Degoutin. Evolutionary, constructive and hybrid procedures for the bi-objective set packing problem. *European Journal of Operational Research*, 204:206–217, 2010.

Y. Deng and J.F. Bard. A reactive GRASP with path relinking for capacitated clustering. *Journal of Heuristics*, 17:119–152, 2011.

Y. Deng, J.F. Bard, G.R. Chacon, and J. Stuber. Scheduling back-end operations in semiconductor manufacturing. *IEEE Transactions on Semiconductor Manufacturing*, 23:210–220, 2010.

S. Dharan and A.S. Nair. Biclustering of gene expression data using reactive greedy randomized adaptive search procedure. *BMC Bioinformatics*, 10 (Suppl 1):S27, 2009.

R. Diestel. *Graph theory*. Springer, New York, 2010.

N. Dodd. Slow annealing versus multiple fast annealing runs: An empirical investigation. *Parallel Computing*, 16:269–272, 1990.

C. Dorea. Stopping rules for a random optimization method. *SIAM Journal on Control and Optimization*, 28:841–850, 1990.

U. Dorndorf and E. Pesch. Fast clustering algorithms. *INFORMS Journal on Computing*, 6:141–153, 1994.

S.E. Dreyfus and R.A. Wagner. The Steiner problem in graphs. *Networks*, 1: 195–201, 1972.

L.M.A. Drummond, L.S. Vianna, M.B. Silva, and L.S. Ochi. Distributed parallel metaheuristics based on GRASP and VNS for solving the traveling purchaser problem. In *Proceedings of the Ninth International Conference on Parallel and Distributed Systems*, pages 257–263, Chungli, 2002. IEEE.

A. Duarte and R. Martí. Tabu search and GRASP for the maximum diversity problem. *European Journal of Operational Research*, 178:71–84, 2007.

A. Duarte, F. Fernández, Á. Sánchez, and A. Sanz. A hierarchical social metaheuristic for the Max-Cut problem. In J. Gottlieb and G.R. Raidl, editors, *Evolutionary computation in combinatorial optimization*, volume 3004 of *Lecture Notes in Computer Science*, pages 84–94. Springer, Berlin, 2004.

A. Duarte, R. Martí, M.G.C. Resende, and R.M.A. Silva. GRASP with path relinking heuristics for the antibandwidth problem. *Networks*, 58:171–189, 2011.

A. Duarte, R. Martí, A. Álvarez, and F. Ángel-Bello. Metaheuristics for the linear ordering problem with cumulative costs. *European Journal of Operational Research*, 216:270–277, 2012.

A. Duarte, J. Sánchez-Oro, M.G.C. Resende, F. Glover, and R. Martí. GRASP with exterior path relinking for differential dispersion minimization. *Information Sciences*, 296:46–60, 2015.

A.R. Duarte, C.C. Ribeiro, and S. Urrutia. A hybrid ILS heuristic to the referee assignment problem with an embedded MIP strategy. In T. Bartz-Beielstein, M.J.B. Aguilera, C. Blum, B. Naujoks, A. Roli, G. Rudolph, and M. Sampels, editors, *Hybrid metaheuristics*, volume 4771 of *Lecture Notes in Computer Science*, pages 82–95. Springer, Berlin, 2007a.

A.R. Duarte, C.C. Ribeiro, S. Urrutia, and E.H. Haeusler. Referee assignment in sports leagues. In E.K. Burke and H. Rudová, editors, *Practice and theory of automated timetabling VI*, volume 3867 of *Lecture Notes in Computer Science*, pages 158–173. Springer, Berlin, 2007b.

C. Duin and S. Voss. The Pilot method: A strategy for heuristic repetition with application to the Steiner problem in graphs. *Networks*, 34:181–191, 1999.

S. Duni Ekşoğlu, P.M. Pardalos, and M.G.C. Resende. Parallel metaheuristics for combinatorial optimization. In R. Corrêa, I. Dutra, M. Fiallos, and F. Gomes, editors, *Models for parallel and distributed computation – Theory, algorithmic techniques and applications*, pages 179–206. Kluwer Academic Publishers, Boston, 2002.

K. Easton, G. Nemhauser, and M.A. Trick. The travelling tournament problem: Description and benchmarks. In T. Walsh, editor, *Principles and practice of constraint programming*, volume 2239 of *Lecture Notes in Computer Science*, pages 580–585. Springer, Berlin, 2001.

J. Edmonds. Paths, trees, and flowers. *Canadian Journal of Mathematics*, 17: 449–467, 1965.

J. Edmonds. Matroids and the greedy algorithm. *Mathematical Programming*, 1: 125–136, 1971.

J. Edmonds. Minimum partition of a matroid in independent subsets. *Journal of Research, National Bureau of Standards*, 69B:67–72, 1975.

H.T. Eikelder, M. Verhoeven, T. Vossen, and E. Aarts. A probabilistic analysis of local search. In I. Osman and J. Kelly, editors, *Metaheuristics: Theory and applications*, pages 605–618. Kluwer Academic Publishers, Boston, 1996.

M. Essafi, X. Delorme, and A. Dolgui. A reactive GRASP and path relinking for balancing reconfigurable transfer lines. *International Journal of Production Research*, 50:5213–5238, 2012.

S. Even, A. Itai, and A. Shamir. On the complexity of timetable and multicommodity flow problems. *SIAM Journal on Computing*, 5:691–703, 1976.

H. Faria Jr., S. Binato, M.G.C. Resende, and D.J. Falcão. Transmission network design by a greedy randomized adaptive path relinking approach. *IEEE Transactions on Power Systems*, 20:43–49, 2005.

T.A. Feo and M.G.C. Resende. A probabilistic heuristic for a computationally difficult set covering problem. *Operations Research Letters*, 8:67–71, 1989.

T.A. Feo and M.G.C. Resende. Greedy randomized adaptive search procedures. *Journal of Global Optimization*, 6:109–133, 1995.

T.A. Feo, M.G.C. Resende, and S.H. Smith. A greedy randomized adaptive search procedure for maximum independent set. Technical report, AT&T Bell Laboratories, 1989.

T.A. Feo, M.G.C. Resende, and S.H. Smith. A greedy randomized adaptive search procedure for maximum independent set. *Operations Research*, 42:860–878, 1994.

P. Festa and M.G.C. Resende. GRASP: An annotated bibliography. In C.C. Ribeiro and P. Hansen, editors, *Essays and surveys in metaheuristics*, pages 325–367. Kluwer Academic Publishers, Boston, 2002.

P. Festa and M.G.C. Resende. An annotated bibliography of GRASP, Part I: Algorithms. *International Transactions in Operational Research*, 16:1–24, 2009a.

P. Festa and M.G.C. Resende. An annotated bibliography of GRASP, Part II: Applications. *International Transactions in Operational Research*, 16, 2009b. 131–172.

P. Festa and M.G.C. Resende. Hybridizations of GRASP with path-relinking. In E-G. Talbi, editor, *Hybrid metaheuristics*, volume 434 of *Studies in Computational Intelligence*, pages 135–155. Springer, New York, 2013.

P. Festa, P.M. Pardalos, M.G.C. Resende, and C.C. Ribeiro. Randomized heuristics for the MAX-CUT problem. *Optimization Methods and Software*, 7:1033–1058, 2002.

P. Festa, P.M. Pardalos, L.S. Pitsoulis, and M.G.C. Resende. GRASP with path-relinking for the weighted maximum satisfiability problem. *Lecture Notes in Computer Science*, 3503:367–379, 2005.

P. Festa, P.M. Pardalos, L.S. Pitsoulis, and M.G.C. Resende. GRASP with path-relinking for the weighted MAXSAT problem. *ACM Journal of Experimental Algorithmics*, 11:1–16, 2006.

M.L. Fisher. The Lagrangean relaxation method for solving integer programming problems. *Management Science*, 50:1861–1871, 2004.

C. Fleurent and F. Glover. Improved constructive multistart strategies for the quadratic assignment problem using adaptive memory. *INFORMS Journal on Computing*, 11:198–204, 1999.

E. Fonseca, R. Fuchsuber, L.F.M. Santos, A. Plastino, and S.L. Martins. Exploring the hybrid metaheuristic DM-GRASP for efficient server replication for reliable multicast. In *International Conference on Metaheuristics and Nature Inspired Computing*, Hammamet, 2008.

B. Fortz and M. Thorup. Increasing Internet capacity using local search. *Computational Optimization and Applications*, 29:13–48, 2000.

A.M. Frieze. Complexity of a 3-dimensional assignment problem. *European Journal of Operational Research*, 13:161–164, 1983.

R.D. Frinhani, R.M. Silva, G.R. Mateus, P. Festa, and M.G.C. Resende. GRASP with path-relinking for data clustering: A case study for biological data. In P.M. Pardalos and S. Rebennack, editors, *Experimental algorithms*, volume 6630 of *Lecture Notes in Computer Science*, pages 410–420. Springer, Berlin, 2011.

K. Fujisawa, M. Fojima, and K. Nakata. Exploiting sparsity in primal-dual interior-point methods for semidefinite programming. *Mathematical Programming*, 79: 235–253, 1997.

K. Fujisawa, M. Fukuda, M. Kojima, and K. Nakata. Numerical evaluation of SDPA (Semidefinite Programming Algorithm). In H. Frenk, K. Roos, T. Terlaky, and S. Zhang, editors, *High performance optimization*, pages 267–301. Springer, Boston, 2000.

L. Gao, Y. Zeng, and A. Dong. An ant colony algorithm for solving Max-cut problem. *Progress in Natural Science*, 18:1173–1178, 2008.

M.R. Garey and D.S. Johnson. Approximation algorithms for combinatorial problems: An annotated bibliography. In J.F. Traub, editor, *Algorithms and complexity: New directions and recent results*, pages 41–52. Academic Press, Orlando, 1976.

M.R. Garey and D.S. Johnson. Strong NP-completeness results: Motivation, examples, and implications. *Journal of the ACM*, 25:499–508, 1978.

M.R. Garey and D.S. Johnson. *Computers and intractability*. Freeman, San Francisco, 1979.

F. Gavril. Algorithms for a maximum clique and a maximum independent set of a circle graph. *Networks*, 3:261–273, 1973.

A. Geist, A. Beguelin, J. Dongarra, W. Jiang, R. Mancheck, and V. Sunderam. *PVM: Parallel virtual machine, A user's guide and tutorial for networked parallel computing*. Scientific and Engineering Computation. MIT Press, Cambridge, 1994.

M. Gendreau and J.-Y. Potvin, editors. *Handbook of metaheuristics*. Springer, New York, 2nd edition, 2010.

J.B. Ghosh. Computational aspects of the maximum diversity problem. *Operations Research Letters*, 19:175–181, 1996.

F. Glover. Tabu Search - Part I. *ORSA Journal on Computing*, 1:190–206, 1989.

F. Glover. Tabu Search - Part II. *ORSA Journal on Computing*, 2:4–32, 1990.

F. Glover. Multilevel tabu search and embedded search neighborhoods for the traveling salesman problem. Technical report, University of Colorado, Boulder, 1991.

F. Glover. Ejection chains, reference structures and alternating path methods for traveling salesman problems. *Discrete Applied Mathematics*, 65:223–253, 1996a.

F. Glover. Tabu search and adaptive memory programing – Advances, applications and challenges. In R.S. Barr, R.V. Helgason, and J.L. Kennington, editors, *Interfaces in computer science and operations research*, pages 1–75. Kluwer Academic Publishers, Boston, 1996b.

F. Glover. Multi-start and strategic oscillation methods – Principles to exploit adaptive memory. In M. Laguna and J.L. González-Velarde, editors, *Computing tools for modeling, optimization and simulation: Interfaces in computer science and operations research*, pages 1–24. Kluwer Academic Publishers, Boston, 2000.

F. Glover. Exterior path relinking for zero-one optimization. *International Journal of Applied Metaheuristic Computing*, 5(3):1–8, 2014.

F. Glover and G. Kochenberger, editors. *Handbook of metaheuristics*. Kluwer Academic Publishers, Boston, 2003.

F. Glover and M. Laguna. *Tabu search*. Kluwer Academic Publishers, Boston, 1997.

F. Glover and A.P. Punnen. The travelling salesman problem: New solvable cases and linkages with the development of approximation algorithms. *Journal of the Operational Research Society*, 48:502–510, 1997.

F. Glover, M. Laguna, and R. Martí. Fundamentals of scatter search and path relinking. *Control and Cybernetics*, 39:653–684, 2000.

F. Glover, M. Laguna, and R. Martí. Scatter search and path relinking: Advances and applications. In F. Glover and G. Kochenberger, editors, *Handbook of metaheuristics*, pages 1–35. Kluwer Academic Publishers, Boston, 2003.

F. Glover, M. Laguna, and R. Martí. Scatter search and path relinking: Foundations and advanced designs. In G.C. Onwubolu and B.V. Babu, editors, *New optimization techniques in engineering*, volume 141 of *Studies in Fuzziness and Soft Computing*, pages 87–100. Springer, Berlin, 2004.

M.X. Goemans and Y. Myung. A catalog of Steiner tree formulations. *Networks*, 23:19–28, 1993.

M.X. Goemans and D.P. Williams. Improved approximation algorithms for Max-Cut and Satisfiability problems using semidefinite programming. *Journal of the ACM*, 42:1115–1145, 1995.

M.X. Goemans and D.P. Williamson. The primal dual method for approximation algorithms and its application to network design problems. In D. Hochbaum, editor, *Approximation algorithms for NP-hard problems*, pages 144–191. PWS Publishing Company, Boston, 1996.

B. Goethals and M.J. Zaki. Advances in frequent itemset mining implementations: Introduction to FIMI03. In B. Goethals and M.J. Zaki, editors, *Proceedings of the IEEE ICDM 2003 Workshop on Frequent Itemset Mining Implementations*, pages 1–12, Melbourne, 2003.

D.E. Goldberg. *Genetic algorithms in search, optimization and machine learning*. Addison-Wesley, Reading, 1989.

O. Goldschmidt and A. Takvorian. An efficient graph planarization two-phase heuristic. *Networks*, 24:69–73, 1994.

M.C. Golumbic. *Algorithmic graph theory and perfect graphs*. Academic Press, New York, 1980.

F.C. Gomes, P. Pardalos, C.S. Oliveira, and M.G.C. Resende. Reactive GRASP with path relinking for channel assignment in mobile phone networks. In *Proceedings of the 5th International Workshop on Discrete Algorithms and Methods for Mobile Computing and Communications*, pages 60–67, Rome, 2001. ACM Press.

G. Grahne and J. Zhu. Efficiently using prefix-trees in mining frequent itemsets, 2003. URL http://bit.ly/1qxiKbl. Last visited on April 16, 2016.

M. Grötschel, L. Lovász, and A. Schrijver. Polynomial algorithms for perfect graphs. *Annals of Discrete Mathematics*, 21:325–356, 1984.

A.L. Guedes, F.D. Moura Neto, and G.M. Platt. Double Azeotropy: Calculations with Newton-like methods and continuous GRASP (C-GRASP). *International Journal of Mathematical Modelling and Numerical Optimisation*, 2:387–404, 2011.

G. Gutin and A.P. Punnen, editors. *The traveling salesman problem and its variations*. Kluwer Academic Publishers, Boston, 2002.

S.L. Hakimi. Steiner's problem in graphs and its applications. *Networks*, 1:113–133, 1971.

J. Han, J. Pei, and Y. Yin. Mining frequent patterns without candidate generation. In *Proceedings of the 2000 ACM SIGMOD International Conference on Management of Data*, pages 1–12, Dallas, 2000. ACM.

J. Han, M. Kamber, and J. Pei. *Data mining: Concepts and techniques*. Morgan Kaufmann Publishers, San Francisco, 3rd edition, 2011.

P. Hansen. The steepest ascent mildest descent heuristic for combinatorial programming. In *Proceedings of the Congress on Numerical Methods in Combinatorial Optimization*, pages 70–145, Capri, 1986.

P. Hansen and L. Kaufman. A primal-dual algorithm for the three-dimensional assignment problem. *Cahiers du CERO*, 15:327–336, 1973.

P. Hansen and N. Mladenović. An introduction to variable neighbourhood search. In S Voss, S. Martello, I.H. Osman, and C. Roucairol, editors, *Metaheuristics: Advances and trends in local search procedures for optimization*, pages 433–458. Kluwer Academic Publishers, Boston, 1999.

P. Hansen and N. Mladenović. Developments of variable neighborhood search. In C.C. Ribeiro and P. Hansen, editors, *Essays and surveys in metaheuristics*, pages 415–439. Kluwer Academic Publishers, Boston, 2002.

P. Hansen and N. Mladenović. Variable neighborhood search. In F. Glover and G. Kochenberger, editors, *Handbook of metaheuristics*, pages 145–184. Kluwer Academic Publishers, Boston, 2003.

J.P. Hart and A.W. Shogan. Semi-greedy heuristics: An empirical study. *Operations Research Letters*, 6:107–114, 1987.

W.E. Hart. Sequential stopping rules for random optimization methods with applications to multistart local search. *SIAM Journal on Optimization*, 9:270–290, 1998.

M.M. Hassan and G.L. Hogg. A review of graph theory applications to the facilities layout problem. *Omega*, 15:291–300, 1987.

M. Held and R.M. Karp. The traveling-salesman problem and minimum spanning trees. *Operations Research*, 18:1138–1162, 1970.

M. Held and R.M. Karp. The traveling-salesman problem and minimum spanning trees: Part II. *Mathematical Programming*, 1:6–25, 1971.

M. Held, P. Wolfe, and H.P. Crowder. Validation of subgradient optimization. *Mathematical Programming*, 6:62–88, 1974.

C. Helmberg and F. Rendl. A spectral bundle method for semidefinite programming. *SIAM Journal on Optimization*, 10:673–696, 2000.

A.J. Higgins, S. Hajkowicz, and E. Bui. A multi-objective model for environmental investment decision making. *Computers & Operations Research*, 35:253–266, 2008.

M.J. Hirsch. *GRASP-based heuristics for continuous global optimization problems.* PhD thesis, Department of Industrial and Systems Engineering, University of Florida, Gainesville, 2006.

M.J. Hirsch, P.M. Pardalos, and M.G.C. Resende. Sensor registration in a sensor network by continuous GRASP. In *IEEE Conference on Military Communications*, pages 501–506, Washington, DC, 2006.

M.J. Hirsch, C.N. Meneses, P.M. Pardalos, M.A. Ragle, and M.G.C. Resende. A continuous GRASP to determine the relationship between drugs and adverse reactions. In O. Seref, O. Erhun Kundakcioglu, and P.M. Pardalos, editors, *Data mining, systems analysis and optimization in biomedicine*, volume 953 of *AIP Conference Proceedings*, pages 106–121. Springer, 2007a.

M.J. Hirsch, C.N. Meneses, P.M. Pardalos, and M.G.C. Resende. Global optimization by continuous GRASP. *Optimization Letters*, 1:201–212, 2007b.

M.J. Hirsch, P.M. Pardalos, and M.G.C. Resende. Solving systems of nonlinear equations with continuous GRASP. *Nonlinear Analysis: Real World Applications*, 10:2000–2006, 2009.

M.J. Hirsch, P.M. Pardalos, and M.G.C. Resende. Speeding up continuous GRASP. *European Journal of Operational Research*, 205:507–521, 2010.

M.J. Hirsch, P.M. Pardalos, and M.G.C. Resende. Correspondence of projected 3D points and lines using a continuous GRASP. *International Transactions in Operational Research*, 18:493–511, 2011.

M.J. Hirsch, H. Ortiz-Pena, and C. Eck. Cooperative tracking of multiple targets by a team of autonomous UAVs. *International Journal of Operations Research and Information Systems*, 3:53–73, 2012.

J.H. Holland. *Adaptation in natural and artificial systems: An introductory analysis with applications to biology, control, and artificial intelligence.* University of Michigan Press, Ann Arbor, 1975.

S. Homer and M. Peinado. Two distributed memory parallel approximation algorithms for Max-Cut. *Journal of Parallel and Distributed Computing*, 1:48–61, 1997.

H.H. Hoos. On the run-time behaviour of stochastic local search algorithms for SAT. In *Proceedings of the Sixteenth National Conference on Artificial Intelligence*, pages 661–666, Orlando, 1999. American Association for Artificial Intelligence.

H.H. Hoos and T. Stützle. On the empirical evaluation of Las Vegas algorithms - Position paper. Technical report, Computer Science Department, University of British Columbia, Vancouver, 1998a.

H.H. Hoos and T. Stützle. Evaluating Las Vegas algorithms – Pitfalls and remedies. In *Proceedings of the 14th Conference on Uncertainty in Artificial Intelligence*, pages 238–245, Madison, 1998b.

H.H. Hoos and T. Stützle. Some surprising regularities in the behaviour of stochastic local search. In M. Maher and J.-F. Puget, editors, *Principles and practice of constraint programming*, volume 1520 of *Lecture Notes in Computer Science*, page 470. Springer, Berlin, 1998c.

H.H. Hoos and T. Stützle. Towards a characterisation of the behaviour of stochastic local search algorithms for SAT. *Artificial Intelligence*, 112:213–232, 1999.

H.H. Hoos and T. Stützle. *Stochastic local search: Foundations and applications*. Elsevier, New York, 2005.

F.K. Hwang, D.S. Richards, and P. Winter. *The Steiner tree problem*. North-Holland, Amsterdam, 1992.

E. Hyytiä and J. Virtamo. Wavelength assignment and routing in WDM networks. In *Proceedings of the Fourteenth Nordic Teletraffic Seminar NTS-14*, pages 31–40, Lyngby, 1998.

C. Ishida, A. Pozo, E. Goldbarg, and M. Goldbarg. Multiobjective optimization and rule learning: Subselection algorithm or meta-heuristic algorithm? In N. Nedjah, L.M. Mourelle, and J. Kacprzyk, editors, *Innovative applications in data mining*, pages 47–70. Springer, Berlin, 2009.

R. Jain. *The art of computer systems performance analysis: Techniques for experimental design, measurement, simulation, and modeling*. Wiley, New York, 1991.

V. Jarník. O jistém problému minimálním. *Práce Moravské Přírodovědecké Společnosti*, 6:57–63, 1930.

D.S. Johnson. *Near-optimal bin-packing algorithms*. PhD thesis, Massachusetts Institute of Technology, Cambridge, 1973.

D.S. Johnson. Approximation algorithms for combinatorial problems. *Journal of Computer and System Sciences*, 9:256–278, 1974.

E.H. Kampke, J.E.C. Arroyo, and A.G. Santos. Reactive GRASP with path relinking for solving parallel machines scheduling problem with resource-assignable sequence dependent setup times. In *Proceedings of the World Congress on Nature and Biologically Inspired Computing*, pages 924–929, Coimbatore, 2009. IEEE.

R.M. Karp. Reducibility among combinatorial problems. In R.E. Miller and J.W. Thatcher, editors, *Complexity of computer computations*. Plenum Press, New York, 1972.

R.M. Karp. On the computational complexity of combinatorial problems. *Networks*, 5:45–68, 1975.

H. Kautz, E. Horvitz, Y. Ruan, C. Gomes, and B. Selman. Dynamic restart policies. In *Proceedings of the Eighteenth National Conference on Artificial intelligence*, pages 674–681, Edmonton, 2002. American Association for Artificial Intelligence.

G. Kendall, S. Knust, C.C. Ribeiro, and S. Urrutia. Scheduling in sports: An annotated bibliography. *Computers & Operations Research*, 37:1–19, 2010.

B.W. Kernighan and S. Lin. An efficient heuristic procedure for partitioning graphs. *Bell System Technical Journal*, 49:291–307, 1970.

R.K. Kincaid. Good solutions to discrete noxious location problems via metaheuristics. *Annals of Operations Research*, 40:265–281, 1992.

S. Kirkpatrick, C.D. Gelatt Jr., and M.P. Vecchi. Optimization by simulated annealing. *Science*, 220(4598):671–680, 1983.

A. Kitnick. Frances Stark: Text after text. *Parkett*, 93:66–71, 2013.

G.A. Kochenberger, B.A. McCarl, and F.P. Wyman. A heuristic for general integer programming. *Decision Sciences*, 5:36–41, 1974.

G.A. Kochenberger, J.-K. Hao, Z. Lu, H. Wang, and F. Glover. Solving large scale Max Cut problems via tabu search. *Journal of Heuristics*, 19:565–571, 2013.

M.R. Krom. The decision problem for a class of first-order formulas in which all disjunctions are binary. *Zeitschrift für Mathematische Logik und Grundlagen der Mathematik*, 13:15–20, 1967.

J.B. Kruskal. On the shortest spanning subtree of a graph and the traveling salesman problem. *Proceedings of the American Mathematical Society*, 7:48–â50, 1956.

K. Kuratowski. Sur le problème des courbes gauches en topologie. *Fundamenta Mathematicae*, 15:271–283, 1930.

N. Labadi, C. Prins, and M. Reghioui. GRASP with path relinking for the capacitated arc routing problem with time windows. In A. Fink and F. Rothlauf, editors, *Advances in computational intelligence in transport, logistics, and supply chain management*, pages 111–135. Springer, Berlin, 2008.

M. Laguna and F. Glover. Bandwidth packing: A tabu search approach. *Management Science*, 39:492–500, 1993.

M. Laguna and R. Martí. GRASP and path relinking for 2-layer straight line crossing minimization. *INFORMS Journal on Computing*, 11:44–52, 1999.

M. Laguna, J.P. Kelly, J.L. González-Velarde, and F. Glover. Tabu search for multilevel generalized assignment problem. *European Journal of Operational Research*, 82:176–189, 1995.

E.L. Lawler. *Combinatorial optimization: Networks and matroids.* Holt, Rinehart and Winston, New York, 1976.

E.L. Lawler, J.K. Lenstra, A.H.G. Rinnooy Kan, and D.B. Shmoys, editors. *The traveling salesman problem: A guided tour of combinatorial optimization.* John Wiley & Sons, New York, 1985.

L.J. LeBlanc, J. Chifflet, and P. Mahey. Packet routing in telecommunication networks with path and flow restrictions. *INFORMS Journal on Computing*, 11: 188–197, 1999.

J.K. Lenstra and A.H.G. Rinnooy Kan. Computational complexity of discrete optimization problems. *Annals of Discrete Mathematics*, 4:121–140, 1979.

R. De Leone, P. Festa, and E. Marchitto. Solving a bus driver scheduling problem with randomized multistart heuristics. *International Transactions in Operational Research*, 18:707–727, 2011.

O. Leue. Methoden zur Lösung dreidimensionaler Zuordnungsprobleme. *Angewandte Informatik*, 14:154–162, 1972.

H. Li and D. Landa-Silva. An elitist GRASP metaheuristic for the multi-objective quadratic assignment problem. In M. Ehrgott, C.M. Fonseca, X. Gandibleux, J.-K. Hao, and M. Sevaux, editors, *Evolutionary multi-criterion optimization*, volume 5467 of *Lecture Notes in Computer Science*, pages 481–494. Springer, Berlin, 2009.

Y. Li, P.M. Pardalos, and M.G.C. Resende. A greedy randomized adaptive search procedure for the quadratic assignment problem. In P.M. Pardalos and H. Wolkowicz, editors, *Quadratic assignment and related problems*, volume 16 of *DIMACS Series in Discrete Mathematics and Theoretical Computer Science*, pages 237–261. American Mathematical Society, Providence, 1994.

S. Lin. Computer solutions of the traveling salesman problem. *Bell System Technical Journal*, 44:2245–2260, 1965.

S. Lin and B.W. Kernighan. An effective heuristic algorithm for the traveling-salesman problem. *Operations Research*, 21:498–516, 1973.

P.C. Liu and R.C. Geldmacher. On the deletion of nonplanar edges of a graph. In *Proceedings of the 10th Southeastern Conference on Combinatorics, Graph Theory and Computing*, pages 727–738, Boca Raton, 1977.

H.R. Lourenço, O. Martin, and T. Stützle. Iterated local search. In F. Glover and G. Kochenberger, editors, *Handbook of metaheuristics*, pages 321–353. Kluwer Academic Publishers, Boston, 2003.

L. Lovász. On the Shannon capacity of a graph. *IEEE Transactions on Information Theory*, IT-25:1–7, 1979.

M. Luby, A. Sinclair, and D. Zuckerman. Optimal speedup of Las Vegas algorithms. *Information Processing Letters*, 47:173–180, 1993.

M. Luis, S. Salhi, and G. Nagy. A guided reactive GRASP for the capacitated multi-source Weber problem. *Computers & Operations Research*, 38:1014–1024, 2011.

D.G. Macharet, A.A. Neto, V.F. da Camara Neto, and M.F.M. Campos. Nonholonomic path planning optimization for Dubins' vehicles. In *2011 IEEE International Conference on Robotics and Automation*, pages 4208–4213, Shanghai, 2011. IEEE.

N. Maculan. The Steiner problem in graphs. *Annals of Discrete Mathematics*, 31: 182–212, 1987.

C.L.B. Maia, R.A.F. Carmo, F.G. Freitas, G.A.L. Campos, and J.T. Souza. Automated test case prioritization with reactive GRASP. *Advances in Software Engineering*, 2010, 2010. doi: 10.1155/2010/428521. Article ID 428521.

P. Manohar, D. Manjunath, and R.K. Shevgaonkar. Routing and wavelength assignment in optical networks from edge disjoint path algorithms. *IEEE Communications Letters*, 5:211–213, 2002.

S. Martello and P. Toth. *Knapsack problems: Algorithms and computer implementations*. John Wiley & Sons, New York, 1990.

R. Martí and F. Sandoya. GRASP and path relinking for the equitable dispersion problem. *Computers & Operations Research*, 40:3091–3099, 2013.

R. Martí, A. Duarte, and M. Laguna. Advanced scatter search for the MAX-CUT problem. *INFORMS Journal on Computing*, 21:26–38, 2009.

R. Martí, J.L. González-Velarde, and A. Duarte. Heuristics for the bi-objective path dissimilarity problem. *Computers & Operations Research*, 36:2905–2912, 2009.

R. Martí, M.G.C. Resende, and C.C. Ribeiro. Multi-start methods for combinatorial optimization. *European Journal of Operational Research*, 226:1–8, 2013a.

R. Martí, M.G.C. Resende, and C.C. Ribeiro. Special issue of Computers & Operations Research: GRASP with path relinking: Developments and applications. *Computers & Operations Research*, 40:3080, 2013b.

R. Martí, V. Campos, M.G.C. Resende, and A. Duarte. Multiobjective GRASP with path relinking. *European Journal of Operational Research*, 240:54–71, 2015.

B. Martin, X. Gandibleux, and L. Granvilliers. Continuous-GRASP revisited. In P. Siarry, editor, *Heuristics: Theory and applications*, chapter 1. Nova Science Publishers, Hauppauge, 2013.

O. Martin and S.W. Otto. Combining simulated annealing with local search heuristics. *Annals of Operations Research*, 63:57–75, 1996.

O. Martin, S.W. Otto, and E.W. Felten. Large-step Markov chains for the traveling salesman problem. *Complex Systems*, 5:299–326, 1991.

S.L. Martins, C.C. Ribeiro, and M.C. Souza. A parallel GRASP for the Steiner problem in graphs. In A. Ferreira, J. Rolim, H. Simon, and S.-H. Teng, editors, *Solving irregularly structured problems in parallel*, volume 1457 of *Lecture Notes in Computer Science*, pages 285–297. Springer, Berlin, 1998.

S.L. Martins, P.M. Pardalos, M.G.C. Resende, and C.C. Ribeiro. Greedy randomized adaptive search procedures for the Steiner problem in graphs. In P.M. Pardalos, S. Rajasejaran, and J. Rolim, editors, *Randomization methods in algorithmic design*, volume 43 of *DIMACS Series in Discrete Mathematics and Theoretical Computer Science*, pages 133–145. American Mathematical Society, Providence, 1999.

S.L. Martins, P.M. Pardalos, M.G.C. Resende, and C.C. Ribeiro. A parallel GRASP for the Steiner tree problem in graphs using a hybrid local search strategy. *Journal of Global Optimization*, 17:267–283, 2000.

S.L. Martins, C.C. Ribeiro, and I. Rosseti. Applications and parallel implementations of metaheuristics in network design and routing. In S. Manandhar, J. Austin, U. Desai, Y. Oyanagi, and A.K. Talukder, editors, *Applied computing*, volume 3285 of *Lecture Notes in Computer Science*, pages 205–213. Springer, Berlin, 2004.

G.R. Mateus, M.G.C. Resende, and R.M.A. Silva. GRASP with path-relinking for the generalized quadratic assignment problem. *Journal of Heuristics*, 17: 527–565, 2011.

K. Melhorn. A faster approximation algorithm for the Steiner problem in graphs. *Information Processing Letters*, 27:125–128, 1988.

B. Melián, M. Laguna, and J.A. Moreno-Pérez. Capacity expansion of fiber optic networks with WDM systems: Problem formulation and comparative analysis. *Computers & Operations Research*, 31:461–472, 2004.

M. Mestria, L.S. Ochi, and S.L. Martins. GRASP with path relinking for the symmetric Euclidean clustered traveling salesman problem. *Computers & Operations Research*, 40:3218–3229, 2013.

Z. Michalewicz. *Genetic algorithms + Data structures = Evolution programs*. Springer, Berlin, 1996.

W. Michelis, E.H.L. Aarts, and J. Korst. *Theoretical aspects of local search*. Springer, Berlin, 2007.

N. Mladenović and P. Hansen. Variable neighborhood search. *Computers & Operations Research*, 24:1097–1100, 1997.

R.E.N. Moraes and C.C. Ribeiro. Power optimization in ad hoc wireless network topology control with biconnectivity requirements. *Computers & Operations Research*, 40:3188–3196, 2013.

L.F. Morán-Mirabal, J.L. González-Velarde, and M.G.C. Resende. Automatic tuning of GRASP with evolutionary path-relinking. In M.J. Blesa, C. Blum, P. Festa, A. Roli, and M. Sampels, editors, *Hybrid metaheuristics*, volume 7919 of *Lecture Notes in Computer Science*, pages 62–77. Springer, Berlin, 2013a.

L.F. Morán-Mirabal, J.L. González-Velarde, M.G.C. Resende, and R.M.A. Silva. Randomized heuristics for handover minimization in mobility networks. *Journal of Heuristics*, 19:845–880, 2013b.

L.F. Morán-Mirabal, J.L. González-Velarde, and M.G.C. Resende. Randomized heuristics for the family traveling salesperson problem. *International Transactions in Operational Research*, 21:41–57, 2014.

R.A. Murphey, P.M. Pardalos, and L.S. Pitsoulis. A parallel GRASP for the data association multidimensional assignment problem. In P.M. Pardalos, editor, *Parallel processing of discrete problems*, volume 106 of *The IMA Volumes in Mathematics and Its Applications*, pages 159–180. Springer, New York, 1998.

J.F. Muth and G.L. Thompson. *Industrial scheduling*. Prentice-Hall, Boston, 1963.

P. Mutzel. *The maximum planar subgraph problem*. PhD thesis, Universität zu Köln, Cologne, 1994.

M.C.V. Nascimento and L. Pitsoulis. Community detection by modularity maximization using GRASP with path relinking. *Computers & Operations Research*, 40:3121–3131, 2013.

M.C.V. Nascimento, M.G.C. Resende, and F.M.B. Toledo. GRASP heuristic with path-relinking for the multi-plant capacitated lot sizing problem. *European Journal of Operational Research*, 200:747–754, 2010.

G.L. Nemhauser and L.A. Wolsey. *Integer and combinatorial optimization*. Wiley, New York, 1988.

V.-P. Nguyen, C. Prins, and C. Prodhon. Solving the two-echelon location routing problem by a GRASP reinforced by a learning process and path relinking. *European Journal of Operational Research*, 216:113–126, 2012.

N.J. Nilsson. *Problem-solving methods in artificial intelligence*. McGraw-Hill, New York, 1971.

N.J. Nilsson. *Principles of artificial intelligence*. Springer, Berlin, 1982.

H. Nishimura and S. Kuroda, editors. *A lost mathematician, Takeo Nakasawa. The forgotten father of matroid theory*. Birkhäuser Verlag, Basel, 2009.

T.F. Noronha and C.C. Ribeiro. Routing and wavelength assignment by partition coloring. *European Journal of Operational Research*, 171:797–810, 2006.

E. Nowicki and C. Smutnicki. An advanced tabu search algorithm for the job shop problem. *Journal of Scheduling*, 8:145–159, 2005.

C.A. Oliveira, P.M. Pardalos, and M.G.C. Resende. GRASP with path-relinking for the quadratic assignment problem. In C.C. Ribeiro and S.L. Martins, editors, *Experimental and efficient algorithms*, volume 3059, pages 356–368. Springer, Berlin, 2004.

S. Orlando, P. Palmerini, and R. Perego. Adaptive and resource-aware mining of frequent sets. In *Proceedings of the 2002 IEEE International Conference on Data Mining*, pages 338–345, Maebashi City, 2002. IEEE.

C. Orsenigo and C. Vercellis. Bayesian stopping rules for greedy randomized procedures. *Journal of Global Optimization*, 36:365–377, 2006.

L. Osborne and B. Gillett. A comparison of two simulated annealing algorithms applied to the directed Steiner problem on networks. *ORSA Journal on Computing*, 3:213–225, 1991.

A. Ouorou, P. Mahey, and J.P. Vial. A survey of algorithms for convex multicommodity flow problems. *Management Science*, 46:126–147, 2000.

A.V.F. Pacheco, , G.M. Ribeiro, and G.R. Mauri. A GRASP with path-relinking for the workover rig scheduling problem. *International Journal of Natural Computing Research*, 1:1–14, 2010.

G. Palubeckis. Multistart tabu search strategies for the unconstrained binary quadratic optimization problem. *Annals of Operations Research*, 131:259–282, 2004.

C.H. Papadimitriou. *Computational complexity*. Addison-Wesley, Reading, 1994.

C.H. Papadimitriou and K. Steiglitz. *Combinatorial optimization: Algorithms and complexity*. Prentice Hall, Englewood Cliffs, 1982.

P.M. Pardalos and L.S. Pitsoulis. *Nonlinear assignment problems: Algorithms and applications*. Kluwer Academic Publishers, Boston, 2000.

P.M. Pardalos and J. Xue. The maximum clique problem. *Journal of Global Optimization*, 4:301–328, 1994.

P.M. Pardalos, L.S. Pitsoulis, and M.G.C. Resende. A parallel GRASP implementation for the quadratic assignment problem. In A. Ferreira and J. Rolim, editors, *Parallel algorithms for irregular problems: State of the art*, pages 115–133. Kluwer Academic Publishers, Boston, 1995.

P.M. Pardalos, L.S. Pitsoulis, and M.G.C. Resende. A parallel GRASP for MAX-SAT problems. In J. Waśniewski, J. Dongarra, K. Madsen, and D. Olesen, editors, *Applied parallel computing industrial computation and optimization*, volume 1184 of *Lecture Notes in Computer Science*, pages 575–585. Springer, Berlin, 1996.

M. Parker and J. Ryan. A column generation algorithm for bandwidth packing. *Telecommunication Systems*, 2:185–195, 1994.

F. Parreño, R. Alvarez-Valdes, J.M. Tamarit, and J.F. Oliveira. A maximal-space algorithm for the container loading problem. *INFORMS Journal on Computing*, 20:412–422, 2008.

R.A. Patterson, H. Pirkul, and E. Rolland. A memory adaptive reasoning technique for solving the capacitated minimum spanning tree problem. *Journal of Heuristics*, 5:159–180, 1999.

J. Pearl. *Heuristics: Intelligent search strategies for computer problem solving*. Addison-Wesley, Reading, 1985.

O. Pedrola, M. Ruiz, L. Velasco, D. Careglio, O. González de Dios, and J. Comellas. A GRASP with path-relinking heuristic for the survivable IP/MPLS-over-WSON multi-layer network optimization problem. *Computers & Operations Research*, 40:3174–3187, 2013.

M. Pérez, F. Almeida, and J.M. Moreno-Vega. A hybrid GRASP-path relinking algorithm for the capacitated $p$-hub median problem. In M.J. Blesa, C. Blum, A. Roli, and M. Sampels, editors, *Hybrid metaheuristics*, volume 3636 of *Lecture Notes in Computer Science*, pages 142–153. Springer, Berlin, 2005.

E. Pesch and F. Glover. TSP ejection chains. *Discrete Applied Mathematics*, 76: 165–181, 1997.

L.S. Pessoa, M.G.C. Resende, and C.C. Ribeiro. Experiments with the LAGRASP heuristic for set $k$-covering. *Optimization Letters*, 5:407–419, 2011.

L.S. Pessoa, M.G.C. Resende, and C.C. Ribeiro. A hybrid Lagrangean heuristic with GRASP and path-relinking for set $k$-covering. *Computers & Operations Research*, 40:3132–3146, 2013.

W.P. Pierskalla. The tri-substitution method for the three-multidimensional assignment problem. *Journal of the Canadian Operational Research Society*, 5:71–81, 1967.

W.P. Pierskalla. The multidimensional assignment problem. *Operations Research*, 16:422–431, 1968.

R.Y. Pinter. Optimal layer assignment for interconnect. *Advances in VLSI and Computer Systems*, 1:123–137, 1984.

L.S. Pitsoulis. *Topics in matroid theory*. SpringerBriefs in Optimization. Springer, 2014.

L.S. Pitsoulis and M.G.C. Resende. Greedy randomized adaptive search procedures. In P.M. Pardalos and M.G.C. Resende, editors, *Handbook of applied optimization*, pages 168–183. Oxford University Press, New York, 2002.

A. Plastino, E.R. Fonseca, R. Fuchshuber, S.L. Martins, A.A. Freitas, M. Luis, and S. Salhi. A hybrid data mining metaheuristic for the $p$-median problem. In H. Park, S. Parthasarathy, H. Liu, and Z. Obradovic, editors, *Proceedings of the 9th SIAM International Conference on Data Mining*, pages 305–316, Sparks, 2009. SIAM.

A. Plastino, R. Fuchshuber, S.L. Martins, A.A. Freitas, and S. Salhi. A hybrid data mining metaheuristic for the $p$-median problem. *Statistical Analysis and Data Mining*, 4:313–335, 2011.

A. Plastino, H. Barbalho, L.F.M. Santos, R. Fuchshuber, and S.L. Martins. Adaptive and multi-mining versions of the DM-GRASP hybrid metaheuristic. *Journal of Heuristics*, 20:39–74, 2014.

S. Poljak and Z. Tuza. Maximum cuts and largest bipartite subgraphs. In W. Cook, L. Lovász, and P. Seymour, editors, *Papers from the special year on Combinatorial Optimization*, volume 20 of *DIMACS Series in Discrete Mathematics and Theoretical Computer Science*, pages 181–244. American Mathematical Society, Providence, 1995.

M. Prais and C.C. Ribeiro. Parameter variation in GRASP implementations. In C.C. Ribeiro and P. Hansen, editors, *Extended Abstracts of the Third Metaheuristics International Conference*, pages 375–380, Angra dos Reis, 1999.

M. Prais and C.C. Ribeiro. Reactive GRASP: An application to a matrix decomposition problem in TDMA traffic assignment. *INFORMS Journal on Computing*, 12:164–176, 2000a.

M. Prais and C.C. Ribeiro. Parameter variation in GRASP procedures. *Investigación Operativa*, 9:1–20, 2000b.

R.C. Prim. Shortest connection networks and some generalizations. *Bell System Technical Journal*, 36:1389–1401, 1957.

C. Prins, C. Prodhon, and R.Wolfler-Calvo. A reactive GRASP and path relinking algorithm for the capacitated location routing problem. In *Proceedings of the International Conference on Industrial Engineering and Systems Management*, Marrakech, 2005. I4E2. ISBN 2-9600532-0-6.

M. Rahmani, M. Rashidinejad, E.M. Carreno, and R.A. Romero. Evolutionary multi-move path-relinking for transmission network expansion planning. In *2010 IEEE Power and Energy Society General Meeting*, pages 1–6, Minneapolis, 2010. IEEE.

R.L. Rardin, R., and Uzsoy. Experimental evaluation of heuristic optimization algorithms: A tutorial. *Journal of Heuristics*, 7:261–304, 2001.

C. Reeves and J.E. Rowe. *Genetic algorithms: Principles and perspectives*. Springer, Berlin, 2002.

C.R. Reeves. *Modern heuristic techniques for combinatorial problems*. Blackwell, London, 1993.

C. Rego. Relaxed tours and path ejections for the traveling salesman problem. *European Journal of Operational Research*, 106:522–538, 1998.

C. Rego and F. Glover. Local search and metaheuristics. In G. Gutin and A.P. Punnen, editors, *The traveling salesman problem and its variations*, pages 309–368. Kluwer Academic Publishers, Boston, 2002.

P.P. Repoussis, C.D. Tarantilis, and G. Ioannou. A hybrid metaheuristic for a real life vehicle routing problem. In T. Boyanov, S. Dimova, K. Georgiev, and G. Nikolov, editors, *Numerical methods and applications*, volume 4310 of *Lecture Notes in Computer Science*, pages 247–254. Springer, Berlin, 2007.

L.I.P. Resende and M.G.C. Resende. A GRASP for frame relay permanent virtual circuit routing. In C.C. Ribeiro and P. Hansen, editors, *Extended Abstracts of the III Metaheuristics International Conference*, pages 397–401, Angra dos Reis, 1999.

M.G.C. Resende. Computing approximate solutions of the maximum covering problem using GRASP. *Journal of Heuristics*, 4:161–171, 1998.

M.G.C. Resende. Metaheuristic hybridization with greedy randomized adaptive search procedures. In Zhi-Long Chen and S. Raghavan, editors, *Tutorials in Operations Research*, pages 295–319. INFORMS, 2008.

M.G.C. Resende and T.A. Feo. A GRASP for satisfiability. In D.S. Johnson and M.A. Trick, editors, *Cliques, coloring, and satisfiability: The second DIMACS implementation challenge*, volume 26 of *DIMACS Series in Discrete Mathematics and Theoretical Computer Science*, pages 499–520. American Mathematical Society, Providence, 1996.

M.G.C. Resende and J.L. González-Velarde. GRASP: Procedimientos de búsqueda miope aleatorizado y adaptativo. *Inteligencia Artificial*, 19:61–76, 2003.

M.G.C. Resende and C.C. Ribeiro. A GRASP for graph planarization. *Networks*, 29:173–189, 1997.

M.G.C. Resende and C.C. Ribeiro. Graph planarization. In C. Floudas and P.M. Pardalos, editors, *Encyclopedia of optimization*, volume 2, pages 368–373. Kluwer Academic Publishers, Boston, 2001.

M.G.C. Resende and C.C. Ribeiro. A GRASP with path-relinking for private virtual circuit routing. *Networks*, 41:104–114, 2003a.

M.G.C. Resende and C.C. Ribeiro. Greedy randomized adaptive search procedures. In F. Glover and G. Kochenberger, editors, *Handbook of metaheuristics*, pages 219–249. Kluwer Academic Publishers, Boston, 2003b.

M.G.C. Resende and C.C. Ribeiro. GRASP with path-relinking: Recent advances and applications. In T. Ibaraki, K. Nonobe, and M. Yagiura, editors, *Metaheuristics: Progress as real problem solvers*, pages 29–63. Springer, New York, 2005a.

M.G.C. Resende and C.C. Ribeiro. Parallel greedy randomized adaptive search procedures. In E. Alba, editor, *Parallel metaheuristics: A new class of algorithms*, pages 315–346. Wiley-Interscience, Hoboken, 2005b.

M.G.C. Resende and C.C. Ribeiro. Greedy randomized adaptive search procedures: Advances and applications. In M. Gendreau and J.-Y. Potvin, editors, *Handbook of metaheuristics*, pages 293–319. Springer, New York, 2nd edition, 2010.

M.G.C. Resende and C.C. Ribeiro. Restart strategies for GRASP with path-relinking heuristics. *Optimization Letters*, 5:467–478, 2011.

M.G.C. Resende and C.C. Ribeiro. GRASP: Greedy randomized adaptive search procedures. In E.K. Burke and G. Kendall, editors, *Search methodologies: Introductory tutorials in optimization and decision support systems*, chapter 11, pages 287–312. Springer, New York, 2nd edition, 2014.

M.G.C. Resende and R.M.A. Silva. GRASP: Greedy randomized adaptive search procedures. In J.J. Cochran, L.A. Cox, Jr., P. Keskinocak, J.P. Kharoufeh, and J.C. Smith, editors, *Encyclopedia of operations research and management science*, volume 3, pages 2118–2128. Wiley, New York, 2011.

M.G.C. Resende and R.M.A. Silva. GRASP: Procedimentos de busca gulosos, aleatórios e adaptativos. In H.S. Lopes, L.C.A. Rodrigues, and M.T.A. Steiner, editors, *Meta-heurísticas em pesquisa operacional*, chapter 1, pages 1–20. Omnipax Editora, Curitiba, 2013.

M.G.C. Resende and R.F. Werneck. A hybrid heuristic for the *p*-median problem. *Journal of Heuristics*, 10:59–88, 2004.

M.G.C. Resende and R.F. Werneck. A hybrid multistart heuristic for the uncapacitated facility location problem. *European Journal of Operational Research*, 174: 54–68, 2006.

M.G.C. Resende, P.M. Pardalos, and Y. Li. Algorithm 754: Fortran subroutines for approximate solution of dense quadratic assignment problems using GRASP. *ACM Transactions on Mathematical Software*, 22:104–118, 1996.

M.G.C. Resende, L.S. Pitsoulis, and P.M. Pardalos. Approximate solution of weighted MAX-SAT problems using GRASP. In J. Gu and P.M. Pardalos, editors, *Satisfiability problems*, volume 35 of *DIMACS Series in Discrete Mathematics and Theoretical Computer Science*, pages 393–405. American Mathematical Society, Providence, 1997.

M.G.C. Resende, T.A. Feo, and S.H. Smith. Algorithm 787: Fortran subroutines for approximate solution of maximum independent set problems using GRASP. *ACM Transactions on Mathematical Software*, 24:386–394, 1998.

M.G.C. Resende, L.S. Pitsoulis, and P.M. Pardalos. Fortran subroutines for computing approximate solutions of MAX-SAT problems using GRASP. *Discrete Applied Mathematics*, 100:95–113, 2000.

M.G.C. Resende, R. Martí, M. Gallego, and A. Duarte. GRASP and path relinking for the max-min diversity problem. *Computers & Operations Research*, 37: 498–508, 2010a.

M.G.C. Resende, C.C. Ribeiro, F. Glover, and R. Martí. Scatter search and path-relinking: Fundamentals, advances, and applications. In M. Gendreau and J.-Y. Potvin, editors, *Handbook of metaheuristics*, pages 87–107. Springer, New York, 2nd edition, 2010b.

M.G.C. Resende, G.R. Mateus, and R.M.A. Silva. GRASP: Busca gulosa, aleatorizada e adaptativa. In A. Gaspar-Cunha, R. Takahashi, and C.H. Antunes, editors, *Manual da computação evolutiva e metaheurística*, pages 201–213. Coimbra University Press, Coimbra, 2012.

A.P. Reynolds and B. de la Iglesia. A multi-objective GRASP for partial classification. *Soft Computing*, 13:227–243, 2009.

A.P. Reynolds, D.W. Corne, and B. de la Iglesia. A multiobjective GRASP for rule selection. In *Proceedings of the 11th Annual Conference on Genetic and Evolutionary Computation*, pages 643–650, Montreal, 2009. ACM.

C.C. Ribeiro. GRASP: Une métaheuristique gloutonne et probabiliste. In J. Teghem and M. Pirlot, editors, *Optimisation approchée en recherche opérationnelle*, pages 153–176. Hermès, Paris, 2002.

C.C. Ribeiro. Sports scheduling: Problems and applications. *International Transactions in Operational Research*, 19:201–226, 2012.

C.C. Ribeiro and M.G.C. Resende. Algorithm 797: Fortran subroutines for approximate solution of graph planarization problems using GRASP. *ACM Transactions on Mathematical Software*, 25:341–352, 1999.

C.C. Ribeiro and M.G.C. Resende. Path-relinking intensification methods for stochastic local search algorithms. *Journal of Heuristics*, 18:193–214, 2012.

C.C. Ribeiro and I. Rosseti. A parallel GRASP heuristic for the 2-path network design problem. In B. Monien and R. Feldmann, editors, *Euro-Par 2002 Parallel Processing*, volume 2400 of *Lecture Notes in Computer Science*, pages 922–926. Springer, Berlin, 2002.

C.C. Ribeiro and I. Rosseti. Efficient parallel cooperative implementations of GRASP heuristics. *Parallel Computing*, 33:21–35, 2007.

C.C. Ribeiro and I. Rosseti. Exploiting run time distributions to compare sequential and parallel stochastic local search algorithms. In *Proceedings of the VIII Metaheuristics International Conference*, Hamburg, 2009.

C.C. Ribeiro and I. Rosseti. tttplots-compare: A perl program to compare time-to-target plots or general runtime distributions of randomized algorithms. *Optimization Letters*, 9:601–614, 2015.

C.C. Ribeiro and S. Urrutia. Heuristics for the mirrored traveling tournament problem. *European Journal of Operational Research*, 179:775–787, 2007.

C.C. Ribeiro, E. Uchoa, and R.F. Werneck. A hybrid GRASP with perturbations for the Steiner problem in graphs. *INFORMS Journal on Computing*, 14:228–246, 2002.

C.C. Ribeiro, I. Rosseti, and R. Vallejos. On the use of run time distributions to evaluate and compare stochastic local search algorithms. In T. Stützle, M. Biratari, and H.H. Hoos, editors, *Engineering stochastic local search algorithms*, volume 5752 of *Lecture Notes in Computer Science*, pages 16–30. Springer, Berlin, 2009.

C.C. Ribeiro, I. Rosseti, and R.C. Souza. Effective probabilistic stopping rules for randomized metaheuristics: GRASP implementations. In C.A.C. Coello, editor, *Learning and intelligent optimization*, volume 6683, pages 146–160. Springer, Berlin, 2011.

C.C. Ribeiro, I. Rosseti, and R. Vallejos. Exploiting run time distributions to compare sequential and parallel stochastic local search algorithms. *Journal of Global Optimization*, 54:405–429, 2012.

C.C. Ribeiro, I. Rosseti, and R.C. Souza. Probabilistic stopping rules for GRASP heuristics and extensions. *International Transactions in Operational Research*, 20:301–323, 2013.

M.H.F. Ribeiro, V.F. Trindade, A. Plastino, and S.L. Martins. Hybridization of GRASP metaheuristic with data mining techniques. In *Proceedings of the ECAI Workshop on Hybrid Metaheuristics*, pages 69–78, Valencia, 2004.

M.H.F. Ribeiro, A. Plastino, and S.L. Martins. Hybridization of GRASP metaheuristic with data mining techniques. *Journal of Mathematical Modelling and Algorithms*, 5:23–41, 2006.

R.Z. Ríos-Mercado and E. Fernández. A reactive GRASP for a commercial territory design problem with multiple balancing requirements. *Computers & Operations Research*, 36:755–776, 2009.

Y. Rochat and É. Taillard. Probabilistic diversification and intensification in local search for vehicle routing. *Journal of Heuristics*, 1:147–167, 1995.

F.J. Rodriguez, C. Blum, C. García-Martínez, and M. Lozano. GRASP with path-relinking for the non-identical parallel machine scheduling problem with minimising total weighted completion times. *Annals of Operations Research*, 201:383–401, 2012.

D.P. Ronconi and L.R.S. Henriques. Some heuristic algorithms for total tardiness minimization in a flowshop with blocking. *Omega*, 37:272–281, 2009.

I. Rosseti. *Sequential and parallel strategies of GRASP with path-relinking for the 2-path network design problem*. PhD thesis, Department of Computer Science, Pontifical Catholic University of Rio de Janeiro, Rio de Janeiro, 2003. In Portuguese.

B. Roy and B. Sussmann. Les problèmes d'ordonnancement avec contraintes disjonctives. Technical Report Note DS no. 9 bis, SEMA, Montrouge, 1964.

M.A. Salazar-Aguilar, R.Z. Ríos-Mercado, and J.L. González-Velarde. GRASP strategies for a bi-objective commercial territory design problem. *Journal of Heuristics*, 19:179–200, 2013.

J. Santamaría, O. Cordón, S. Damas, R. Martí, and R.J. Palma. GRASP & evolutionary path relinking for medical image registration based on point matching. In *2010 IEEE Congress on Evolutionary Computation*, pages 1–8. IEEE, 2010.

J. Santamaría, O. Cordón, S. Damas, R. Martí, and R.J. Palma. GRASP and path relinking hybridizations for the point matching-based image registration problem. *Journal of Heuristics*, 18:169–192, 2012.

D. Santos, A. de Sousa, and F. Alvelos. A hybrid column generation with GRASP and path relinking for the network load balancing problem. *Computers & Operations Research*, 40:3147–3158, 2013.

L.F. Santos, M.H.F. Ribeiro, A. Plastino, and S.L. Martins. A hybrid GRASP with data mining for the maximum diversity problem. In M.J. Blesa, C. Blum, A. Roli, and M. Sampels, editors, *Hybrid metaheuristics*, volume 3636 of *Lecture Notes in Computer Science*, pages 116–127. Springer, Berlin, 2005.

L.F. Santos, C.V. Albuquerque, S.L. Martins, and A. Plastino. A hybrid GRASP with data mining for efficient server replication for reliable multicast. In *Proceedings of the 49th Annual IEEE GLOBECOM Technical Conference*, pages 1–6, San Francisco, 2006. IEEE. doi: 10.1109/GLOCOM.2006.246.

L.F. Santos, S.L. Martins, and A. Plastino. Applications of the DM-GRASP heuristic: A survey. *International Transactions on Operational Research*, 15:387–416, 2008.

M. Sarrafzadeh and D. Lee. A new approach to topological via minimization. *IEEE Transactions on Computer-Aided Design*, 8:890–900, 1989.

M. Scaparra and R. Church. A GRASP and path relinking heuristic for rural road network development. *Journal of Heuristics*, 11:89–108, 2005.

A. Scholl, R. Klein, and W. Domschke. Pattern based vocabulary building for effectively sequencing mixed-model assembly lines. *Journal of Heuristics*, 4:359–381, 1998.

A. Schrijver. *Theory of linear and integer programming*. Wiley, New York, 1986.

B. Selman, H.A. Kautz, and B. Cohen. Noise strategies for improving local search. In *Proceedings of the Twelfth National Conference on Artificial Intelligence*, pages 337–343, Seattle, 1994. American Association for Artificial Intelligence.

S. Senju and Y. Toyoda. An approach to linear programming with 0-1 variables. *Management Science*, 15:196–207, 1968.

K. Seo, S. Hyun, and Y.-H. Kim. A spanning tree-based encoding of the MAX CUT problem for evolutionary search. In C.A.C. Coello, V. Cutello, K. Deb, S. Forrest, G. Nicosia, and M. Pavone, editors, *Parallel problem solving from nature - Part I*, volume 7491 of *Lecture Notes in Computer Science*, pages 510–518. Springer, Berlin, 2012.

I.V. Sergienko, V.P. Shilo, and V.A. Roshchin. Optimization parallelizing for discrete programming problems. *Cybernetics and Systems Analysis*, 40:184–189, 2004.

F.S. Serifoglu and G. Ulusoy. Multiprocessor task scheduling in multistage hybrid flow-shops: A genetic algorithm approach. *Journal of the Operational Research Society*, 55:504–512, 2004.

N.Z. Shor. Quadratic optimization problems. *Soviet Journal of Computer and Systems Science*, 25:1–11, 1987.

O.V. Shylo, T. Middelkoop, and P.M. Pardalos. Restart strategies in optimization: Parallel and serial cases. *Parallel Computing*, 37:60–68, 2011a.

O.V. Shylo, O.A. Prokopyev, and J. Rajgopal. On algorithm portfolios and restart strategies. *Operations Research Letters*, 39:49–52, 2011b.

C.-C. Shyur and U.-E. Wen. Optimizing the system of virtual paths by tabu search. *European Journal of Operational Research*, 129:650–662, 2001.

F. Silva and D. Serra. Locating emergency services with different priorities: The priority queuing covering location problem. *Journal of the Operational Research Society*, 59:1229–1238, 2007.

G.C. Silva, L.S. Ochi, and S.L. Martins. Experimental comparison of greedy randomized adaptive search procedures for the maximum diversity problem. In C.C. Ribeiro and S.L. Martins, editors, *Experimental and efficient algorithms*, volume 3059 of *Lecture Notes in Computer Science*, pages 498–512. Springer, Berlin, 2004.

G.C. Silva, M.R.Q. de Andrade, L.S. Ochi, S.L. Martins, and A. Plastino. New heuristics for the maximum diversity problem. *Journal of Heuristics*, 13: 315–336, 2007.

R.M.A. Silva, M.G.C. Resende, P.M. Pardalos, and M.J. Hirsch. A Python/C library for bound-constrained global optimization with continuous GRASP. *Optimization Letters*, 7:967–984, 2013a.

R.M.A. Silva, M.G.C. Resende, P.M. Pardalos, G.R. Mateus, and G. de Tomi. GRASP with path-relinking for facility layout. In B.I. Goldengorin, V.A. Kalyagin, and P.M. Pardalos, editors, *Models, algorithms, and technologies for network analysis*, volume 59 of *Springer Proceedings in Mathematics & Statistics*, pages 175–190. Springer, Berlin, 2013b.

M. Snir, S. Otto, S. Huss-Lederman, D. Walker, and J. Dongarra. *MPI – The complete reference, Volume 1 – The MPI core*. MIT Press, Cambridge, 1998.

K. Sörensen. Metaheuristics – The metaphor exposed. *International Transactions in Operational Research*, 22:1–16, 2015.

K. Sörensen and P. Schittekat. Statistical analysis of distance-based path relinking for the capacitated vehicle routing problem. *Computers & Operations Research*, 40:3197–3205, 2013.

M.C. Souza, C. Duhamel, and C.C. Ribeiro. A GRASP heuristic for the capacitated minimum spanning tree problem using a memory-based local search strategy. In M.G.C. Resende and J. Souza, editors, *Metaheuristics: Computer decision-making*, pages 627–658. Kluwer Academic Publishers, Boston, 2004.

C.S. Sung and S.K. Park. An algorithm for configuring embedded networks in reconfigurable telecommunication networks. *Telecommunication Systems*, 4: 241–271, 1995.

E.D. Taillard. Robust taboo search for the quadratic assignment problem. *Parallel Computing*, 17:443–455, 1991.

H. Takahashi and A. Matsuyama. An approximate solution for the Steiner problem in graphs. *Mathematica Japonica*, 24:573–577, 1980.

E.-G. Talbi. *Metaheuristics: From design to implementation*. Wiley, New York, 2009.

R. Tamassia and G. Di Battista. Automatic graph drawing and readability of diagrams. *IEEE Transactions on Systems, Man, and Cybernetics*, 18:61–79, 1988.

F.L. Usberti, P.M. França, and A.L.M. França. GRASP with evolutionary path-relinking for the capacitated arc routing problem. *Computers & Operations Research*, 40:3206–3217, 2013.

P.J.M. van Laarhoven and E. Aarts. *Simulated annealing: Theory and applications*. Kluwer Academic Publishers, Boston, 1987.

V.V. Vazirani. *Approximation algorithms*. Springer, Berlin, 2001.

M.G.A. Verhoeven and E.H.L. Aarts. Parallel local search. *Journal of Heuristics*, 1:43–66, 1995.

D.S. Vianna and J.E.C. Arroyo. A GRASP algorithm for the multi-objective knapsack problem. In *Proceedings of the 24th International Conference of the Chilean Computer Science Society*, pages 69–75, Arica, 2004. IEEE.

J.X. Vianna Neto, D.L.A. Bernert, and L.S. Coelho. Continuous GRASP algorithm applied to economic dispatch problem of thermal units. In *Proceedings of the 13th Brazilian Congress of Thermal Sciences and Engineering*, Uberlandia, 2010.

J.G. Villegas. *Vehicle routing problems with trailers*. PhD thesis, Universite de Technologie de Troyes, Troyes, 2010.

J.G. Villegas, C. Prins, C. Prodhon, A.L. Medaglia, and N. Velasco. A GRASP with evolutionary path relinking for the truck and trailer routing problem. *Computers & Operations Research*, 38:1319–1334, 2011.

M. Vlach. Branch and bound method for the three index assignment problem. *Ekonomicko-Mathematický Obzor*, 3:181–191, 1967.

S. Voss. Steiner's problem in graphs: Heuristic methods. *Discrete Applied Mathematics*, 40:45–72, 1992.

S. Voss. Heuristics for nonlinear assignment problems. In P.M. Pardalos and L.S. Pitsoulis, editors, *Nonlinear assignment problems: Algorithms and applications*, pages 175–215. Kluwer Academic Publishers, Boston, 2000.

S. Voss, A. Fink, and C. Duin. Looking ahead with the Pilot method. *Annals of Operations Research*, 136:285–302, 2005.

D.B. West. *Introduction to graph theory*. Pearson, 2001.

H. Whitney. On the abstract properties of linear dependence. *American Journal of Mathematics*, 57:509–533, 1935.

D.P. Williamson and D.B. Shmoys. *The design of approximation algorithms*. Cambridge University Press, New York, 2011.

P. Winter. Steiner problem in networks: A survey. *Networks*, 17:129–167, 1987.

I.H. Witten, E. Frank, and M.A. Hall. *Data mining: Practical machine learning tools and techniques*. Morgan Kaufmann, San Francisco, 3rd edition, 2011.

L.A. Wolsey. *Integer programming*. Wiley, New York, 1998.

Q. Wu and J.-K. Hao. A memetic approach for the Max-Cut problem. In C.A.C. Coello, V. Cutello, K. Deb, S. Forrest, G. Nicosia, and M. Pavone, editors, *Parallel problem solving from nature - Part II*, volume 7492 of *Lecture Notes in Computer Science*, pages 297–306. Springer, Berlin, 2012.

F.P. Wyman. Binary programming: A occasion rule for selecting optimal vs. heuristic techniques. *The Computer Journal*, 16:135–140, 1973.

M. Yagiura and T. Ibaraki. Local search. In P.M. Pardalos and M.G.C. Resende, editors, *Handbook of applied optimization*, pages 104–123. Oxford University Press, 2002.

M. Yagiura, T. Ibaraki, and F. Glover. An ejection chain approach for the generalized assignment problem. *INFORMS Journal on Computing*, 16:133–151, 2004.

M. Yannakakis. Computational complexity. In E.H.L. Aarts and J.K. Lenstra, editors, *Local search in combinatorial optimization*, chapter 2, pages 19–55. Wiley, Chichester, 2007.

J.R. Yee and F.Y.S. Lin. A routing algorithm for virtual circuit data networks with multiple sessions per O-D pair. *Networks*, 22:185–208, 1992.

# Index

© Springer Science+Business Media New York 2016
M.G.C. Resende, C.C. Ribeiro, *Optimization by GRASP*,
DOI 10.1007/978-1-4939-6530-4